GALILEO IN CONTEXT

edited by JÜRGEN RENN

Max-Planck-Institut für Wissenschaftsgeschichte, Berlin

CAMBRIDGE
UNIVERSITY PRESS

CAMBRIDGE UNIVERSITY PRESS
Cambridge, New York, Melbourne, Madrid, Cape Town,
Singapore, São Paulo, Delhi, Tokyo, Mexico City

Cambridge University Press
The Edinburgh Building, Cambridge CB2 8RU, UK

Published in the United States of America by Cambridge University Press, New York

www.cambridge.org
Information on this title: www.cambridge.org/9780521001038

First published 2001
Re-issued 2011

A catalogue record for this publication is available from the British Library

ISBN 978-0-521-00103-8 Paperback

Title page illustration: Analysis of the composition of the ink
of Galileo's worksheet folio 107r (MS.Gal.72) by means
of Particle Induced X-ray Emission, performed at the Ipstituto
Nazionale di Fisica Nucleare, Florence; for more
information see 'Hunting the White Elephant' in this volume

Galileo in Context

Edited by Jürgen Renn

Lindy Divarci, Editorial Assistant

CONTENTS

Galileo in Context: An Engineer-Scientist, Artist, and Courtier at the Origins of Classical Science

Andrea (in the door): Unhappy is the land that breeds no hero.
Galileo: No, Andrea. Unhappy is the land that needs a hero.
(Life of Galileo. Bertold Brecht)

The present volume documents recent attempts to explore the science of Galileo Galilei beyond its traditional perception as an isolated pioneering achievement into the intellectual, cultural, and social contexts that made it possible and that shaped it substantially. Three such contexts are singled out as having been of paramount importance for the genesis of Galilean science: the context of the engineer-scientists in which Galileo grew up and which provided his physics with much of its experiential basis; the closely related context of art which provided him not only with a model for his career as a courtier but also with the techniques of visual representation that he employed in his astronomical work; and finally, the context of contemporary power structures (including their ideological component), comprising those of the church as well as those of the courts and of the emerging scientific community. These structures determined not only Galileo's career but also the ways in which scientific information was produced, organized, and communicated in early modern Europe.

Several of the essays build on recent in-depth studies of Galileo's contexts that attempted also to develop new historiographical approaches. They range from an analysis of his relation to the church in terms of power-knowledge structures via a cultural anthropology of science under the conditions of patronage, and an examination of the formative role of representational techniques for scientific thinking, to a study of the knowledge structures common to the thinking of Galileo and his contemporaries and characteristic of "preclassical mechanics." These different perspectives are brought together here to show that, far from excluding each other, they in fact give rise to a surprisingly coherent new picture challenging the entrenched views of Galileo as a hero of science. That such a challenge might actually succeed in affecting the traditional image of Galileo is, however, rather unlikely, given the regularity with which historical research on Galileo tends to fall into oblivion under the spell of the Galileo myth.

Of course, the Galileo myth keeps changing with the changing images of science, but what has remained is that, for more than three hundred years, his science and life have served as archetypes for the scientific enterprise: he is still widely recognized as the lonely founding hero of modern science who introduced the scientific method and, in defending it, became a victim of the repression of science by the Catholic Church. Yet concurrently, Galileo's science and life have also been an object of ever more extensive scholarly studies. When opening modern textbooks however, be they of physics or of the history of science, not to speak of encyclopedias or popular biographies, there can be little doubt: Galileo, the myth, has remained largely untouched by scholarly insights into his historical situation and role. Conversely, scholarly literature has often failed to dissociate itself from the myth and has hence unquestioningly accepted the paradigmatic role ascribed to Galileo. The specific contexts of Galileo's life and science hence often come into play only as attenuating or reinforcing factors in the development of this paradigmatic role and not as elements that make this development understandable in the first place. By focusing on a model scientist, scholars hoped to attain universally valid insights into the functioning of science and into its conflicts with power, independently of the specificity of the historical situation. Remarkably, this is not only the case for the older literature extolling the virtues of Galileo's experimental method, but also for more recent heterodox discussions in which the trustworthiness of his procedures is severely criticized — but still with the aim of showing the problematic character of scientific reasoning in general. Even the highly specialized recent Galilean scholarship is under the spell of the Galileo myth to the extent that research questions such as the sources of his scientific method, or the crucial experiments by which he supposedly made his decisive discoveries are pursued. Obviously, such questions presuppose that Galileo indeed introduced a novel scientific method guiding his research and that he indeed made crucial discoveries in the sense of later classical physics, issues on which some of the contributions to this volume throw a new light.

A volume on "Galileo in context" must challenge the notion of context as well as the traditional image of Galileo, if it is to undermine the Galileo myth. As long as putting Galileo's science into its historical contexts means only identifying "influences" or "conditions" affecting his actions and thinking and does not mean re-examining the traditional epistemological understanding of the cognitive core of the scientific enterprise, the Galileo myth will continue to haunt scholarship dedicated to early modern science. Instead of studying contexts in order to determine the supposedly decisive factors of Galileo's life and science, it seems more enlightening to take Galileo rather as a probe for exploring a *cultural system of knowledge*, that is, the shared knowledge of the time with its social structures of transmission and dissemination, its material representations, and its cognitive organization. The contexts of Galileo's science would thus no longer have to be interpreted as pointing to competing explanatory frameworks emphasizing for instance either social or cognitive factors of the development of knowledge, but rather as layers of the historical reality from which this cultural system of knowledge would have to be reconstructed. Thus, truly putting Galileo's achievements into their historic contexts requires building up an epistemological framework wherein different contexts, whether referring to the social, material or cognitive dimensions of science are no longer fragmented and played off against each other but rather can be integrated so that the question "which context is relevant and why?" becomes answerable.

The present volume offers clues suggesting that a synthesis of different approaches to the question of the contexts of Galileo's science can actually succeed. The essays in this volume show Galileo not as a singular figure but as representative of the groups of actors who shaped the Scientific Revolution of the early modern period, ranging from engineer-scientists such as Guidobaldo del Monte, via philosophers such as Pierre Gassendi, to artists such as Ludovico Cigoli. This volume not only points to a great variety in the specific problems addressed by these actors, an equally great variety of the approaches taken in coping with them, and of the individual fates of these actors with regard to the success or failure of their contributions. It also suggests that the emergence and dissemination of the new sciences of Galileo's times were shaped by constraints that turn out to be surprisingly similar in spite of their individual variety. In fact, all of these actors were not only bound by similar social structures but also had to confront an array of *shared bodies and images of knowledge*, to use the terminology of Yehuda Elkana. The shared bodies and images of knowledge constituting the intellectual resources of early modern science ranged from the heritage of Aristotelian physics and ancient mechanics to the drawing techniques of contemporary art and engineering. The historical actors exploited these resources in their struggle with the *challenging objects* of the time, whether these were represented by the new technological achievements or by the newly discovered celestial phenomena. Taken together, these constraints and common challenges constitute the boundary conditions of a cultural system of knowledge which has to be reconstructed before one can truly understand individual intellectual trajectories such as that of Galileo as "science in context."

Approaches that focus on Galileo as an archetypical figure of modern science paradoxically lose sight of precisely what one might call "*the Galilean moment*" in the history of science. This Galilean moment may indeed be associated with a subversive power of knowledge, not in the sense of a clash between modern rationality and ancient dogmas, but in the sense of an explosion of technical and scientific knowledge so powerful that it unavoidably became central also to the symbolic politics of the time, deeply ingrained in canonized views of the natural world. It is the great paradox of the Galilean moment that this explosion of knowledge was not initiated by the birth of a new form of rationality but rather by the conflictual, yet productive, encounter of traditional bodies of knowledge such as the practical knowledge of the engineers and the theoretical tradition of the universities. This encounter was, at least in part, triggered by the great practical ventures of the time, from intercontinental navigation to large-scale engineering projects. It took place within a setting where the advancement of science was regulated by mechanisms to which its intellectual achievements were, taken by themselves, of only limited relevance, being promoted or suppressed within a patronage system. Nevertheless, the impact of the resulting integration of heterogeneous strands of knowledge on the advancement of science was so powerful that it eventually led not only to the revision of traditional conceptual systems, but also to the creation of new social structures for the production and dissemination of scientific knowledge. Hence the role of "the social context" of scientific development would be underestimated were it considered solely as an external framing condition for a specific subculture of society. Instead, the development of scientific knowledge in the early modern era must be understood as an essential part of societal dynamics itself, namely as the self-reflection of an increasingly knowledge-based society.

Around the turn of the last century, there have been remarkable attempts to identify the specificity of Galileo's historical situation, beginning with Antonio Favaro's *Edizione Nazionale* of Galileo's works. By making the early, evidently Aristotelian writings of Galileo as well as his vast correspondence available, insights into Galileo's intellectual debts with regard to his antique and medieval predecessors, as well as to his contemporaries became inescapable. Among the first scholars to draw consequences from such insights, albeit in quite different ways, were Raffaello Caverni and Emil Wohlwill. They seriously confronted the challenge of making sense, not only of Galileo's major works such as the *Dialogue on the Two Great World Systems* and the *Discorsi on Two New Sciences*, but also of the numerous documents — unpublished manuscripts as well as letters — that show Galileo at work as one among many of his contemporaries struggling with the authority of Scholasticism, while still thinking in terms of Aristotelian notions (or polemically defending his inventions against competitors, occasionally maintaining his own priority only with the help of false pretenses). Caverni, in his six-volume *Storia del metodo sperimentale in Italia (History of the Experimental Method in Italy)*, and Wohlwill, in his two-volume biography *Galilei und sein Kampf für die Copernicanische Lehre (Galileo and His Battle for the Copernican System)* as well as in several of his book-length papers, were the first historians to examine Galileo's work also with an eye to his failure to attain crucial conceptual breakthroughs such as a general principle of inertia. Galileo's thinking thus emerged as much closer to that of his predecessors and contemporaries than the Galileo myth would have it, an insight that was also confirmed by the extensive studies of Pierre Duhem, Anneliese Meier, and Eduard Jan Dijksterhuis, in particular in his neglected masterpiece *Val en worp (Fall and Projection)*. It is a singular fact characterizing the present state of Galileo studies that thousands of pages written by authors circa 1900 on Galileo's science had no substantial impact — either on the public perception of Galileo or on the specialized scholarly literature.

The unassailable character of the Galileo myth has many roots, among them the disciplinary splintering of studies dealing with the development of scientific knowledge. In particular, the analysis of the emergence of early modern science at large, in its nature an undertaking involving epistemological, historical, and sociological dimensions, has been, to a considerable extent, pursued separately from the study of Galilean science in the sense of a highly specialized sub-discipline of the history of science, comprising careful editions, detailed commentaries, and subtle interpretations of historical sources. For example, in the early twentieth century, historians such as Edgar Zilsel and Leonard Olschki drew attention to the context of contemporary engineer-scientists and to their role in the genesis of modern science. Their studies began to shed some light on the structural character-istics of early modern science, emphasizing its dependence on the social and material working conditions of engineer-scientists and, in particular, its roots in the social and cognitive integration of various traditions of knowledge. But these early pioneers succeeded only to a limited degree in linking their questions with a detailed examination of the historical material that was gradually becoming available through editions, commentaries, and specialized studies. As a conse-quence of the disciplinary separation of philological, historical, philosophical, and other approaches, this material thus remained largely unexploited for answering theoretical questions related to the emergence of early modern science.

Galilean studies in the philosophical tradition often exploited historical sources merely as a quarry from which to pick and choose, instead of systematically confronting their theoretical claims with the wealth of extant sources. In fact, even philosophers who extensively discussed Galileo, as did Natorp, Cassirer, or Hus-serl, hardly took the results of the historical research on Galileo by Wohlwill and others as a serious challenge to their philosophical views. Cassirer, for instance, claimed that Galileo deduced the principle of inertia in his *Discorsi* and, with only a few words, rejected objections based on Wohlwill's detailed historical research to this interpretation as being philosophically irrelevant and as illuminating "only the historical difficulties" presenting themselves to the achievement of the new insight. Koyré claimed that Galileo's mathematical Platonism was crucial for the success of the new science on motion, denying the relevance and even the existence of experiments on motion performed by Galileo. He effectively ignored the numerous references to experiments in Galileo's published and unpublished writings and concentrated his textual analysis instead on those passages supporting his opinion. The overwhelming richness of sources on Galileo's science may have even appeared as irrelevant to Koyré since his primary aim was not to reconstruct a sequence of historical events. He attempted instead to identify mental attitudes characteristic of the historical actors. This approach was possibly related to that of contemporary students of "collective representations" such as Durkheim or Lévy-Bruhl, as has been recently suggested by Paola Zambelli. In fact, however, the "collective" of Koyré's actual historical studies of Galileo's science remains essentially restricted, not only to a small intellectual elite, as Yehuda Elkana sees it, but actually to a single individual, Galileo himself. Koyré's identification of mental attitudes, such as his characterization of Galileo as a physical thinker in contrast to that of Descartes as a mathematical thinker, thus fails to achieve the historical contextualization of Galileo's science for which he is often credited.

Given the one-sidedness of Koyré's interpretation and his highly selective treatment of historical sources, it was not difficult for another towering figure of Galileo scholarship in the twentieth century, Stillman Drake, to challenge this interpretation and to identify documents favoring instead his view of Galileo as the first modern experimental physicist. Projecting his own opposition to philosophy on Galileo, Drake saw himself as a pioneer in opening up the study of Galileo as a working scientist, a claim that is justified by his extensive studies of Galileo's manuscripts on mechanics and numerous other contributions. But Drake's publications hardly take into account the substantial earlier research on Galileo's science, both in history and philosophy. Drake's own preconceived opinions, less reflective than those of the philosophers, shaped his historical work all the more strongly, as may be illustrated by occasional misleading quotations, twisted translations, or suitably arranged cut-and-paste editions that can be found alongside the lasting achievements in Drake's work. While his translations and editions considerably widened the scope of the historical documentation available to the English-speaking world, his leading role in Galileo scholarship contributed at the same time to the oblivion of the earlier historical and philosophical research, which seemed to be superseded.

In Galileo scholarship, the opposition between Koyré and Drake has taken on an almost archetypical role in shaping the historical questions, the controversial issues, as well as the literary style of many contributions, even critical ones. Nevertheless, this limitation of the intellectual horizon of Galileo studies has, over the years, gradually been undermined by the research of scholars such as Thomas B. Settle, Pierre Souffrin, and Winifred Wisan, who also saw the necessity of looking back to Favaro, Wohlwill, and Caverni. The widened scope of Galileo scholarship is, in fact, becoming visible in recent work such as that collected in *The Cambridge Companion to Galileo*, edited by Peter Machamer. There still remains, however, a considerable gap between systematic epistemological questions, for instance concerning the role of shared knowledge for the emergence of early modern science, and thorough historical research, identifying such shared knowledge with as much empirical rigor as historians expect from reconstructions of Galileo's individual contributions. Closing this gap will make it necessary to overcome traditional boundaries of specialization by means of new forms of collaboration between scholars and new ways of making historical documents available, in particular by exploiting the potential of the Internet. Meanwhile, it is the aim of the present volume to survey recent approaches to Galileo scholarship that, particularly when taken together, offer perspectives on the potential outcome of such a joint effort, that is, an *historical epistemology* of early modern science.

The essays of the first section, "The Context of the Practitioners: Mechanics and its New Objects," deal with the relationships between practical and theoretical knowledge in the emergence of classical mechanics. They show that neither the reliance on experiments, nor the continuity of theoretical traditions, nor the social context, taken by themselves, sufficiently account for the eventual success of Galilean science. The studies of the first section rather suggest that *challenging objects* which entered the intellectual horizon of the new engineer-scientists from outside the dominating academic traditions triggered the transformation of scholastic physical concepts towards classical mechanics. The origin of these objects in the accumulated *shared knowledge* of the practitioners and engineers of the time reveals their role as important and irreducible mediatory instances between early modern science and its social and technological context.

The essays of the second section, "The Context of the Artists: Astronomy and its New Representations," deal with another mediatory instance between early modern science and its social and technical contexts, *visual representations*. They also illustrate the extent to which the distinction between the history of science and the history of art is an artificial one when it comes to the early modern period. Artists and engineer-scientists not only shared similar career-patterns in the fragile social environment of patronage but also a common curriculum of learning that equipped them with similar techniques for addressing similar problems, the challenges of design involved in practical tasks such as those of architecture and the challenges of visual representation when confronted with the new experience of the age. The essays of the second section open a wide field of questions, worthy of being followed up in future studies: Which precisely were Galileo's artistic tools and in which tradition do they stand? How did these artistic traditions affect the perception and representation of the objects of his science? What was the function of Galileo's artistic production for his social role and its advancement, so similar to that of contemporary artists? How did Galileo's representations of the moon contribute to the dissemination and acceptance of his scientific results and their intellectual provocation? And more generally: Which role do visual representations play as mediatory instances between observation and theoretical convictions?

The essays of the third section, "The Contexts of Church, Patrons, and Colleagues: New Science and Traditional Power Structures," show how social, material, and cognitive factors act together in shaping the collective processes of the production, dissemination, and transmission of knowledge. By focussing on the dissemination and transmission of Galileo's contributions, the studies of this section provide general insights into the dynamics of the *cultural system of knowledge* constituting early modern science. As the essays of this section indicate, this dynamics is characterized by an economy of credit and disclosure due to the patronage system of early modern science, by the determining role of the Church in the institutionalization of teaching and learning, and by the potential of early modern science to undermine the dominant worldview of the Church. Early modern science results from an integration of disparate contributions, such as the theoretical knowledge of Scholasticism and the practical knowledge of the engineer-scientists, into the emerging framework of classical science.

The essays of this section make it clear to what extent such an *integration of knowledge* and its results are dependent on the historically contingent infrastructure of the shared knowledge of the time.

In order to encourage modern readers to make use of the forgotten treasures of Galileo scholarship from the turn of the last century, the present volume includes as an appendix essays in English translation by Favaro, Caverni, and Wohlwill which form part of a controversy on the origins of Galileo's great achievements in mechanics, traditionally identified with the discovery of the law of free fall and of the parabolic shape of the projectile trajectory. Naturally, the three essays can hardly provide more than a glimpse into the wealth of sources, contexts, and interpretations offered by the three masters of Galileo studies. The essays here published for the first time in English translation are prefaced by biographical introductions placing the pioneering works of Favaro, Caverni, and Wohlwill within the context of the historical scholarship of their period.

The roots of this volume go back to a workshop on new trends in Galileo scholarship, held in January 1996 at the ETH Zurich, and organized by Yehuda Elkana and Helga Nowotny. Without their encouragement this volume would not have been realized. Giuseppe Castagnetti played a crucial role in editing the Appendix. Support which helped to complete this volume was furthermore offered by Jochen Büttner, Peter Damerow, Lorraine Daston, Gideon Freudenthal, Wolfgang Lefèvre, Simone Rieger, Urs Schoepflin, Petra Schröter, Matteo Valleriani and other colleagues from the Max Planck Institute for the History of Science in Berlin.

Jürgen Renn
Max-Planck-Institut für Wissenschaftsgeschichte, Berlin

1. The Context of the Practioners: Mechanics and its New Objects

WOLFGANG LEFÈVRE

Galileo Engineer: Art and Modern Science

The Argument

In spite of Koyré's conclusions, there are sufficient reasons to claim that Galileo, and with him the beginnings of classical mechanics in early modern times, was closely related to practical mechanics. It is, however, not completely clear how, and to what extent, practitioners and engineers could have had a part in shaping the modern sciences. By comparing the beginnings of modern dynamics with the beginnings of statics in Antiquity, and in particular with Archimedes — whose rediscovery in the sixteenth century was of great consequence — I will focus on the question of which devices played a comparable role in dynamics to that of the lever and balance in statics. I will also examine where these devices came from. In this way, I will show that the entire world of mechanics of that time — "high" and "low," practical and theoretical — was of significance for shaping classical mechanics and that a specific relationship between art and science was and is constitutive for modern sciences.

Koyré's Provocation

In 1943, Alexandre Koyré wrote: "The Cartesian and Galilean science has, of course, been of extreme importance for the engineer and the technician; ultimately it has produced a technical revolution. Yet it was created and developed neither by engineers nor technicians, but by theorists and philosophers" (Koyré 1943a, 401 n. 5). In the same essay: "Their [Galileo's and Descartes'] science is made ... by men who seldom built or made anything more real than a theory" (ibid., 401).

Against these and other similar arguments by Koyré, it is not difficult to show that Galileo pertains just to the tradition of the Italian engineers of the Renaissance. The facts are so well known that it will be sufficient only to list some points:

- his training in mathematics in the early 1580s by the mathematician and engineer Ostilio Ricci (see Galilei 1890–1909 XIX, 36; see also Masotti 1977), who is said to have been a pupil of Tartaglia (Drake 1978, 3) and taught later mathematics at the Accademia del Disegno in Florence, founded, on instruction of Cosimo I, Duke (then) of Florence, by Vasari, an educational institution for artists and engineers like Gresham College in London in the seventeenth century;

- his lectures on practical mathematics (see Galilei 1890–1909 XIX, 149–158) — fortification,[1] surveying, mechanics, optics, use of the sector[2] etc. — given in Padua in addition to his regular teaching activities at the university;
- his running his own workshop in Padua, which was not so much of use for performing experiments but served as a workshop for manufacturing instruments[3] that he either invented or developed in a special way;
- his successful application for a patent (*privilegio*) of the *Signoria* of Venice for a device for raising water in 1593–94 (see Galilei 1890–1909 XIX, 126–129);
- his varied activities as an inventor, which prompted Leonardo Olschki to write: "One has to imagine that every one of Galileo's discoveries in physics and astronomy is closely connected with any instrument which was either invented or modified in a special way by him" (Olschki 1927, 140);
- his occupation with engineering problems — pumps, regulation of rivers, fortification etc. — all his life;
- his function as mathematician at the Medicean court in which capacity he had to supervise all suggestions of important engineering projects.

Finally, there are indications that Galileo saw himself within the tradition of the Italian engineers of the Renaissance: His last and perhaps most important book bears the title *Discorsi e dimonstrazioni mathematiche intorno a due nuove scienze attenenti alla Mecanica & i movimenti locali*. Alluding thus to Nicolo Tartaglia's *La Nuova Scientia*, it seems to me that Galileo himself ranked his *Discorsi* as within the tradition of treatises that are known as vernacular engineering literature of early modern times.

All these facts were of course known to Koyré as well, whose writings on Galileo are later than those of Olschki or of Zilsel.[4] Conversely, it was known to Olschki and Zilsel that there was not only accordance between men like Tartaglia and Galileo and common practitioners and engineers, but there was also distance and even open conflicts. What Koyré wanted to deny was that it was possible to gain anything for a true understanding of the modern sciences — and these sciences are of course at issue when Galileo is the topic — by studying, as their context, the world of craftsmen and engineers, the "tradition of the workshops" (see Mittelstraß 1970, 175 ff). According to Koyré, the modern sciences resulted from a radical turn of philosophical paradigms, that is, from the replacement of a view of nature in the tradition of Aristotle — seen as bound to sense perceptions and to everyday life concepts — by a mathematical one in the tradition of Plato.

Even if little convinced of Koyré's claims and rather inclined to follow Olschki,

---

[1] See also Galileo, "Breve instruzione all'architettura militare" (Galilei 1890–1909 II, 15–75), and Galileo, "Trattato di fortificazione" (ibid., 77–146).

[2] See also Galileo, "Il compasso geometrico e militare" (Galilei 1890–1909 II, 343–361), and Galileo, "Le operazioni del compasso geometrico e militare" (ibid., 363–424).

[3] See above all the bookkeeping accounts regarding "L'officina di strumenti mathematici in Padova" (Galilei 1890–1909 XIX, 131–149).

[4] Koyré mentions Edgar Zilsel's *The Social Roots of Science* (1942) as well as Olschki's book (see Koyré 1943a, 401 n. 6).

one has to admit that Koyré's claims remain provoking as long as it is not really clear what is meant by the opposite claim that the emergence of the modern sciences becomes only intelligible when seen in the context of the world of craftsmen and engineers, of the tradition of the workshops. As discussed extensively elsewhere (see Lefèvre 1978, especially Part I), it seems clear that one possible meaning drops out in advance — the meaning that the modern sciences can be derived from needs or bottlenecks of the technology of early modern times.

Though not on the level of regular (non-agricultural) production, which was almost completely performed by a smoothly-functioning system of crafts and artisanship, there were in fact serious problems and bottlenecks in the few exceptional areas of "high tech" production. The problem of regulating the water level in mines of some depth with water pumps is one such example. Such problems were certainly stimulating for theoretically-interested people but it is clear that the sciences of that time were not able to contribute much to the solution of such problems: Contrary to Koyré's position, the world of production gained almost no benefit from the modern sciences before the nineteenth century.[5] Furthermore and perhaps more importantly, it did (and does) not depend on the practical urgency of a problem whether it was the subject of theoretical investigations with fruitful consequences for the sciences or not. That problems which occur in the high tech areas of production are rather unsuitable candidates in this respect is impressively shown by Leonardo da Vinci's admirable inquiries into mechanical problems. The fact that these inquiries were apparently of rather limited consequence for scientific mechanics might be due to the fact that he treated these problems with the nearly unreduced complexity that confronted the men of praxis.

But can we then expect more from the world of craftsmen and engineers, from the tradition of the workshops, than that, at best, it contributed among other factors to a favorable and stimulating climate for the development of the modern sciences? At any rate, it seems not yet sufficiently clear how and to what extent the sphere of practitioners and engineers could play an important role in shaping the modern sciences.[6] In order to examine these questions further within the surroundings of Galileo's life and work, I will start by calling to mind one of the achievements that has earned Galileo a place as one of the founding heroes of the modern sciences.

---

[5] This applies even to fields like construction, irrigation, and draining, or military architecture where engineers made wide use of geometry, statics, etc., i.e. of sciences, but very rarely of modern sciences. Achievements like the eighteenth-century moon tables, so desperately needed for navigation, were indeed genuinely the fruits of modern science, viz. of Newton's moon theory, but at the same time rather exceptions.

[6] For a more recent "Interim Assessment" of the debate on the tradition of the workshops and the emergence of early modern science, see Cohen 1994, 345ff.

Dynamics — "An Entirely Modern Science"

Galileo's contribution to the foundation of modern dynamics proved to be a decisive step toward the emergence of the modern sciences.[7] Considering the impact he actually had on the development of modern scientific mechanics, one even can say — *cum grano salis* — that his contribution consists precisely in his derivation of the law of free fall and of the projectile trajectory in vacuum. This assessment of his achievements is of course anachronistic. From the perspective of the later, fully developed scientific mechanics, the criterion for judging significance is which of his theories — as reformulated as they may be — was incorporated into classical mechanics. On the other hand it should also be recognized that the decisive event of the seventeenth-century scientific revolution in the field of physics was in fact the origin of modern dynamics. The concepts of these new dynamics enabled the generation of Huygens, Leibniz, and Newton to lay the foundation stone of the building of scientific mechanics in its specific modern shape.

Ernst Mach regarded dynamics as "entirely a modern science." "The mechanical speculations of the ancients, particularly of the Greeks, related wholly to statics. Only in mostly unsuccessful paths, does their thinking extend into dynamics" (Mach 1989, 151). It actually seems that the physicist Mach was unable to take seriously natural philosophy in the tradition of Aristotle, with its statements on natural and forced motion, heavy and light bodies, etc. Koyré, conversely, too solid an historian of ideas not to recognize the legitimacy of the Aristotelian dynamics, emphasized that these dynamics are — perhaps with the exception of the theory of projection (see Koyré 1943a, 411) — much more plausible for understanding everyday life than modern dynamics. Nevertheless, both Mach and Koyré agree on what is most relevant for us: among the theories of mechanics before Galileo's time, only the field of statics and hydrostatics theories can be assessed as scientific from a modern point of view. This is by no means true in the field of dynamics.[8] For Mach and Koyré, there exists no previous history of dynamics in the Middle Ages[9] or in Antiquity, whereas both establish such a

---

[7] Damerow et al. have shown that Galileo himself remained within the limits of pre-classical mechanics (see Damerow et. al. 1992).

[8] Seen from the perspective of modern mechanics, a rigid distinction between statics and dynamics is artificial. To a certain extent, it is even artificial with respect to the history of mechanics from Antiquity to the early modern era. Within the mechanics tradition of that period, we find not only classical topics of statics, but also of dynamics. In addition to treatments of these statics topics without any application of concepts of movement or force (above all Archimedes), we also find treatments of these topics which make use of dynamics concepts, such as the pseudo-Aristotelian treatise on *Mechanical Problems*. The distinction I want to make here (borrowing from Mach) is the following: Whereas certain classical topics of statics, like the lever, were treated in a modern way in Antiquity by men like Archimedes, essential topics of dynamics — free fall, impact, projection etc. — were handled in a manner known from Aristotle's *Physica*. When I speak of statics or dynamics in this article, I thus always refer only to different topics and not to absolutely separated scientific fields.

[9] Koyré was familiar with the writings of Duhem on the Parisian nominalists of the fourteenth century (see Koyré 1943a, 406).

history for modern statics beginning in the Hellenistic period with Archimedes. Koyré came to see Archimedes and Galileo, though separated by almost two thousand years, as a kind of twin figure personifying the true founding hero of the modern sciences:

> the new, Galilean, physics is a geometry of motion, just as the physics of his true master, the *divus Archimedes*, was a geometry of rest. ... Motion is subjected to number; that is something which even the greatest of the old Platonists, the superhuman Archimedes himself, did not know, something which was left to discover to ... the Platonist Galileo Galilei. (Koyré 1943b, 347f.)

Postponing all objections prompted by these statements, I want to single out two of them with which I agree:

- Modern dynamics, which goes back to Galileo, is a genuine novelty of modern times, whereas modern statics goes back to Archimedes.
- The way in which Galileo treated problems of dynamics is — in principle, regarding the type of treatment — comparable with the way in which Archimedes treated problems of statics.

Thus, it may be appropriate to have a short look back to the mechanics of Archimedes.

Archimedes — A Scientific Engineer

I will start by recalling some well known things. Although it is today a sub-discipline of physics, and became the archetype of physics for the generations after Galileo, mechanics did not belong to physics from Antiquity until the time of Galileo (see Hoykaas 1963) when it was conceived of as knowledge about devices and machines. This knowledge, according to a traditional understanding, which Galileo still had to criticize,[10] was thought of as a means of outwitting nature. Accordingly, it did not make sense to expect that one could gain knowledge about nature by investigating those devices. To complete the picture, making it more complicated at the same time, we have to add that, on the other hand, certain topics that are treated today within a sub-discipline of scientific mechanics were regarded as belonging to physics since the time of Aristotle, namely — as previously mentioned — dynamical topics. The situation is thus the following: Topics that now belong to the realm of statics or hydrostatics as sub-disciplines of mechanics were then the subject matter of "mechanics" in the sense of knowledge — not about nature, but — of art; whereas topics that now belong to the realm of dynamics as a sub-discipline of mechanics were then the subject matter of "physics," in the sense of an all-encompassing natural philosophy.

---

[10] See the introduction of Galileo's *Le meccaniche* in Galilei 1960, 147f. (Galilei 1890–1909 II, 155f.).

With respect to Archimedes, one consequence of this situation was that, until the time of Galileo, his writings on statics and hydrostatics were not considered to be writings on physics, as was the case for writings like the *Collectiones mathematicae* of Pappus or the *Mechanica* of Heron of Alexandria. They were counted as "mechanics" in the sense of a doctrine on art. Accordingly, only from an anachronistic point of view, could Koyré claim that Archimedes obtained his achievements in statics and hydrostatics because of a mathematical view of nature. The old distinction between mechanics and physics brings our attention to the real content of Archimedes' mechanical writings: theories on certain tools, instruments, simple machines, and devices — on devices the installation, adjustment, and application of which sometimes exceeded the capacities of regular craftsmen and were thus the business of experts who were later called engineers in the West. Archimedes, for Koyré "the greatest of the old Platonists," was honored in Antiquity as the most outstanding engineer whose inventions became entangled by an interweaving of legends not easily deciphered. Even his biography seems to show that the engineer Archimedes preceded the mathematician (see Schneider 1979).

Of course, skills and abilities in the field of technology do not explain those in the theoretical field. There is no gradual transition from the practical experiences and knowledge of the engineer Archimedes to the theoretical form of knowledge his writings on mechanics demonstrate. Compared to engineering treatises like the *De Architectura Libri Decem* of Vitruvius which pass on practical knowledge and experiences, these writings of Archimedes mark a qualitative difference — a difference which is comparable to that between experienced and skilled mathematical practitioners and theoretical mathematicians. In arithmetic and geometry, the decisive step which led in Greek Antiquity to systems of knowledge with a deductive structure was made by reflecting on the possibilities of acting with the symbol systems of the time — means of counting or constructions of spatial relations, respectively. A basic prerequisite of this reflection was an external, i.e. non mental, representation of the elaborated characters, structures, and laws of these actions with symbols, in this case a representation in the medium of the colloquial but written language (see Damerow and Lefèvre 1998, 88f.). The theoretical achievements of Archimedes in the field of mechanics can be understood analogously. They resulted from the reflection on the possibilities of acting with certain material means; not of symbolic operations, but with those of technical operations. In this case too, an external representation of the elaborated characters, structures, and laws of these actions with technical tools was a basic prerequisite of this reflection, but this time Archimedes could use the already available theoretical mathematics for this representation. These mathematical procedures were doubtless of great significance for his success. The precondition for such an application of mathematical tools as a means of theoretical representation and of deduction in statics was, however, that there were devices and mechanical arrangements by reflection and measurement of which the quantities and the kind of relationships expressed by such laws can be obtained.

In the case of Archimedes, we know which technical devices he used to develop his theoretical statics; he developed them on the so-called simple machines (*potentiae staticae*), especially the balance (lever). If we agree with Koyré that Galileo's theories of dynamics are in principle similar to Archimedes' theories of statics, the question becomes what devices did men like Galileo then use to develop modern dynamics in early modern times?

New Problems in Dynamics

Western Europe got to know of the preserved mechanical writings of Archimedes — *On the equilibrium of planes* and *On floating bodies* — in the thirteenth century through the translations of William of Moerbeke.[11] One cannot speak of a real appropriation and assimilation of these writings, however, before the sixteenth century. Their first publication as printed text in 1543 (Archimedes 1543), an event of serious consequence, was the act of the self-educated engineer and mathematician Nicolo Tartaglia. It can be stated generally that the sixteenth-century revival of the classical tradition of mechanics was not in the first place the concern and work of natural philosophy at the universities or of the humanist movement,[12] but of laymen in classics, namely of engineers who were interested in theoretical questions.[13] This confirms once more that the traditional distinction between mechanics and physics was still valid in the sixteenth century.

It is striking to observe that this appropriation of traditions of classical mechanics by theoretically interested engineers was genuine, original, and creative from its beginnings. They tried to apply it to solving problems like impact, momentum, free fall, and projection etc., to problems that went beyond the limits of statics and hydrostatics passed on from Antiquity. It does not mean that these engineers aspired to develop a new treatment of the traditional problems of dynamics within the natural philosophy taught at the universities. Rather, they were striving to solve the dynamics problems which occurred within their practical occupation as engineers, and only in consequence of that did it happen that they were sometimes forced to discuss theories from academic natural philosophy.

This connection between the independent appropriation of the classical mechanics traditions by sixteenth-century engineers, the application of this inherit-

---

[11] Among the lost writings of Archimedes was one with the title *On balance*. For its presumed contents, see above all Knorr 1982.

[12] One has to add immediately, however, that the humanist movement of fifteenth-century Italy provided the ground on which laymen like Tartaglia could base, namely the collection of the antique texts in several libraries (see Rose 1975, particularly chap. 2).

[13] Within this paper, we have to pass over the short and isolated renaissance of studies in statics in the thirteenth century centered on writings attributed to Jordanus Nemorarius (see Moody and Clagett 1960). It may be of interest to observe that the rediscovery of these sources in the sixteenth century was the result of editorial activities of those engineer figures of the Renaissance, of Apianus (*Liber Jordani Nemorarii ... de ponderibus ... Petro Apiano ...* Nürnberg 1533) and again of Tartaglia (*Iordani opusculum de ponderositate Nicolai Tartaleae ...* Venice 1565).

ance to problems which arose from their practical occupation, and occasional examinations of dynamics theories from academic natural philosophy is obvious, for instance, in the case of projectile trajectory, a central problem for engineers from Tartaglia to Galileo. No theoretical investigation of this problem could ultimately avoid dealing with key concepts from the doctrine of motion in the tradition of academic natural philosophy — concepts like heaviness, lightness, natural versus forced motion, the Aristotelian theory of free fall, the theory of projection in the Aristotelian tradition as well as in the tradition of the *impetus* physics, etc. But theoretically interested engineers discussed these traditional problems and concepts of motion in light of their new questions, freed these from traditional connections to the old conceptual framework, and put them in new ones, discussing, for instance, the Aristotelian propositions regarding free fall within the conceptual framework of hydrostatics (see Galilei 1960, 35ff.; Galilei 1890–1909 I, 271ff.).

The practical context of the new dynamic questions that these engineers raised is well known: ballistics, water pumps, transmissions, etc. The technical revolution of early modern times constituted the general background to these questions, and their outcome is known as well: These engineers dealt with and developed the topics in dynamics, which had previously been treated within traditional natural philosophy, in a way so new that within an interval of only one hundred years — Tartaglia's *Nuova Scientia* appeared in 1537, Galileo's *Discorsi* in 1638 — they produced the preconditions for the development of modern classical mechanics in the second half of the seventeenth century. Although the practical context of the new questions in dynamics and the new ways of their treatment are clear, at least in principle, the exact contribution of this context to the emergence of the new mechanics remains unclear.

Devices of Production and of Scientific Research

It seems obvious that the mere fact that these engineers had motives[14] for being interested in problems in dynamics explains very little. The desire to solve a problem does not constitute a sufficient reason or basis for actually solving it. The Archimedean attitude of these men does not explain much more either. To be able to treat problems in dynamics in an Archimedean way requires not only highly developed mathematical competence, but, above all, suitable technical arrangements with which the problems can be investigated. Thus, we have to examine the devices men like Galileo used to perform their research, the type of devices they used, and investigate where they came from.

These questions seem even more natural in the case of dynamics than in statics.

---

[14] And perhaps not only professional ones, but also motives which were connected with the contemporary struggles of world views.

It is obviously much more difficult to investigate dynamic phenomena like projection, free fall, impact, etc. than those of statics. One need only think of the swiftness of such phenomena and consider in addition how poor — compared to today — the instruments of that era were, for measuring short intervals of time, for example. Galileo's inventive spirit in working with such difficulties has long prompted special admiration among Galileo scholars; one example is his idea of investigating the law of free fall by letting spheres roll down an inclined plane (see Galilei 1974, 169f. [Galilei 1890–1909 VIII, 212f.]).

Galileo's fame as a founding hero of modern sciences rests not least on the fact that he was one of the first scientists who performed experiments. In contrast to the time when Koyré dominated the research on Galileo, today almost nobody doubts that Galileo performed experiments.[15] Instead, nowadays the opposite danger exists — in seeing Galileo in the light of the narrow and absolutely ahistorical view of experimentation developed by the traditional philosophy of science. According to this view, experiments are only carried out in order to test theories, and the exploratory character of experimentation is largely ignored.[16] It does not seem clear to me, for example, whether the usual emphasis on the genius of Galileo's experiments with the inclined plane is misleading or not. It is known that investigations concerning motions on inclined planes were undertaken at that time which had nothing to do with the problem of free fall — Galileo's own earlier investigations on "ratios of motions of the same body over various inclined planes" (*de proportionibus motuum eiusdem mobilis super diversa plana inclinata*), for instance, were not only connected with problems of statics but were, moreover, pursued within the conceptual framework of statics (Galilei 1960, 63ff. [Galilei 1890–1909 I, 296ff.]; see Drake 1978, 23ff.). Thus, it would be interesting to know the significance of the experiences and insights gained through these earlier studies for his later experimental research on free fall reported in the *Discorsi* (see Galilei 1974, 169f. [Galilei 1890–1909 VIII, 212f.]). Inventions require discoveries, and the question of which technical arrangements are suitable to treat a given mechanical problem in an Archimedean way is not primarily a question of brilliant inventions but of discoveries which require a close familiarity with the available technologies of the time.

This point is valid not only for finding suitable arrangements in order to investigate an already clear-cut problem, but also for finding the problems themselves. Whether problems can be treated in an Archimedean way depends on the technologies available in an era; the technical possibilities of a given time shape the specific form in which problems can be investigated. Thus, finding the problems

---

[15] Concerning the efforts to reconstruct Galileo's experiments, see above all the writings of Thomas B. Settle. The English version of Settle 1995 appeared in 1996 as a preprint of the Max-Planck Institut für Wissenschaftsgeschichte, Berlin.

[16] Jürgen Renn et al. ("Hunting the White Elephant," in this volume) present new insights into Galileo's experimental practice which may necessitate rethinking, among other things, the usual understanding of established or supposed experiments of Galileo as theory testing key experiments (see, for instance, Drake 1978, 127ff.).

themselves is also a question of discovery that requires familiarity with the available technologies of the time.

The example of the pendulum might be helpful in determining whether problems can be treated in an Archimedean way in order to illustrate the significance of familiarity with technology for the modern sciences. The pendulum is so suitable a device for investigating dynamics problems, it can seem that it had been invented for this purpose. Galileo discovered the pendulum law at the beginning of his career.[17] The anecdote that he discovered it by observing the swinging chandeliers of the cathedral of Pisa has long been ranked among the fables of science. I am not sure that it is such a great advance to assume that experiments his father Vincenzio performed in connection with theoretical questions of music led the son to his discovery (Drake 1978, 21). In any case, the old fable with the chandeliers had at least one advantage: It makes us wonder why the characteristics of pendulum motion were not noticed much earlier. Such swinging objects can be found in civilizations much less developed than Tuscany at the turn of the sixteenth to the seventeenth century. What other prerequisites were needed to discover the characteristics of pendulum motion? It is striking that the pendulum law was not established before pendulums were used as elements of certain machines — a use that can not be traced back to any document older than Leonardo da Vinci (see Feldhaus 1970, column 1218; Usher 1988, 310)[18] — and which was first publicized through Jacques Besson's *Theatrum instrumentorum et machinarum* (1569 and 1578),[19] — only a few years before Galileo's discovery (see White 1966, 108) (fig. 1). Inversely, a new practical utilization of the pendulum emerged apparently in close vicinity of Galileo's investigations — its application as time-keeper.[20]

---

[17] It is not always clear what is meant by Galileo's discovery of the pendulum law: the simple law that the swinging time depends solely on the length of the thread or the (wrong) isochronism law, i.e. the assumption that a pendulum traverses equal arcs in equal times. The latter "law" is an implicit main topic of the Third Day of the *Discorsi* and can be traced back at least to 1602 (see Galileo's letter to Guidobaldo del Monte from November 29, 1602, in Galilei 1890–1909 X, 97–100). Earlier remarks of Galileo on the pendulum focus on the time pendulums of different weight take to come to rest — see Galileo Galilei: Memoranda on Motion in Drake and Drabkin 1969, 383 (memorandum 20a) (Galilei 1890–1909 I, 413.); see also Galileo Galilei: *De motu*. Galilei 1960, 108 (Galilei 1890–1909 I, 335). For the purposes of this article, it is of less interest which aspect of the pendulum was actually studied by Galileo at a certain time than the very fact that he made the pendulum an object of theoretical investigations which, to my knowledge, was never done before.

[18] The trebuchet (leverage artillery), a medieval ballistic device which makes use of a lever to which a counterweight is attached, may well have been inspiring for the heavy pendulum as it occurs with Leonardo and later with Besson since it was still known in early modern times. It is, however, no precursor of those machines with heavy pendulums because, in contrast to the latter, it does not use the reciprocating movement of pendulums. About the origins of the trebuchet, see Huuri 1941; Needham 1976; Hill 1973; Hansen 1992. See also White 1962, 102f.

[19] An Italian edition of Besson's *Theatrum* with the title *Il teatro de gli istrumenti* appeared in 1582. For the machines with heavy pendulums as elements displayed in Besson's *Theatrum* (see Beck 1899, 191ff.). Later, as mathematician at the Florentine court, Galileo had to evaluate a proposition for a machine which apparently suggested the use of a heavy pendulum as a device for "accumulating power" in the way of fly-wheels (see Galilei 1890–1909 VIII, 571–584).

[20] The *pulsilogium*, a pulse-clock that used a pendulum, was first described by the Venetian physician Santorio Santorio in 1602 (*Methodi vitandorum errorum qui in arte medica contingunt*. Venice). Santorio was a friend of Guidobaldo del Monte and acquainted with Galileo who, as

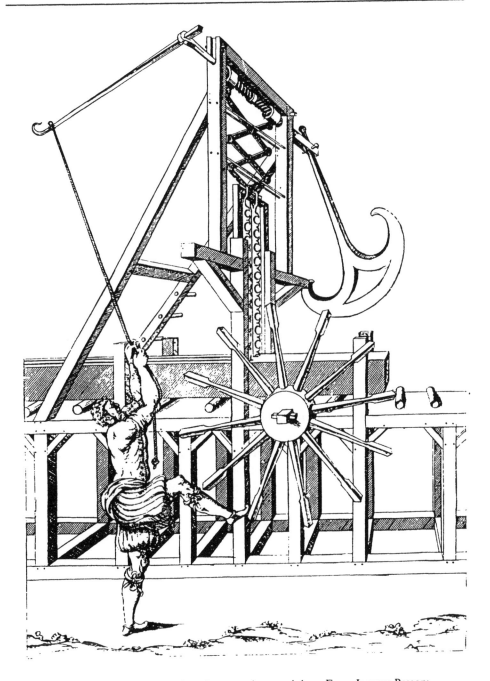

Fig. 1. Sawing-machine with a heavy anchor-pendulum. From Jacques Besson: *Theatrum instrumentorum et machinarum*. Lugdunum 1578, Plate 14. Courtesy Niedersaechsische Staats- und Landesbibliothek Goettingen.

Was that a mere coincidence or was there a systematic connection between the practical utilization of the properties of pendulums by engineers or physicians and the establishment of these characteristics in a theoretical way? Without knowing more about the particulars — for instance, about engineers' concepts of pendulums — one cannot dare to give a definitive answer in this case. It is generally clear, however, that the properties of a device like the pendulum cannot be discovered by mere observation. That is the systematic reason why anecdotes like the one told about Galileo or Newton and the falling apple are implausible — all other faults aside. The properties of objects can only determined by practical experience proving what is possible and what impossible. Compared to observation, it constitutes an entirely different basis for conceptualizations if those properties are already exploited in certain technical arrangements — whatever the practitioners actually think.

What is interesting here is therefore not that the first attempts to frame the pendulum law were undertaken just a few years after the first use of the pendulum as an element of machines. It is not important in the first place how long it had been used practically in devices like the swinging anchors in Besson's machines and not merely as a possible subject of observation like swinging lamps. The devices whose role for modern dynamics can be compared with that of balances for statics in Antiquity need not necessarily be new inventions in the early modern era. As in the case of the inclined plane, even devices which were well-known in Antiquity could be of significance for the beginnings of modern dynamics. What is decisive is rather that they actually were such subjects of practical use and experience. The experiences gained by practical use are not only an obvious prerequisite for developing theories of such devices, but also for detecting questions of theoretical interest that can be studied by means of them. Stressing this significance of technologies for scientific conceptualizations, Lynn White coined the illuminating phrase: "Art has always been a highly selective mirror of nature" (White 1966, 110).[21]

Machines, devices, and techniques using certain material objects have had this significance for modern sciences since the sixteenth and seventeenth century. The role of technology and art deserves special attention at their beginnings, however, when they were still far away from the stage of institutionalized research enterprises with a systematic praxis of experimentation. To a certain extent, this role can be described as compensating for the lack of systematic experimentation. The differ-

---

mentioned above, wrote at the end of the same year a letter to Guidobaldo del Monte about the alleged isochronism of pendulums. Mitchell 1892 and more recently Bedini 1991, 7f., have assumed that Santorio's instrument has to be regarded as a fruit of Galileo's experiments with pendulums. It seems to me no less plausible to assume that Santorio's instrument — which, of course, only made use of the comparative simple relationship between length of the thread and swinging time — was a source of inspiration for Galileo (for Santorio, see Grmek 1977).

[21] Peter Ruben has suggested that the epistemological theory of mirroring, notorious in Lenin's version, might lead to fruitful consequences if it is separated from a sensualistic understanding and if the "mirrors" are conceived of as material entities which serve as material means of thinking, entities which are constructed historically and hence are subject to historical change (see Ruben 1978).

ence between an experiment and a technical operation within the realm of production consists not in a divergence with respect to the devices and materials used but in the divergence of the purposes. As an extreme case, even the same physical procedure with the same technical arrangement can be both: an experiment when it is performed in order to gain knowledge, or an act of production if its goal is the use of the produced effect. Thus, there is no principled impediment why men with theoretical questions could not study technical processes carried out in workshops, mines, or factories as if they were performed to gain insights. The most important role the technical revolution of early modern times and the tradition of the workshops played for the modern scientific mechanics might therefore have consisted in unintentionally creating devices that were suitable for finding and investigating questions in dynamics and that can be regarded as prototypes of experimental arrangements.

Low and High Mechanics

Galileo combined his investigations on bodies descending inclined planes and on the motion of pendulums by attempting to reconstruct the falling arc of the pendulum by approximating its segments through an infinite number of infinitely small inclined planes of decending. In our context it is not of great interest that the result of this endeavor, namely the isochronism of the circular pendulum, proved to be wrong, as Huygens would show. Such combinations themselves deserve attention here. What kind of intelligence enables one to contrive such subtle combinations? How and where does one get the idea of investigating pendulum motion by means of inclined planes? Can that be ascribed alone to the mathematical competence of Galileo or does it testify at the same time to the capacity of a more practical imagination that inventive engineers have at their disposal. Olschki went so far as to suppose that Galileo's scientific achievements cannot be understood properly without taking into account a specific ability which we can perhaps call engineering heuristics. "His [Galileo's] technical genius is the essential prerequisite for the scientific experiments that first shaped the true dimensions of his theoretical originality" (Olschki 1927, 140).

Even if this statement of Olschki seems exaggerated, it gains its significance from the background of the shown role of technologies and art in the early modern era. For without something like "technical genius," Galileo might have been unable to realize the possibilities for theoretical mechanics which the arts of his time provided. The personal abilities of the protagonists, and not least their interest in and their understanding of technical arrangements, constituted an important part of the prerequisites for the realization of these possibilities. They gained their significance, conversely, by a world of mechanics which, due to effective methods of communication, had become a connected space of experience by the sixteenth century, reaching from the activities and skills of practitioners without any intellectual ambitions to educated and exceptionally creative engineers, and even to engineering scientists like Galileo. Here, we only need call to mind well known general factors for this development: the continued economical prosperity of the Northern Italian states; the commerce and trade among them as well as with other parts of Europe; new means of communication such as printed books and the use of the vernacular language in writings on professional and theoretical topics; new institutions of education such as the Accademia del Disegno, and courts as centers of communication between experts in different fields, etc. J. A. Bennett has shown for the world of "mathematics" in seventeenth century England — reaching from practitioners like sea captains or surveyors who only used mathematical instruments to theoretical mathematicians and natural philosophers — that particularly the producers of mathematical instruments functioned as an intermediary center of this world, rendering it a realm of exchange between "low" and "high" mathematics (see Bennett 1986, 1–28). It would be desirable to have an accordingly close investigation of the special functions of the different types inhabiting the world of mechanics in Northern Italy, The Netherlands, and England at the turn of the sixteenth to the seventeenth century.

In this paper, I only want to draw our attention to the existence of this world of mechanics not unified by theory — it was probably not before the nineteenth century that even scientific mechanics itself became a truly coherent theoretical building — but which constituted nevertheless a connected realm of experience, a realm within which the established theories were not more than a few scattered islands. The theoretical achievements of men like Galileo are only properly understandable by recognizing that they were based on the whole realm of mechanics of the time — on the "low" mechanics no less than on the "high." Not only specifically scientific strategies of gaining empirical knowledge like experiments, but those strategies together with experiences and reflections on devices and procedures used in the world of production led to the threshold of classical mechanics.[22]

Nature and Art

The creative appropriation of the classical mechanics tradition as well as the development of the real starting points of modern dynamics in the sixteenth and in the first decades of the seventeenth century was — as I tried to show — the work of men who must be thought of no less as engineers than as scientists. It does not diminish the fame of Galileo when we state that in this respect, no categorical difference can be established between him and Tartaglia, Benedetti, or Guidobaldo del Monte. There are no convincing reasons for calling the one a scientist and the other an engineer. But, of course, we have to distinguish between such theoretically

---

[22] The same was shown for chemistry at the turn from the seventeenth to the eighteenth century by Ursula Klein (see Klein 1994).

committed engineers on the one hand and engineers and practitioners — not to mention common craftsmen — on the other who never thought of solving problems in any other than a pragmatic if not a traditional way. The dialogues in Tartaglia's *Quesiti et inventioni diverse* (1546) are excellent documents for showing the differences, misunderstandings, and the latent or open tensions between men of such a pragmatic bent and theoretically interested ones like Tartaglia (see Olschki 1927, 81f.). The latter do not cease to be engineers only because of their success in treating some mechanical problems in a theoretical manner comparable to that of men like Archimedes in Antiquity.

What makes it difficult to consider men like Galileo engineers is perhaps not primarily the fact that they produced truly scientific writings; the difficulty might lay elsewhere. It could be that we are not used to recognizing that the theories of these men — and *a fortiori* it is valid in the case of theories seen as part of the history of modern mechanics — are theories about technical arrangements and procedures. Moreover, we have to recognize them as theories about technical arrangements that were most significant within the sphere of production — including military — whereas arrangements set up for research purposes, i.e. experimental arrangements, played a more minor role. Galileo's famous *Discorsi* is the classic example in this respect. Used to considering the modern sciences, the sciences based on experimentation, to be physics in the Aristotelian sense of theories about nature as such, we are a little embarrassed when the beginnings of these sciences in the early modern era make it clear that their statements are statements about nature mediated by our technical intervention. In other words, theories that show nature "mirrored" by art.

The truly important replacement of deeply rooted philosophical paradigms that accompanied the birth of modern sciences was not an exchange of an Aristotelian view of nature for a Platonic one. Rather, through the activities of men like Tartaglia or Galileo, the fundamental change consisted of transforming "mechanics," traditionally conceived of as mere knowledge about art, into the paradigm for physics.

References

Archimedes. 1543. *Opera Archimedis ... per Nicolaum Tartaleam ... in luce posita.* Venice.

Beck, Theodor. 1899. *Beiträge zur Geschichte des Maschinenbaues.* Berlin: Julius Springer.

Bedini, Silvio A. 1991. *The Pulse of Time.* Firenze: Leo. S. Olschki.

Bennett, J. A. 1986. "The Mechanics' Philosophy." *History of Science* IV/1:1–28.

Cohen, H. Floris. 1994. *The Scientific Revolution — A Historiographical Inquiry.* Chicago and London: University of Chicago Press.

Damerow, Peter, Gideon Freudenthal, Peter McLaughlin, and Jürgen Renn. 1992. *Exploring the Limits of Preclassical Mechanics. A Study of Conceptual Development in Early Modern Science: Free Fall and Compounded Motion in the Work of Descartes, Galileo, and Beeckman.* New York: Springer.

Damerow, Peter and Wolfgang Lefèvre. 1998. "Wissenssysteme im geschichtlichen Wandel." In *Enzyklopädie der Psychologie* (Themenbereich C — Serie II — Band 6), edited by F. Klix and H. Spada, 77–113. Göttingen: Hogrefe.

Drake, Stillman and I. E. Drabkin. 1969. *Mechanics in Sixteenth-Century Italy.* Madison: University of Wisconsin Press.

Drake, Stillman. 1978. *Galileo at Work: His Scientific Biography.* Chicago: University of Chicago Press.

Feldhaus, F. M. 1970. *Die Technik — Ein Lexikon.* Wiesbaden: Heinz Moos.

Galilei, Galileo. [1890–1909] 1964–66. *Le opere di Galileo; nuova ristampa della edizione nazionale,* edited by Antonio Favaro. Florence: Barbéra.

——. 1960. *On Motion and On Mechanics.* Translated by Stillman Drake and I. E. Drabkin. Madison: University of Wisconsin Press.

——. 1974. *Two New Sciences.* Translated by Stillman Drake. Madison: University of Wisconsin Press.

Grmek, M.D. 1977. "Santorio Santorio." *Dictionary of Scientific Biography* XII:101–103.

Hansen, Peter V. 1992. "Experimental Reconstruction of a Medieval Trébuchet." *Acta Archeologica* LXIII:189–208.

Hill, Donald. 1973. "Trebuchets." *Viator* IV:99–116.

Hoykaas. R. 1963. *Das Verhältnis von Physik und Mechanik in historischer Hinsicht.* Wiesbaden: Franz Steiner.

Huuri, Kalervo. 1941. *Zur Geschichte des mittelalterlichen Geschützwesens aus orientalischen Quellen.* Helsinki.

Klein, Ursula. 1994. *Verbindung und Affinität.* Basel: Birkhäuser.

Knorr, Wilbur Richard. 1982. *Ancient Sources of the Medieval Tradition of Mechanics.* Firenze: Istituto e Museo di Storia della Scienza

Koyré, Alexandre. 1943a. "Galileo and Plato." *Journal of the History of Ideas* IV/4:400–428.

——. 1943b. "Galileo and the Scientific Revolution of the XVIIth Century." *Philosophical Review* LII/4 :333–348.

Lefèvre, Wolfgang. 1978. *Naturtheorie und Produktionsweise*. Darmstadt and Neuwied: Luchterhand.

Mach, Ernst. 1989. *The Science of Mechanics*. La Salle: Open Court.

Masotti, Arnaldo. 1977. "Ostilio Ricci." *Dictionary of Scientific Biography* XI: 405 f.

Mitchell, S. Weir. 1892. "The Early History of Instrumental Precision in Medicine." *Transactions of the Congress of American Physicians and Surgeons.* New Haven: Tuttle, Morehouse and Taylor.

Mittelstraß, Jürgen. 1970. *Neuzeit und Aufklärung. Studien zur Entstehung der neuzeitlichen Wissenschaft und Philosophie.* Berlin and New York: de Gruyter.

Moody, Ernest A., and Marshall Clagett, eds. 1960. *The Medieval Science of Weights.* Madison: University of Wisconsin Press.

Needham, Joseph. 1976. "Chinas trebuchets, manned and counterweighted." In *On Pre-Modern Technology and Science*, edited by B. S. Hall and D. C. West, 107–145. Malibu.

Olschki, Leonardo. 1927. *Galilei und seine Zeit.* Halle: Max Niemeyer.

Rose, Paul Lawrence. 1975. *The Italian Renaissance of Mathematics.* Genève: Librairie Droz.

Ruben, Peter. 1978. "Wissenschaft als allgemeine Arbeit." In Peter Ruben, *Dialektik und Arbeit der Philosophie*, 9–51. Köln: Pahl-Rugenstein.

Schneider, Ivo. 1979. *Archimedes — Ingenieur, Naturwissenschaftler und Mathematiker.* Darmstadt: Wissenschaftliche Buchgesellschaft.

Settle, Thomas B. 1995. "La rete degli esperimenti Galileiani." In P. Bozzi, C. Maccagni, C. Olivieri, and T. B. Settle. *Galileo e la scienza sperimentale*, edited by M. Baldo Ceolin, 11–62. Padova.

Usher, Abbott P. 1988. *A History of Mechanical Inventions.* New York: Dover.

White, Lynn Jr. 1962. *Medieval Technology and Social Change.* Oxford: Oxford University Press.

——. 1966. "Pumps and Pendula: Galileo and Technology." In *Galileo Reappraised*, edited by Carlo L. Golino, 96–110. Berkeley and Los Angeles: University of California Press.

Max Planck Institute for the History of Science

JÜRGEN RENN, PETER DAMEROW, AND SIMONE RIEGER
WITH AN APPENDIX BY DOMENICO GIULINI

Hunting the White Elephant: When and How did Galileo Discover the Law of Fall?

In *The Stolen White Elephant* Mark Twain tells the story of a white elephant, a gift of the King of Siam to Queen Victoria of England, which somehow got lost in New York on its way to England. An impressive army of highly qualified detectives swarmed out over the whole country in search of the lost treasure. And after a short time an abundance of optimistic reports with precise observations were returned by every detective giving evidence that the elephant must shortly before have been at the very place he had chosen for his investigation. Although one elephant could never have been strolling around at the same time at such different places over a vast area, and in spite of the fact that the elephant, wounded by a bullet, lay dead the whole time in the cellar of the police headquarters, the detectives were highly praised by the public for their professional and effective execution of their task.

The Argument

We present a number of findings concerning Galileo's major discoveries which question both the methods and the results of dating his achievements by common historiographic criteria. The dating of Galileo's discoveries is, however, not our primary concern. This paper is intended to contribute to a critical reexamination of the notion of discovery from the point of view of historical epistemology. We claim that the puzzling course of Galileo's discoveries is not an exceptional comedy of errors but rather illustrates the normal way in which scientific progress is achieved. We argue that scientific knowledge generally develops not as a sequence of independent discoveries accumulating to a new body of knowledge but rather as a network of interdependent activities which only as a whole makes the individual steps understandable as meaningful "discoveries."

Introduction

The present paper deals with sources documenting Galileo's work and that of his contemporaries on issues that have always been a focus of studies in the history of science, the discovery of the law of fall and of the parabolic shape of the projectile trajectory, discoveries which mark, according to common understanding, the origins of classical mechanics. In spite of being the subject of more than a century of historical research, the question of when and how Galileo made his major discoveries is, however, still only insufficiently answered. It is generally assumed that he must have found the law of fall around the year 1604 and that he discovered the parabolic shape of the trajectory of projectile motion only several years later. There is, however, no such agreement concerning the question of how he arrived at these achievements. In particular, it is still controversial whether he found these laws primarily by empirical observations or by theoretical speculation.

As far as the date of the discovery of the parabolic trajectory is concerned, we shall show that as early as 1592 Galileo and Guidobaldo del Monte jointly performed an experiment on an inclined plane, which Galileo later in the *Discorsi* described as producing a parabolic trajectory. The law of fall, according to our reconstruction, was merely a trivial consequence of the recognition of the parabolic shape of the trajectory.

We argue further that at first this experiment was not an exciting or dramatic event at all for Galileo. In particular, that the experiment suggested the parabolic shape of the trajectory, implying the law of fall, seems initially not to have been significant to him. However, Galileo did immediately consider it remarkable that the experiment suggested a symmetry of natural and violent motion which had no foundation in the prevailing Aristotelian theory of nature that he largely shared. Even after he had grasped the parabolic shape of the trajectory and its most plausible explanation by the law of fall, he did not realize for a considerable time the further theoretical implications of the experiment. Only much later did he passionately maintain that the discovery of the parabolic shape of the projectile trajectory was his most important breakthrough. And finally in the *Discorsi* he claimed priority for himself in performing this experiment and explaining its outcome, the parabolic shape of the curve.

Galileo's conviction that the shape of the projectile trajectory is parabolic was based on dynamical arguments. These arguments allowed him to infer that the trajectory is curved in the same way as the catenary, although from a modern point of view the arguments must be regarded as fallacious. Furthermore, we shall show that Galileo, when he checked by means of a hanging chain the validity of his claim that the trajectory and the catenary are both parabolic, he arrived at the correct result that the curve of the hanging chain deviates considerably from a parabola. Nevertheless, he stuck to his invalid argument and tried to find an indubitable proof of the parabolic shape of both curves. Finally, he became convinced that he had found such a proof and intended to make it a core topic of the *Fifth Day* of the *Discorsi*. It was only his declining health that prevented him from realizing this plan, and the *Fifth Day* remained unwritten.

The interpretation of clues which seem to indicate a discovery or a new idea presupposes an indisputable answer to two questions. Firstly, what does it mean to say that someone has discovered something? Secondly, how can such a discovery be identified using clues provided by the available sources? We claim that the answer to these questions concerning the nature of discoveries is not at all obvious. On the contrary, as will become clear, premature answers to these questions have led historians of science astray.

As a consequence, it has, first of all, been widely neglected that Galileo saw a close connection between the parabolic trajectory and the catenary, the curve of a hanging chain. Secondly, explicit hints given by Galileo in his publications and his correspondence concerning when and how he became convinced of the parabolic shape of both curves have been disregarded or have been considered as unreliable because they seemingly did not fit into an alleged scheme of his work.

In the following we address the canonical issue of the discovery of the law of fall and the origins of classical mechanics from a different perspective, in which the notion of discovery is not taken for granted but rather itself becomes a subject of reflection. Consequently, sections dealing with details of Galileo's scholarly activities will be complemented by sections which either investigate these activities in the broader social and historical context, such as the challenges of technology to engineer-scientists, or pursue questions of historical epistemology, such as the question of whether experiments or theoretical reflections were the foundation of Galileo's judgments, beliefs, and research strategies.

The paper begins with a brief account of the canonical viewpoint of Galileo's major discoveries (*The Standard Dating of the Discovery of the Law of Fall and of the Parabolic Trajectory*) and then turns to an issue neglected in the standard accounts, Galileo's preoccupation with the catenary (*The Neglected Issue: Trajectory and Hanging Chain*). Two sections are devoted to the evidence that testifies to the significance of this neglected issue throughout Galileo's life (*Evidence I: Galileo Using Hanging Chains* and *Evidence II: Viviani's Addition to the Discorsi and Guidobaldo's Protocol of an Experiment*). The role of a particular theoretical context for Galileo's interpretation of his findings is analyzed in the following two sections (*How Can the Aristotelian View of Projectile Motion Account for a Symmetrical Trajectory?* and *Decomposing the Trajectory—Neutral Motion and the Law of Fall*). The next section returns to the historiographical aspect of dating Galileo's discoveries and questions the standard dating without, however, reducing it to the alternative attribution of a different date to the discovery of the parabolic shape of the projectile trajectory and of the law of fall (*Dating Guidobaldo's Protocol*). In order to illuminate Galileo's own perspective on his discoveries, the next three sections discuss the impact of the changing contexts of Galileo's work, from the scholarly environment of Pisan philosophy and mathematics via the intellectual challenges of early modern engineering technology, to the open horizons of the intellectual elites of Venice, one of the industrial and commercial centers of the early modern world (*Guidobaldo del Monte as an Engineer-Scientist, Galileo in the Footsteps of Guidobaldo del Monte*, and *Galileo and Paolo Sarpi—Towards a New Science of Motion*). Having developed a complex picture of the contexts of Galileo's discoveries and his understanding of them, the paper turns to the question of the mutual influence of theory and experiment which led to his final research project on the shape of the trajectory and the law of fall (*Did Galileo Trust the Dynamical Argument?—Galileo the Experimenter* and *Did Galileo Trust his Experiments?—Galileo's First Attempt of a Proof*). The next section elaborates on the observation of how unsatisfactorily open an individual scientific biography usually ends, in the case of Galileo's biography in an unpublished proof of an erroneous theorem representing what he considered to be the keystone of his new theory of motion (*Returning to the Dynamical Argument—The Final Proof*). The final section turns again from the historiographic perspective to the broader issue of the paper (*When and How Did Galileo Discover the Law of Fall?*). In a first appendix four letters documenting the changing contexts of Galileo's work are presented in English translation. In a second appendix Galileo's claims concerning his discoveries are reexamined from the perspective of modern physics.

The Standard Dating of the Discovery of the Law of Fall and of the Parabolic Trajectory

The question of when and how Galileo made his celebrated discoveries of the law of fall and of the parabolic shape of the projectile trajectory has been extensively discussed in the last one hundred years by historians of science. Contrary to the testimony of Galileo's disciple Viviani,[1] who ascribed such discoveries already to the young Galileo, it is widely accepted today that these discoveries date into the late Paduan period. Most writers date the discovery of the law of fall to the year 1604 and assume that Galileo discovered the parabolic shape of the projectile trajectory even some years later. In his influential *Galileo Studies*, Alexandre Koyré lapidarily affirmed:

> The law of falling bodies—the first law of classical physics—was formulated by Galileo in 1604. (Koyré 1966, 83)

This dating is primarily based on the few contemporary documents by Galileo himself which provide clues to his knowledge of the law of fall and the form of the trajectory. In particular, two letters by Galileo stand out because of the testimony they offer to his knowledge at certain precise dates. The first letter is directed to Paolo Sarpi and dated 16 October 1604 (Galilei 1890-1909, X: 116); it provides clear evidence that, at this point in time, Galileo knew the law of fall. The second letter is directed to Antonio dei Medici and dated 11 February 1609 (Galilei 1890-1909, X: 228). It shows that Galileo, by that time, knew that projectiles reaching the same height take the same time to fall down, a property that fol-

---

[1] According to Viviani, Galileo discovered the isochronism of the pendulum already as a student in Pisa around 1583; see Viviani's letter to Leopoldo dei Medici, Galilei 1890-1909, XIX: 648; see also his biography, Galilei 1890-1909, XIX: 603. He claims furthermore that Galileo performed experiments on free fall already between 1589 and 1592 when he was professor in Pisa; see Viviani's biography of Galileo, Galilei 1890-1909, XIX: 606.

lows, from a modern point of view, from the decomposition of the parabolic trajectory into its horizontally uniform and vertically accelerated components.

Of course, these documents provide at most a *terminus ante quem* for Galileo's discoveries. The hesitation of modern Galileo scholars to follow Viviani's testimony and to accept an earlier date is, partly at least, due to the study of Galileo's early manuscript, presumably kept in a folder labelled *de motu antiquiora scripta mea* in the following simply referred to as *De Motu*, which in spite of its anti-Aristotelian tendency shows him as deeply influenced by the categories and assumptions of medieval Aristotelian physics.

The cornerstones 1604 and 1609 define a scaffolding for more or less speculative stories about what really happened at those times. It is, indeed, customary to find these dates in reconstructions of the sequence of Galileo's discoveries, even though the claims of what made up these discoveries widely diverge among different authors – some even maintain that the law of fall was only discovered in 1609 and the parabolic trajectory perhaps even later. As an example we quote the succinct reconstruction of the sequence of Galileo's discoveries by the influential German historian of science Friedrich Klemm:

> Attempts showing him that the difference in the speed of fall of bodies of the same size with different specific weights becomes smaller the more the medium is diluted, suggested to Galileo around 1600 to assume that the speed of fall in vacuum is going to be of equal magnitude for all bodies. (...)
>
> The next step is now to give up all considerations about the cause of the growing speed of fall and to limit himself to treating mathematically the motion of fall taking place for all bodies in a vacuum in the same way. In 1604 Galileo makes the assumption, convinced that in nature everything is constituted as simply as possible: the speed of fall in vacuum increases with the distance of fall traversed. This approach leads him into contradictions. Finally in 1609 he comes to recognize the process of fall as a *uniformly* accelerated motion, that is, he comes to the insight that the velocity grows with time. Starting from here he obtains by using developments of the late middle ages: the distances are to each other as the squares of the times. This again gives him the possibility to verify the hypothetical approach $v = a \cdot t$ by the experiment (the fall trench experiment). (Klemm 1964, 79-80)

This reconstruction is obviously guided by the supposition that Galileo proceeded methodically and step-by-step from one discovery to the next. Naturally, such a view of Galileo's progress must remain speculative as long as it is not supported by a detailed analysis of contemporary sources.

Fortunately, an extensive collection of Galileo's notes on his research on mechanics has survived and is now kept in the Galilean Collection of the National Library in Florence, the so-called *Codex Ms. Gal. 72*. However, these notes are neither dated nor preserved in their original order of composition. The most influ-

ential interpretations of these notes are those going back to the publications of
Stillman Drake who devoted a great deal of his research to the study of these
notes. Drake derived his interpretations in particular from his attempts at a chronological ordering of the pages of that manuscript. He writes:

> Arranged in their order of composition and considered together with theo-
> rems found on other pages or in the text of *Two New Sciences*, those notes
> tell a story of the origin of modern physical science. It is not the story on
> which historians of science were generally agreed in 1974, nor did I then
> foresee the extent to which that would in time be modified by Galileo's
> working papers. (Galilei 1989, vii, part from the introduction to the second
> edition of Drake's translation of the *Two New Sciences*)

Drake's claim that his story differs from the story on which historians of science
were generally agreed may be true for his reconstructions of the process of Galileo's discoveries which have changed over time. Contrary to such claims, however, the dates which he derived for the major steps do not differ very much from
what has been commonly accepted. Elsewhere he writes:

> From the beginning of his professional career, Galileo's main interest was in
> problems of motion and of mechanics. His first treatise on motion (pre-
> served in manuscript but left unpublished by him) belongs probably to the
> year 1590, when he held the chair of mathematics at the University of Pisa.
> It was followed by a treatise on mechanics begun at the University of Padua
> in 1593 and probably put into its posthumously published form around
> 1600. Galileo made some interesting discoveries concerning fall along arcs
> and chords of vertical circles in 1602, and two years later he hit upon the law
> of acceleration in free fall. With the aid of this law he developed a large
> number of theorems concerning motion along inclined planes, mainly in
> 1607-8. Toward the end of 1608 he confirmed by ingenious and precise
> experiments an idea he had long held: that horizontal motion would con-
> tinue uniformly in the absence of external resistance. These experiments led
> him at once to the parabolic trajectory of projectiles. Meanwhile he had
> been at work on a theory of the breaking strength of beams, which seems
> also to have been virtually completed in 1608. (Galilei 1974, ix)[1]

But even after Drake had extensively studied Galileo's working papers, and
after repeatedly changing his views on Galileo's discoveries, he accommodated
the dating of Galileo's manuscripts to the standard dating rather than the other

---

[1]   Drake explains how he himself came to three different versions of the process of Galileo's
discovery of the law of fall. The third version he finally adopted in this article was not the last one.
His very last one is contained in the preface to his edition of the Discorsi. This version is essentially
based on a fancyful interpretation of folio 189 of Ms. Gal. 72 (Drake 1973). We shall publish elsewhere a reconstruction of this folio which will show that Drake's interpretation of this folio cannot
be maintained.

way around. In his latest book, Drake gives a detailed time-table of Galileo's activities and achievements (Drake 1990, XIIIf); according to this time-table, in 1602 Galileo "begins studies of long pendulums and motion on inclines", in 1604 he "discovers the law of pendulums from careful timings; finds the law of fall," and only in 1608 he "discovers the parabolic trajectory by measurements."

It thus emerges as a peculiarity of recent Galileo historiography that the dates at which Galileo supposedly made his major discoveries have remained largely unchallenged, in spite of the relatively weak direct evidence available for this dating. This peculiarity is all the more surprising as the dating and the sequence of Galileo's discoveries have been extensively and controversially discussed in the older Galileo literature around the turn of the century. The possibility that Galileo first discovered the parabolic form of the trajectory and only then the law of fall was, for instance, seriously considered by Emil Wohlwill, together with the possibility that the law of fall was discovered much earlier than is now commonly assumed (Wohlwill 1993, I: 144-162 and Wohlwill 1899).[1] But Emil Wohlwill's substantial contributions had as little impact on the Galileo studies of the last fifty years as those of his eminent Italian contemporaries, Antonio Favaro and Raffaello Caverni. As we will see in the following, there are good reasons to take up the debate where it was left a century ago.

The Neglected Issue: Trajectory and Hanging Chain

Concerning the sources from which Galileo derived his major discoveries there is much less agreement among recent historians of science than concerning the dating. The assumptions about his sources range in fact from pure empirical evidence achieved exclusively by means of careful experimentation and precise measurements, on the one hand, to predominantly theoretical speculation in direct continuation of scholastic traditions only scarcely supported by empirical demonstrations, on the other hand.[2]

In spite of the wide range of different reconstructions of the discovery process,[3] however, a simple fact has nearly been completely neglected both by the older

---

[1] In the latter article Wohlwill still argues that the law of fall was discovered shortly before he wrote his letter to Paolo Sarpi in 1604 in which the law is explicitly mentioned. In the first volume of his final work on Galileo published shortly before his death, however, he developed an ingenious argument for a quite different dating. Based on the sources available at that time and, in particular, based on the first publication of excerpts from Galileo's notes by Caverni and Favaro, he developed a reconstruction of Galileo's discovery which he qualified as "only the most probable" way of how Galileo might have found the law of fall; this reconstruction is essentially coherent with what is presented in the following. The new evidence provided here shows that Wohlwill's conjecture was much more sound than the interpretations which are currently *en vogue*.

[2] In discussions, these positions are still represented prototypically by Drake, on the one side, and Koyré, on the other side.

[3] For the literature on this subject, see the references in section 3.3.1 of Damerow, Freudenthal, McLaughlin, and Renn 1992; see in particular, the ingenious reconstruction of Galileo's inclined plane experiment, described in Settle 1961.

and the more recent literature: for Galileo, there exists a close connection between the parabolic trajectory and the catenary, that is, the curve of hanging chains. This neglect is all the more astonishing as the connection is explicitly made a subject of discussion in his final word on the matter, the *Discorsi*. In the course of the discussions of the *Second Day*, Galileo's spokesman Salviati describes two methods of drawing a parabola, one of them involving the trajectory of a projected body, the other one using a hanging chain:

> There are many ways of drawing such lines, of which two are speedier than the rest; I shall tell these to you. One is really marvelous, for by this method, in less time than someone else can draw finely with a compass on paper four or six circles of different sizes, I can draw thirty or forty parabolic lines no less fine, exact, and neat than the circumferences of those circles. I use an exquisitely round bronze ball, no larger than a nut; this is rolled [*tirata*] on a metal mirror held not vertically but somewhat tilted, so that the ball in motion runs over it and presses it lightly. In moving, it leaves a parabolic line, very thin, and smoothly traced. This [parabola] will be wider or narrower, according as the ball is rolled higher or lower. From this, we have clear and sensible experience that the motion of projectiles is made along parabolic lines, an effect first observed by our friend, who also gives a demonstration of it. We shall all see this in his book on motion at the next [*primo*] meeting. To describe parabolas in this way, the ball must be somewhat warmed and moistened by manipulating it in the hand, so that the traces it will leave shall be more apparent on the mirror.

> The other way to draw on the prism the line we seek is to fix two nails in a wall in a horizontal line, separated by double the width of the rectangle in which we wish to draw the semiparabola. From these two nails hang a fine chain, of such length that its curve [*sacca*] will extend over the length of the prism. This chain curves in a parabolic shape, so that if we mark points on the wall along the path of the chain, we shall have drawn a full parabola. By means of a perpendicular hung from the center between the two nails, this will be divided into equal parts. (Galilei 1974, 142f)

At a prominent place, the end of the *Fourth Day* of the *Discorsi*, Galileo returns to what he considered a surprising fact, the parabolic shape of the catenary:

> *Salviati.* (...) But I wish to cause you wonder and delight together by telling you that the cord thus hung, whether much or little stretched, bends in a line that is very close to parabolic. The similarity is so great that if you draw a parabolic line in a vertical plane surface but upside down—that is, with the vertex down and the base parallel to the horizontal—and then hang a little chain from the extremities of the base of the parabola thus drawn, you will see by slackening the little chain now more and now less, that it curves and

adapts itself to the parabola; and the agreement will be the closer, the less curved and the more extended the parabola drawn shall be. In parabolas described with an elevation of less than 45°, the chain will go almost exactly along the parabola.

Sagredo. Then with a chain wrought very fine, one might speedily mark out many parabolic lines on a plane surface.

Salviati. That can be done, and with no little utility, as I am about to tell you. (Galilei 1974, 256)

However, before Salviati can fulfill his promise of further explanations, he is interrupted by Simplicio who asked for a demonstration of the impossibility to stretch a rope or chain into a perfectly straight line before he continues. Unfortunately, we do not get the end of the story. About three pages later the *Discorsi* end abruptly because, as we know and will further discuss later, Galileo was not able to finish his book as planned. We are asked to wait for the unwritten part when Simplicio, satisfied with the answer, tried to bring up the issue again:

Simplicio. I am fully satisfied. And now Salviati, in agreement with his promise, shall explain to us the utility that may be drawn from the little chain, and afterward give us those speculations made by our Author about the force of impact.

Salviati. Sufficient to this day is our having occupied ourselves in the contemplations now finished. The time is rather late, and will not, by a large margin, allow us to explain the matters you mention; so let us defer that meeting to another and more suitable time. (Galilei 1974, 259)

The question of what Galileo through his spokesman Salviati intended to say about the hanging chain deserves attention not only for the sake of the curiosity of the issue but also for a systematic reason. Galileo's claim that the curve of a hanging rope or chain comes arbitrarily close to a parabola is obviously wrong. A correct mathematical representation of the catenary presupposes, as we know today, the knowledge of hyperbolic functions.[1] Galileo and his contemporaries had no chance to derive, and not even to mathematically describe the shape that a hanging chain assumes. In fact, the actual catenary even deviates considerably from a parabola as long as the distance between the two suspension points of the chain is not much greater than the vertical distance between the suspension points

---

[1] The catenary is represented by the equation

$$y = h \cdot \cosh\frac{x}{h}$$

with a parameter h depending on the horizontal tension (which is constant for a specific catenary) and on the derivative of the volume of the chain or rope considered as a function of its length (a ratio which is also assumed to be constant over the length of the chain or rope). For an extensive discussion of the catenary see Appendix B.

and the lowest point of the hanging chain. But if the distance between the two suspension points substantially exceeds the vertical distance between the suspension points and the lowest point of the chain, the parabola is a reasonable approximation of the catenary.

Was it this approximation that Galileo had in mind when he identified the catenary with the parabola? There is overwhelming evidence that this is not the case. In the finished parts of the *Discorsi* he clearly pointed out that he assumed not a fortuitous but a substantial relation between the parabolic trajectory and the catenary.

> *Salviati.* Well, Sagredo, in this matter of the rope, you may cease to marvel at the strangeness of the effect, since you have a proof of it; and if we consider well, perhaps we shall find some relation between this event of the rope and that of the projectile [fired horizontally].

> The curvature of the line of the horizontal projectile seems to derive from two forces, of which one (that of the projector) drives it horizontally, while the other (that of its own heaviness) draws it straight down. In drawing the rope, there is [likewise] the force of that which pulls it horizontally, and also that of the weight of the rope itself, which naturally inclines it downward. So these two kinds of events are very similar. (Galilei 1974, 256)

This supposed theoretical relationship between the catenary and the trajectory will be extensively discussed in the following. Previously, however, another question has to be answered: did Galileo really produce, as he claimed, parabolic curves by means of projected bronze balls and hanging chains or did he merely invent these stories in order to give his idea of a theoretical connection between catenary and trajectory a lively illustration?

Evidence I: Galileo Using Hanging Chains

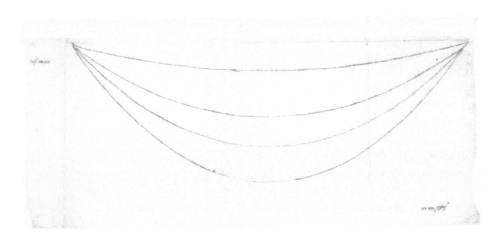

Figure 1. Ms. Gal. 72, folio 41/42 with curves produced by means of a hanging chain

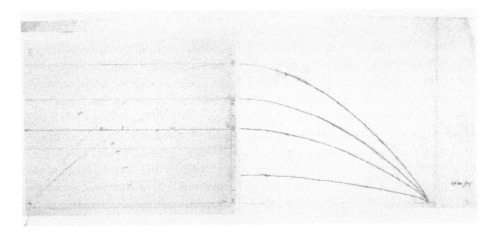

Figure 2. Ms. Gal. 72, folio 41/42 (folio in the back) used as a template for drawing projectile trajectories on folio 113 (folio in front)

Fortunately, this question can easily be answered in the case of the first method of drawing parabolic lines he describes. Among Galileo's notes on motion, Ms. Gal. 72, there is a folded sheet of rough paper designated as folio 41/42 (see figure 1) that has obviously been used for drawing catenaries as Galileo described it in the *Discorsi*. The sheet was fixed to the wall by means of two nails; the holes in the sheet of paper, through which the nails were driven, are still visible.[1] Chains of different length where fixed at these two nails and their shape was copied to the paper by means of some needles. Finally, using this perforated sheet, the resulting curves were copied by letting ink seep through the little holes pierced into the paper along the hanging chains.

Another folio page, 113 recto, shows a drawing containing curves which represent projectile trajectories of oblique gun shots projected under various angles. The curves consist of ink dots which are joined by faint lines. In addition to these representations of trajectories, the folio page contains several drawn or scratched auxiliary lines, such as straight lines representing the directions of the shots or the levels of their maximum heights.[2]

A comparison of the curves representing the trajectories has shown that they fit precisely the template represented by folio 41/42 (see figure 2). Thus folio 113 is a preserved example of the application of the very technique of drawing supposedly parabolic curves by means of a hanging chain which Galileo describes in the *Discorsi*.

Evidence II: Viviani's Addition to the *Discorsi* and Guidobaldo's Protocol of an Experiment

In the case of the second method for drawing parabolas which Galileo in the *Discorsi* claims to have been using, that is, the generation of parabolas by rolling a ball over an inclined mirror, the evidence of the truth of his claim is somewhat more indirect. Galileo's copy of the first edition of the *Discorsi* contains numerous corrections, notes, and additions mostly by the hand of his disciple Vincenzio Viviani. These notes were probably added in part still by Galileo himself for a revised edition, but apparently written by Viviani (and possibly by other disciples) because of Galileo's progressive blindness during the last years of his life. Unfortunately, Galileo's *Discorsi* were never published with the revisions according to these notes, most likely because it is impossible to distinguish which of

---

[1] The distance between the two suspension points is 443 mm. According to two notes, one on the left and the other on the right side of the curves, "total amplitude 465," this distance was measured by Galileo as 465, probably indicating 465 "points." Thus, the size of the unit used by Galileo is 0.95 mm.

[2] Several uninked construction lines can be found on the folio page 113r. An analysis of these lines has provided evidence that a basic unit of exactly the same size was used on this folio as on folio 41/42. A set of parallel lines can be identified which are drawn vertically to the baseline of the parabolic trajectory in equal distances of precisely 15 "points" measured in the basic unit of folio 41/42. The total distance measured along the baseline from the origin of the shots to the vertical representing the target is divided by the parallel lines precisely into 16 parts of 15 "points" each. Furthermore, an arc of a circle with a radius of 100 "points" can also be identified. At the vertical target line a number "100" is written exactly 100 "points" above the base line. But not only the small unit is common to both folios. Assuming that the distance between the parallel lines defines a higher unit of 15 "points," the "total amplitude 465" on the template folio 41/42 measures precisely 31 of these units.

these notes were authorized by Galileo and which of them were inserted by Viviani on his own account only after Galileo's death. In any case, these notes provide striking insights into consequences of Galileo's *Discorsi* which were either implicit but insufficiently expressed in the printed text of the first edition or could be achieved immediately by an elaboration of his practical and theoretical achievements.

With regard to the present problem of whether or not Galileo really used the second method of drawing parabolas, the inspection of his copy of the first edition of the *Discorsi* provided a surprise. On a sheet of paper[1] inserted close to Galileo's description of the second method for drawing parabolas one finds two curves which show the typical characteristics of such a method: the indications of the bouncing of the ball at the beginning and the slight deformation of the parabola at the end due to friction (see figure 3).[2] If one could be certain that it was Galileo himself who produced these parabolas, it would be thus clear that the description in the *Discorsi* refers to an experiment he had actually performed himself.

The claim that Galileo used this method receives strong confirmation by the analysis of another manuscript, although this manuscript is definitely not written by Galileo himself but by Guidobaldo del Monte, a correspondent, benefactor, and a close associate of Galileo in his early research on mechanics. This document nevertheless provides additional evidence and even allows the conclusion, as we will show, that Galileo was familiar with this method of drawing parabolas already long before he wrote the *Discorsi*.

At the end of a notebook of Guidobaldo there are two drawings which possibly depict an inclined plane used for such an experiment, together with a protocol which perfectly resembles by the description of Galileo's second method mentioned in the *Discorsi* (see figures 4 and 5). A closer inspection of Guidobaldo's drawings shows that they actually represent a roof which may have offered a convenient setting ready-at-hand for originally trying out a method similar to that described by Galileo on a scale comparable to that of ballistics, the usual context in which projectile motion was considered at that time.[3]

---

[1]   Folio page 90v of Galileo's copy of the *Discorsi*, Galilei after 1638.
[2]   This judgment is based on a careful repetition of the experiment under controlled conditions with the help of equipment designed by Henrik Haak and realized by the workshop of the Fritz Haber Institute of the Max Planck Society. Henrik Haak has furthermore assisted us in the reproduction of the historical experiment; the results will be the subject of a forthcoming publication.
[3]   Henrik Haak, who constructed the apparatus for our reproduction of the historical experiment, has directed our attention to the fact that the inclined planes depicted by Guidobaldo immediately before and almost immediately behind the protocol represent a roof construction.

Figure 3. Sheet of paper found in Galileo's copy of the first edition of the Discorsi (Ms. Gal. 79, folio page 90 verso) containing two parabolic curves generated by an inked ball thrown along an inclined plane, inserted near to the place where this method is described

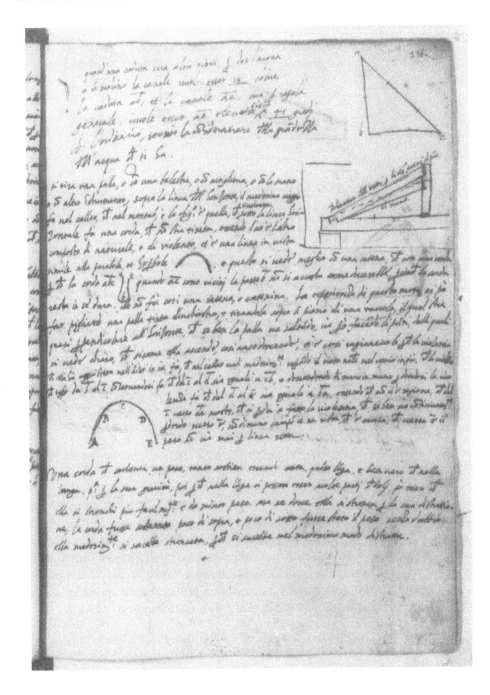

Figure 4. Guidobaldo's protocol of the projectile trajectory experiment

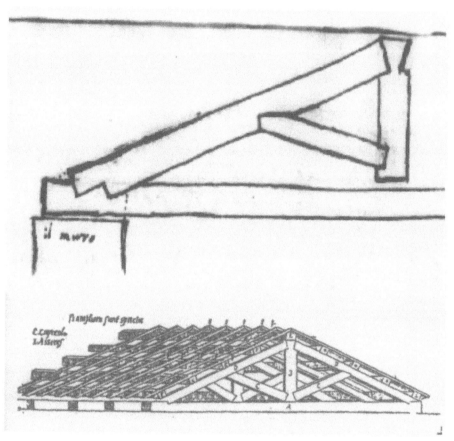

Figure 5. Guidobaldo's sketch (del Monte ca. 1587-1592, 237) and a contemporary representation
of a roof with "capriata" (illustration by Palladio, reproduced from Tampone 1996, 71)

Guidobaldo's protocol not only describes precisely the experimental setting but
also reports results, such as the symmetry of the generated curve and the close
relation to the curve of a hanging chain, that can be deduced from the observation
that the curves in both cases result from the same configuration of forces. The
protocol begins with a summary of consequences that can be drawn from the outcome of the experiment, followed by a description of the method applied. It ends
with a theoretical interpretation of the symmetry of the trajectory:

> If one throws a ball with a catapult or with artillery or by hand or by some
> other instrument above the horizontal line, it will take the same path in fall-
> ing as in rising, and the shape is that which, when inverted under the horizon, a
> rope makes which is not pulled, both being composed of the natural and the
> forced, and it is a line which in appearance is similar to a parabola and
>
> hyperbola ⌒ . And this can be seen better with a chain than with a

rope, since [in the case of] the rope abc  , when ac are close to each other, the part b does not approach as it should because the rope remains hard in itself, while a chain or a little chain does not behave in this way. The experiment of this movement can be made by taking a ball colored with ink, and throwing it over a plane of a table which is almost perpendicular to the horizontal.

Although the ball bounces along, yet it makes points as it goes, from which one can clearly see that as it rises so it descends, and it is reasonable this way, since the violence it has acquired in its ascent operates so that in falling it overcomes, in the same way, the natural movement in coming down so that the violence that overcame [the path] from b to c, conserving itself, operates so that from c to d [the path] is equal to cb, and the violence which is gradually lessening when descending operates so that from d to e [the path] is equal to ba, since there is no reason from c towards de that shows that the violence is lost at all, which, although it lessens continually towards e, yet there remains a sufficient amount of it, which is the cause that the weight never travels in a straight line towards e.

The similarity of this protocol of Guidobaldo's experiment and Galileo's description of his second method to draw parabolas raises, of course, the question of what relation exists between these two reports. Did Guidobaldo and Galileo independently make the same observation? If not, who of them did the experiment and who only heard of or reproduced it? We will show in the following that not only are both referring to the same experiment, but that, moreover, Galileo was even present when this experiment was performed. First, however, the issue has to be analyzed in some more detail. (del Monte ca. 1587-1592, 236)[1]

---

[1] A transcription of the text has been first published by Libri 1838, IV: 397f. Its significance for dating Galileo's early work on motion was first recognized by Fredette 1969. The experiment described by Guidobaldo has been extensively discussed in Naylor 1974. Naylor concludes that Galileo could not have been convinced by the outcome of this experiment alone of the parabolic shape of the trajectory and that it was only in 1607 that he arrived at this conviction. Naylor thus agrees with the standard dating of this discovery, a conclusion that we will attempt to refute in the following. The crucial significance of the experiment for Galileo's discovery of the law of fall was first suggested by Damerow, Freudenthal, McLaughlin, and Renn 1992, 336f.

How Can the Aristotelian View of Projectile Motion Account for a Symmetrical Trajectory?

When has the technique of tracing the trajectory of a ball, which both Galileo and Guidobaldo described, been developed? A first clue to the answer to this question is provided by the outcome of Guidobaldo's experiment itself. This outcome, as it is reported in Guidobaldo's note, was in one respect incompatible with the view of projectile motion prevailing at the time of the young Galileo. In the Aristotelian tradition, projectile motion was conceived of as resulting from the contrariety of natural and violent motion, the latter according to medieval tradition acting through an impetus impressed by the mover into the moving body. According to this understanding of projectile motion, the trajectory cannot be symmetrical because the motion of the projectile is determined at the beginning and at the end by quite different causes. At the beginning it is dominated by the impetus impressed into the projectile, at the end by its natural motion towards the center of the earth.

At the time of Galileo, this tradition was primarily represented by Tartaglia's systematic treatise on artillery, his *Nova Scientia* published in 1537 (Tartaglia 1984). In this treatise, he struggled with the problem that Aristotelian dynamics could not be satisfactorily brought into accordance with the knowledge of the practitioners on projectile motion at that time. The Aristotelian distinction between natural and violent motion seemed to be promising as part of an axiomatic foundation of a theory of projectile motion, perfectly represented by the axiomatic exposition in the first book of Tartaglia's *Nova Scientia*. On the other hand, this foundation did not even provide a definite answer to such a simple question as whether or not the projectile trajectory has any straight part. Tartaglia knew very well that any shot systematically deviated from the target in the line of vision and he was perfectly able to explain this phenomenon:

> Truly no violent trajectory or motion of uniformly heavy bodies outside the perpendicular of the horizon can have any part that is perfectly straight, because of the weight residing in that body, which continually acts on it and draws it towards the center of the world. (Drake and Drabkin 1969, 84)[1]

But in contradiction to this assumption he also assumed:

> Every violent trajectory or motion of uniformly heavy bodies outside the perpendicular of the horizon will always be partly straight and partly curved,

---

[1] Tartaglia is already in his *Nova Scientia* very explicit in this point (see the footnote by the editors, see also Damerow, Freudenthal, McLaughlin, and Renn 1992, 144f). Nevertheless, this aspect of Tartaglia's theory is still widely neglected. It seemingly does not fit into the simple scheme of historical explanation which associates Aristotelian dynamics with a preclassical conception of the projectile trajectory and classical physics with its parabolic shape. For a recent example see, for instance, the monograph on Tartaglia by Arend 1998, chapter 4, in particular 174f.

and the curved part will form part of the circumference of a circle. (Drake and Drabkin 1969, 84)

According to Tartaglia's theory, the trajectory of a projectile consists of three parts. It begins with a straight part that is followed by a section of a circle and then ending in a straight vertical line (see figure 6). This form of the trajectory also corresponds to Tartaglia's adaptation of the Aristotelian dynamics to projectile motion in the case of artillery, a case that was, of course, much more complicated than what was traditionally treated in Aristotelian physics.

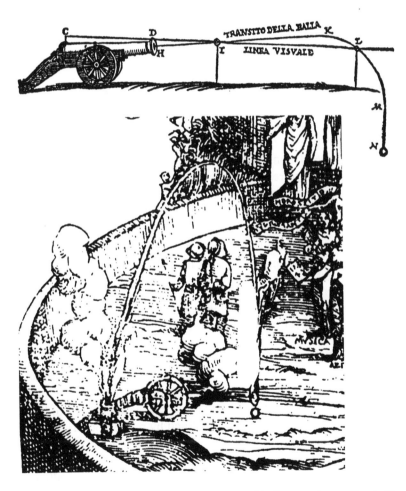

Figure 6. Tartaglia's projectile trajectories according to his theory and according to his practical experience[1]

---

[1] The first figure is taken from N. Tartaglia 1959, the second one from S. Drake and I. E. Drabkin 1969.

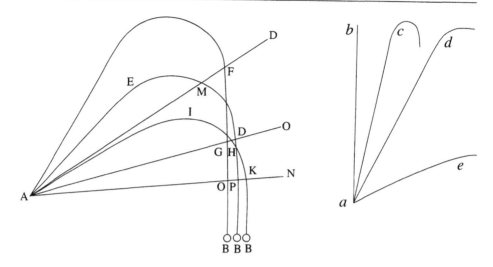

Figure 7. Comparison of Tartaglia's construction of projectile trajectories (left) with Galileo's figure in *De Motu* (right)[1]

The first part of the trajectory was conceived by Tartaglia as reflecting the initially dominant role of the violent motion, whereas the last straight part is in
accord with the eventual dominance of the projectile's weight over the violent
motion and the tendency to reach the center of the earth. The curved middle part
might have been conceived of as a mixed motion compounded of both violent
motion in the original direction and natural motion vertical downward. But, since
violent and natural motion were supposed to be contrary to each other, this conclusion appeared to be impossible to Tartaglia. He claimed instead the curved part
to be exclusively due to violent motion as is the first straight part of the trajectory.
He proved the proposition:

> No uniformly heavy body can go through any interval of time or of space
> with mixed natural and violent motion. (Drake and Drabkin 1969, 80)

Tartaglia's construction of the trajectory was influential throughout the sixteenth
century, although it could not be brought into agreement with the simple explanation for the obvious fact that non-vertical projection is never perfectly straight,
and, in addition, corresponds only roughly to the visual impression of the motion
of projectiles, and certainly could not be justified by precise observations of their
trajectory. The simple shape of Tartaglia's trajectory, however, immediately
allows one to draw a number of conclusions about projectile motion by geometrical reasoning that seemed to be theoretically convincing and practically useful.

It is well known that Galileo originally also adhered to this theory. In his early
manuscript *De Motu*, written about 1590, he contributed to this theory by proving

---

[1] The first figure is taken from Tartaglia 1984, 11. The second figure has been redrawn on the
basis of a microfilm reproduction of the original manuscript.

at the end of his treatise the proposition that objects projected by the same force move farther on a straight line the less acute are the angles they make with the plane of the horizon. At that time, he obviously had no doubt that the traditional view of a straight beginning of the trajectory was correct, adding his own contribution to further developing this theory (see figure 7). He tried to explain the different lengths of the straight parts of the trajectories of bodies projected under different angles by arguing that different amounts of force are impressed into the body according to the different resistances if the angle of projection is varied. In the course of the proof of this proposition, however, he developed Tartaglia's theory further in the direction already taken by Tartaglia himself. In Tartaglia's later publication, the *Quesiti*, he again, possibly under the pressure of Cardano's criticism of his claim that natural and forced motion cannot act simultaneously,[1] stated even more clearly than in his *Nova Scientia* that the trajectory is in no part perfectly straight. Galileo elaborated the theoretical explanation for the curvature of the trajectory given by Tartaglia even further. This theoretical explanation brought him into conflict with an argument he had developed earlier in order to explain acceleration in free fall. Galileo argued that only in the case of vertical projection violent and natural motion due to their contrariety cannot act together at the same time, whereas in the case of oblique projection the trajectory may well be explained by the simultaneous effects of both.

> When a ball is sent up perpendicularly to the horizon, it cannot turn from that course and make its way back over the same straight line, as it must, unless the quality that impels it upward has first disappeared entirely. But this does not happen when the ball is sent up on a line inclined to the horizon. For in that case it is not necessary for the [impressed] projecting force to be entirely used up when the ball begins to be deflected from the straight line. For it is enough that the impetus that impels the body by force keeps it from [returning to] its original point of departure. And this it can accomplish so long as the body moves on a line inclined to the horizon, even though it may be only a little inclined [from the perpendicular] in its motion. For at the time when the ball begins to turn down [from the straight line], its motion is not contrary to the [original] motion in a straight line; and, therefore, the body can change over to the [new] motion without the complete disappearance of the impelling force. But this cannot happen while the body is moving perpendicularly upward, because the line of the downward path is the same as the line of the forced motion. Therefore, whenever in its downward course, the body does not move toward the place from which it was projected by the impressed force, that force permits it to turn downward. For it is sufficient for that force that it keeps the body from returning to the point from which it departed. (Galilei 1960, 113)

---

[1] See Drake and Drabkin 1969, 100-104. On Cardano's criticism see Arend 1998.

The assumption that the curved part of the trajectory of a body projected obliquely may be compounded of violent and natural motion at the same time immediately raises the question of what ratio they might have. Galileo obviously felt it necessary to assume in the case of vertical projection that the impressed force had to disappear entirely before the projected body can turn downward. This, however, contradicts the theory of vertical projection which he developed in the context of his discussion of acceleration of free fall and according to which, at the turning point of a projection directed upwards, impressed force and weight of the body equilibrate each other:

> For a heavy body to be able to be moved upward by force, an impelling force greater than the resisting weight is required; otherwise the resisting weight could not be overcome, and, consequently, the body could not move upward. That is, the body moves upward, provided the impressed motive force is greater than the resisting weight. But since that force, as has been shown, is continuously weakened, it will finally become so diminished that it will no longer overcome the weight of the body and will not impel the body beyond that point. And yet, this does not mean that at the end of the forced motion this impressed force will have been completely destroyed, but merely that it will have been so diminished that it no longer exceeds the weight of the body but is just equal to it. To put it in a word, the force that impels the body upward, which is lightness, will no longer be dominant in the body, but it will have been reduced to parity with the weight of the body. And at that time, in the final moment of the forced motion, the body will be neither heavy nor light. (Galilei 1960, 89)

The uncertainty of Galileo about the problem of how to account for the shape of the trajectory of a projectile in terms of the interaction of violent and natural motion indicates the implicit difficulty of the medieval Aristotelian view of projectile motion mentioned above. This difficulty must have become even more demanding when a symmetrical shape of the trajectory had to be taken into account. Galileo or anybody who performed the experiment recorded in Guidobaldo's protocol or heard about its outcome must have realized immediately that it sheds new light upon the prevailing view of projectile motion. The trajectory cannot be symmetrical unless the impetus determining the first part acts exactly the same way as the natural motion acts in the second part. Hence, the symmetry of the trajectory must have been remarkable to everybody who was familiar with the Aristotelian view of projectile motion. This is probably the reason why Guidobaldo, before he described the experimental setting, started his note by stating this puzzling fact, and then drew attention to an observation that might make it plausible. He compared the constellation of violent force and natural tendency in the case of the trajectory with another case, showing a similar constellation, but exhibiting a perfect and intelligible symmetry: the catenary.

The explanation for the unexpected symmetry suggested by this comparison implies certain assumptions which challenged the medieval Aristotelian theoretical framework even more. Whereas in this tradition violent motion and natural motion were contraries which could not contribute together to one and the same motion the comparison with the catenary requires that they act jointly and in the same way, mutually exchanging their roles when ascending turns into descending. This conclusion is, in fact, drawn in Guidobaldo's protocol.

For the same reason, the prevailing belief that the beginning part and the ending part of the projectile trajectory are straight lines had definitely to be dismissed. From the first moment on the trajectory has to be curved by the weight of the projectile, even though it looks perfectly straight. Accordingly, as Guidobaldo writes, the violence is gradually lessened but is never lost completely.

Summing up: The immediate outcome of the experiment described in Guidobaldo's protocol is the observation that projectile motion has a *symmetrical trajectory*. Although this observation could be explained within the conceptual framework of the medieval Aristotelian tradition, such an explanation indirectly implied characteristics of projectile motion that were specific and unusual at the time of the young Galileo:

- The symmetry of the dynamic situation suggested that the *projectile trajectory and the catenary have the same form.*

- It further implied that the projectile never moves in a perfectly straight line but in a curve determined by two components, the violent one and the natural one. Both have to *act equally while mutually changing their roles* in ascending or descending, respectively.

Decomposing the Trajectory—Neutral Motion and the Law of Fall

There is still another implication of the experiment recorded in Guidobaldo's note which is somewhat more hidden but leads to much more dramatic consequences. For someone mathematically educated like Galileo or Guidobaldo the symmetrical curves of projectile trajectories and hanging chains must have raised immediately the question of whether these curves coincide with any of the well-known curves of ancient or contemporary mathematics. And indeed, Guidobaldo mentioned already in his protocol the hyperbola and the parabola as curves which look similar to the curves generated by the experiment. He must also have been well aware of the fact that in order to ascertain that such a curve fits the trajectory exactly the curve had to be derived from assumptions regarding the forces that determine in an equal way the curves of the trajectory and the catenary.

This, however, was by no means a simple task. It is difficult already to decide which of the two curves, the hyperbola or the parabola, is a more promising candidate, although the curves generated by the same procedure documented by the folio sheet in Galileo's hand copy of the Discorsi mentioned above clearly exclude the hyperbola option. Moreover, the dynamical explanation given in the protocol implies, on closer inspection, that the constantly decreasing ratio between violent and natural motion in descent assumed in the protocol is incompatible with the asymptotic behavior of the hyperbola. It follows, in fact, from this asymptotic behavior that this ratio should approach a constant different from zero. But once the parabola has been chosen, its geometrical properties, well-known since ancient times, suggest certain assumptions about the forces and how they act together. Mathematically trained scientists like Galileo or Guidobaldo will have been able to see that in every point of the trajectory or of the hanging chain the square of the horizontal distance from the highest or lowest point respectively is proportional to the vertical distance. Any assumed relation of the two dimensions of the parabola with the two types of forces, the violent and the natural, leads therefore automatically to statements about how precisely these forces generate the curves of the catenary and the trajectory. This suggests in particular to conceive of the motion of a projectile as being composed of two motions, a uniform horizontal motion and a vertical motion that is first an upwards decelerated and, after having reached the highest point, downwards accelerated motion.

This consideration within the conceptual framework of contemporary Aristotelian thinking does, of course, neither lead to the mechanical explanation of the catenary in classical physics, nor to that of projectile motion. It is not even sufficient to construct a precise analogy between the way in which the forces act together in the case of the catenary and in the case of the trajectory as it is so strongly pointed out both in Guidobaldo's protocol and even more vividly later in Galileo's Discorsi. But if Galileo had ever followed this line of intuitive thought, he would have realized that the assumption of a uniform motion along the horizontal is closely related to an amazing insight he achieved already much earlier when he studied the motion on inclined planes with different inclinations, and that the second motion leads to an extraordinary discovery, that is, what later was called the law of fall.

Galileo hit on the phenomenon of a uniform horizontal motion when he investigated the force of a body moving down differently inclined planes. This force decreases with decreasing inclination of the plane. Thus, for the case of a horizontal plane he proved in the De Motu manuscript:

> A body subject to no external resistance on a plane sloping no matter how little below the horizon will move down [the plane] in natural motion, without the application of any external force. This can be seen in the case of water. And the same body on a plane sloping upward, no matter how little, above the horizon, does not move up [the plane] except by force. And so the conclusion remains that on the horizontal plane itself the motion of the body

is neither natural nor forced. But if its motion is not forced motion, then it can be made to move by the smallest of all possible forces. (Galilei 1960, 66)

This result is obviously incompatible with the Aristotelian dynamic law according to which the velocity of a motion is proportional to the moving force. Later it became a cornerstone of Galileo's theory of projectile motion, but when Galileo first hit on this result he seems to have had trouble reconciling it with the traditional understanding of natural and forced motion. He added into the margin a note indicating that he intended to interpret the unusual horizontal motion as mixed motion in the Aristotelian sense and added a justification of this statement:

> From this it follows that mixed motion does not exist *except circular* [deleted]. For since the forced motion of heavy bodies is away from the center, and their natural motion toward the center, a motion which is partly upward and partly downward cannot be compounded from these two; unless perhaps we should say that such a mixed motion is that which takes place on the circumference of a circle around the center of the universe. But such a motion will be better described as 'neutral' than as 'mixed.' For 'mixed' partakes of both [natural and forced], 'neutral' of neither. (Galilei 1960, 67)[1]

In accordance with this last remark, he then deleted the words "except circular," obviously because he preferred to return to the traditional view held by Tartaglia that motions mixed of natural and forced motions are impossible. From the viewpoint of later classical physics this reluctance to accept the motion of projectiles to be compounded of natural and forced motions appears to have been a major obstacle against generalizing the concept of "neutral" motion to a general concept of inertial motion.

But even without such a generalization the concept of neutral, horizontal motion that is neither natural nor forced paves the way to the law of fall. If the parabolic trajectory is decomposed into a neutral, horizontal motion and a natural, vertical motion, then in order to ensure the symmetry of the trajectory the easiest way is to consider the neutral motion as uniform. Then, however, the geometrical theorem which is the basis of such a decomposition, stating that the vertical distances are proportional to the squares of the horizontal distances from the vertex, implies the proportionality of the vertical distances to the squares of the times represented by the horizontal distances, that is, the law of fall.

---

[1] See also the reference to the whole problem in the *Discorsi*, Galilei 1974, 157-159.

Dating Guidobaldo's Protocol

In view of such implications of Guidobaldo's experiment concerning Galileo's major discoveries, the question of when it was first performed becomes significant. The experiment was, of course, made before Guidobaldo's death in 1607. The previous analysis of possible consequences of the observations reported by Guidobaldo and Galileo show furthermore, on the one hand, that Galileo when he worked at his *De Motu* manuscript around 1590, that is about two years after first contacts between Galileo and Guidobaldo are documented by the correspondence between them, cannot yet have been aware of the outcome of the experiment. On the other hand, it is also evident that Guidobaldo, when he wrote the protocol, was not yet familiar with Galileo's discovery of the parabolic shape of the trajectory. It is also unlikely that he knew at that time already Galileo's law of fall, because otherwise he would have immediately recognized the close relationship between this law and a parabolic shape of the trajectory and, consequently, would not have considered a hyperbolic shape as an alternative. Thus, the time window of possible dates of the protocol ranges from 1590 to 1607 with the qualification that it must have been written before Guidobaldo had any knowledge of Galileo's discoveries of the law of free fall or the parabolic trajectory, in case they had been made already earlier. This latter clue would not help to date Guidobaldo's protocol if the standard dating of Galileo's discoveries to 1604 and 1609 would be correct. However, as we mentioned already, there is strong evidence that Galileo achieved these major results of his work much earlier and that, in fact, these discoveries are closely connected with the experiment described by Guidobaldo.

It seems that Galileo concealed as long as possible the discovery of the parabolic shape of the trajectory, in contrast to the discovery of the law of fall. In the publication of his *Dialogo* in the year 1632 in which he made known the law of fall for the first time, he also included a misleading discussion of the trajectory of a projected body which seems to indicate that even at that time he still had no idea of the true shape of this trajectory (Galilei 1967, 165ff).

There are indications, that he, indeed, consciously kept his discovery a secret. Following a tradition of the time, he occasionally showed his admiration to a close friend by signing in an "Album Amicorum" with an allegorical drawing containing an allusion to an important achievement which could take the form of a riddle (Galilei 1890-1909, XIX: 204).

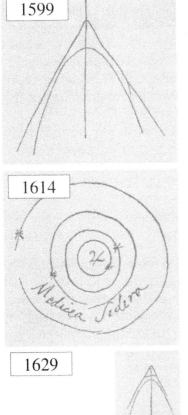

Hoc, Thoma Segete, observantiae et amicitiae
in te meae signum ita perenne servabis,
ut indelebili nota pectori meo virtus infixit tua.
        Galileus Galilei N. Flor.^{us}, Mat.^{rum}
in Academia Pat.^{na} professo (sic),
m. pp.^{a} scripsi Murani, Idib. Augusti 1599.

> This, Thomas Segget, will serve you as a
> sign of my esteem and friendship towards you
> – so durably as your virtue has sticked it to
> my heart by an undestructible mark.

        An. 1614. D. 19 Novembris.
Ut nobili ac generoso studio D. Ernesti Brinctii
rem gratam facerem, Galileus Galileius
Florentinus manu propria scripsi Florentiae.

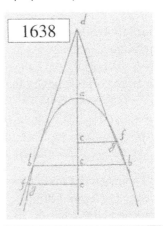

Accedens non conveniam
Galileus Galileius m. p.^{a} scripsi,
die 8^{a} Martii 1629, Florentiae.

> Approaching, I rather not join.

> Figure at the beginning of the
> *Fourth Day* of the *Discorsi*
> used in the proof of the first
> theorem which states that a body
> which is projected horizontally
> describes a semi-parabola

Figure 8. Allegorical drawings of Galileo representing discoveries and a corresponding diagram
in the Discorsi

Four such allegorical drawings used by Galileo are known (see figures 8).[1] One of them expresses his discovery of the satellites of Jupiter and is dated November 19, 1614; the three others each depict a parabola together with its middle axis and its tangents at two symmetrical points, symbolizing a geometrical relation between these tangents and the height of a parabola. The latter figures are similar to the diagrams at the beginning of the *Fourth Day* of his *Discorsi*, representing the parabolic trajectory of a body projected in the direction of the tangent and reaching a maximal height which is one half of the height of the crossing point of the tangents with the middle axis of the symmetrical trajectory. Two of these three figures are dated March 8 and March 20 of the year 1629, the third one is dated as early as August 1599. If, as is most probable, Galileo indeed represented by this allegorical use of a parabola his discovery of the parabolic trajectory, he thus must have made this discovery already earlier than 1599.

It was not until 1632 that certain circumstances, which will presently be discussed, forced him to make his discovery known to the public. This event provides us with a direct statement of Galileo himself about the question of when he made this discovery. When Bonaventura Cavalieri in 1632, shortly after the publication of Galileo's *Dialogo*, published his book *Lo Specchio Ustorio overo Trattato delle Settioni Coniche* on parabolic mirrors, he sent Galileo a letter which contains the following information:

> I have briefly touched the motion of projected bodies by showing that if the resistance of the air is excluded it must take place along a parabola, provided that your principle of the motion of heavy bodies is assumed that their acceleration corresponds to the increase of the odd numbers as they follow each other from one onwards. I declare, however, that I have learned in great parts from you what I touch upon in this matter, at the same time advancing myself a derivation of that principle. (Bonaventura Cavalieri to Galileo, August 31, 1632, Galilei 1890-1909, XIV: 378)

This announcement must have shocked Galileo. In a letter written immediately afterwards to Cesare Marsili, a common friend who lived like Cavalieri in Bologna, he complained:

> I have letters from Father Fra Buonaventura with the news that he had recently given to print a treatise on the burning mirror in which, as he says, he has introduced on an appropriate occasion the theorem and the proof concerning the trajectory of projected bodies in which he explains that it is a parabolic curve. I cannot hide from you, my dear Sir, that this news was any-

---

[1] Another drawing which symbolizes Galileo's discovery of the parabolic shape of the projectile trajectory, and a text, including the date March 20, 1629, which both belong to a hitherto unknown entry in an "Album Amicorum" were recently uncovered as part of a telescope, where they were placed in order to pretend that the telescope was built by Galileo himself and therefore represents a unique and valuable instrument; see Miniati, Greco, Molesini, and Quercioli 1994.

thing but pleasant to me because I see how the first fruits of a study of mine of more than forty years, imparted largely in confidence to the said Father, should now be wrenched from me and how the flower shall now be broken from the glory which I hoped to gain from such long-lasting efforts, since truly what first moved me to speculate about motion was my intention of finding this path which, although once found it is not very hard to demonstrate, still I, who discovered it, know how much labor I spent in finding that conclusion. (Galileo to Cesare Marsili, September 11, 1632, Galilei 1890-1909, XIV: 386)

Galileo received immediate answers both from Marsili and from Cavalieri, written on the same day (Bonaventura Cavalieri to Galileo, September 21, 1632, Galilei 1890-1909, XIV: 395, and Cesare Marsili to Galileo, September 21, 1632, Galilei 1890-1909, XIV: 396). Marsili assured Galileo of Cavalieri's full loyalty. Cavalieri himself expressed his deep concern about Galileo's anger and tried to convince him by a number of different reasons that he did not intend to offend Galileo by his publication. He first claimed that he was uncertain whether the thesis that the trajectory has a parabolic shape entirely corresponds to Galileo's intentions. Then he adduced as an excuse that he was convinced that the thesis had been already widely spread. He furthermore added that he had been uncertain whether the thesis was of any value to Galileo. Finally, he claimed that he had reason to assume that Galileo at that time had published his result already long ago:

I add that I truly thought that you had already somewhere written about it, as I have not been in the lucky situation to have seen all your works, and it has encouraged my belief that I realized how much and how long this doctrine has been circulated already, because Oddi has told me already ten years ago that you have performed experiments about that matter together with Sig.r Guidobaldo del Monte, and that also has made me imprudent so that I have not written you earlier about it, since I believed, in fact, that you do in no way bother about it but would rather be content that one of your disciples would show himself on such a favorable occasion as an adept of your doctrine of which he confesses to have learned it from you. (Bonaventura Cavalieri to Galileo, September 21, 1632, Galilei 1890-1909, XIV: 395)[1]

The conflict about Cavalieri's intended publication of the parabolic shape of the trajectory provides two pieces of information which are highly significant for the question of when and how Galileo really made his discovery.

First, Galileo's claim in his letter to Cesare Marsili makes it conceivable that he had discovered the parabolic shape already around forty years before he wrote this letter, that is, as early as or even earlier than 1592. If this should be true, this

---

[1]   See also the discussion of this correspondence in Wohlwill 1899.

discovery must have been one of the earliest discoveries of Galileo that challenged his *De Motu* theory.[1]

Second, Cavalieri brings Galileo's claim in connection with experiments on projectile motion that Galileo had performed together with Guidobaldo del Monte. The only experiment that is known and fits the account of Muzio Oddi is the one reported in Guidobaldo's protocol. It follows that either Galileo himself must have been present during this experiment or at least have known about it in the case that there were further experiments on projectile motion jointly performed by them.

Cavalieri claims that he had heard about these experiments already ten years earlier, that is around 1622, from Muzio Oddi. Indeed, there is independent evidence confirming Cavalieri's report. It turns out that Cavalieri and Muzio Oddi happened to be both living in the same place, Milan, between 1620 and 1622, that is, just around the time mentioned in the above passage.[2] But Muzio Oddi himself must have recalled these experiments as having been made much earlier, because Guidobaldo del Monte died already in 1607. Muzio Oddi was born and lived – with interruptions – in Urbino. He mentioned that he had been for a short time a disciple of Guidobaldo del Monte. Between 1595 and 1598 he left Urbino to become a military architect in the Bourgogne, with the exception of a period between 1596 and 1597 when he served as an architect in Pesaro. In 1599 he began to get in trouble with the authorities, first in 1599 for illegal fishing and bathing naked in a river, then in 1601 for allegedly stealing from the closet of the Grand Duke; as a consequence he had to flee in the same year from Urbino into the territories of the Venetian Republic. He returned only in 1605 after an amnesty, but soon got again in trouble with the rulers of the city – because of certain favors received from the Grand Duchesse – so that he was arrested again, in the "Rocca di Pesaro." He stayed in prison until he was released in 1610 and finally left Urbino for Milan.[3]

What are the possible dates for occasions on which Muzio Oddi could have heard from Guidobaldo del Monte about the experiments performed together with Galileo? Since, according to his own testimony, he was acquainted with Guidobaldo del Monte already as a disciple, it cannot be excluded that he heard about the experiments even before he left in 1595. The best opportunity must, of course, have been the time between 1596 and 1597, when he worked as an architect in Pesaro itself. A further possibility is the time period between 1598 and

---

[1] It has earlier been assumed that Galileo's claim of such an early discovery of the parabolic shape of the trajectory was exaggerated and that in his letter to Marsili Galileo was actually referring to his treatment of projectile motion in *De Motu*, see Camerota 1992, 79. In the light of the explicit mentioning of the joint experiment with Guidobaldo des Monte in Cavalieri's letter of September 21, 1632, and the precise coincidence of the date of the experiment and Galileo's claim, discussed here, there can be no doubt that he must have said the truth and that the alternative assumption has turned out to be untenable.

[2] For Cavalieri's biography, see Gillispie 1981, for Oddi's biography, see Gamba and Montebelli 1988, chap. IV.

[3] See the short biographical sketch in Gamba and Montebelli 1988, 111-113.

1601 when he stayed again in Urbino, that is, not far away from Pesaro. If we exclude the possibility that Guidobaldo del Monte could have contacted him while he was in prison, the latest date for an encounter is the short stay in Urbino in the year 1605, but given the circumstances of this stay, it is quite unlikely that he should have discussed such experiments with Guidobaldo del Monte at that time. Summing up, the experiments will probably have taken place before 1601, most likely even before 1597. This dating is compatible also with the *terminus ante quem* 1599 suggested by the allegorical drawings mentioned above.

In view of this evidence in favor of an early dating for Galileo's discoveries of the parabolic shape of the trajectory and of the law of fall, it can no longer be excluded that Galileo's own reference to a "study of mine of more than forty years" in his letter of 1632 must actually be taken literally, even though it points at a date for these discoveries as early as 1592.

This circumstantial evidence suggests a close reexamination of the events around the time of 1592 in order to find out whether something special might have happened in this period. It is well known that indeed the year 1592 represents a turning point in Galileo's career. As a result of strong support from Baccio Valori, Consul of the Accademia Fiorentina in 1588 and later representative of Ferdinando I de Medici in the Accademia del Disegno, from Giovanni Vincenzo Pinelli, the head of a group of literary and culturally interested people in Padua, and, in particular from Guidobaldo del Monte, Galileo received, in late 1592, an appointment at the University of Padua (Drake 1987, 32). Earlier in the same year, when Galileo was still desperate about his future and planned a trip to Venice in order to explore his chances of obtaining a position, he received a consoling letter from Guidobaldo who invited him to travel through Monte Baroccio on his way to Venice:

> It also saddens me to see that your Lordship is not treated according to your worth, and even more it saddens me that you are lacking good hope. And if you intend to go to Venice in this summer, I invite you to pass by here so that for my part I will not fail to make any effort I can in order to help and to serve you; because I certainly cannot see you in this state. (Guidobaldo del Monte to Galileo, February 21, 1592, Galilei 1890-1909, X: 47)[1]

Galileo actually travelled twice during this year from Florence to the Venetian republic, the first time probably towards the end of August or in early September in order to receive the appointment (Giovanni Ugoccini to Belisario Vinta, September 21, 1592, Galilei 1890-1909, X: 49), the second time sometime between October and early December when he finally moved to Padua.[2] In the meantime, he had to go back to Florence in order to get permission from the Grand Duke to

---

[1] On the basis of this letter a visit of Galileo with Guidobaldo had been conjectured also by other authors, see e.g. Gamba and Montebelli 1989, 14.

[2] Galileo was finally in Padua by the middle of December as attested to by his correspondence; see Galilei 1890-1909, X: 50ff.

leave Tuscany and take the chair in Padua.[1] There is no reason to doubt that Galileo followed Guidobaldo's invitation on one of these two occasions.

When Galileo arrived at Padua he immediately visited Giovanni Vincenzo Pinelli (Giovanni Vincenzo Pinelli to Galileo, September 3, 1592, Galilei 1890-1909, X: 47) in whose house he also lived for some time in the beginning of his Paduan stay.[2] At some point during Galileo's early stay in the Venetian Republic he must have encountered, possibly in the house of Pinelli, Paolo Sarpi with whom he long afterwards, stayed in close scientific contact. In the notebook of Sarpi under the header "1592" the following entry can be found:

> The projectile not [moving] along the perpendicular to the horizon never moves along a straight line, but along a curve, composed of two straight motions, one natural, and the other one along where the force is directed. The impressed [force] at the beginning is always greater, and, for this reason, the beginning comes very close to the straight line; but the impressed force continues decreasingly and it returns [in a] similar [way] to [that of] the beginning if it [i.e. the impressed force] has the proportion to the natural [force], as the natural one had to it [i.e. the impressed force], and in everything and all the time the descending is similar to the ascending. If, however, the one [i.e. the impressed force] of the projectile expires, the motion finally comes rectilinearly downward; but if (as has been assumed before) it is infinitely divisible and diminishes according to proportional parts, the motion never comes to be a straight line. Hence, the motion of the projectile is compounded by two forces: one of which always remains the same, and the other always decreases.
>
> A similar line is caused by a suspended rope, because its suspension would like to pull each part laterally towards it [i.e. the fixed ends], while it [i.e. the rope] would like to move downwards; therefore, the parts closer to the beginning share more of the lateral [force], and the parts closer to the middle share more of the natural [force], the middle has equal shares of both of them and is the vertex of the figure. (Notes number 537 and 538 in Sarpi 1996, 398f)[3]

In view of the in no way uncertain terms with which Sarpi gives a description of the trajectory of projectile motion that is in flat contradiction with the accepted view, it seems inconceivable that this description should be unrelated to Guidobaldo del Monte's interpretation of the experiment described in his proto-

---

[1] Galileo left Venice to Florence on September 27, 1592; see Giovanni Ugoccini to the Grand Duke of Tuscany, September 26, 1592, Galilei 1890-1909, X: 50.

[2] Benedetto Zorzi to Galileo, December 12, 1592, Galilei 1890-1909, X: 51; see also Favaro 1966, I: 50.

[3] Due to the local calendar in the Venetian republic, the entries of 1592 may include notes up to February 28, 1593 which makes it even more likely that these entries were made at a time when Galileo and Sarpi were already in close contact.

col. Although it is obvious that this note has been written independently of the actual wording of Guidobaldo del Monte's protocol, it corresponds point by point to the at that time unorthodox theoretical assumptions which in this protocol are used to explain the symmetry of the trajectory which was the unexpected outcome of the experiment.

The symmetry is attributed to a symmetry of the dynamical constellation. This dynamical constellation is conceived as paradigmatically represented by a hanging chain. As an implication the trajectory is everywhere conceived as determined by two components, a violent one and a natural one which explains that the trajectory is nowhere perfectly straight. Finally, according to the dynamical interpretation given for the symmetry of the trajectory, violent and natural motion have to mutually exchange their roles in the ascending and the descending part of the trajectory respectively.

Given the fact that direct contact between Guidobaldo del Monte and Paolo Sarpi in 1592 is extremely improbable and that, as has been discussed already above, Galileo claims in his *Discorsi* that he himself had invented the method to trace the trajectory by means of an inclined plane as it is described in Guidobaldo del Monte's protocol, the remarkable correspondence of this protocol and Paolo Sarpi's note strongly suggests that it was nobody else but Galileo himself who communicated the information about the outcome of the experiment and its interpretation to Paolo Sarpi. The striking similarities between Guidobaldo del Monte's protocol and Paolo Sarpi's note find indeed a simple explanation if Galileo on his first trip[1] from Florence to the Venetian republic in 1592 followed the invitation of Guidobaldo del Monte, performed together with him the experiments on the projectile trajectory and afterwards discussed the puzzling results with his new friend Paolo Sarpi, who became one of his most important intellectual companions in the coming years of his work at Padua. At the same time, Galileo's claim in his conflict with Cavalieri in 1632 that his discovery of the parabolic shape of the trajectory of projectile motion reaches back to work done more than 40 years ago as well as Cavalieri's report that he had heard about Galileo's discovery of the parabolic trajectory by means of experiments performed together with Guidobaldo del Monte, turn out to be perfectly justified, whoever of the two had the idea for designing these experiments. Consequently, according to common historiographic criteria the discovery of the parabolic shape of the trajectory of projectile motion has to be dated to the year 1592. It must, in fact, have been one of Galileo's earliest discoveries reported in his latest publication, the *Discorsi*.

In addition to the reoccurrence of the experiment reported in Guidobaldo's notebook in Galileo's *Discorsi* and of its interpretation in the notebook of Paolo

---

[1] It must have been the first trip because as late as January 1593 Guidobaldo was still not informed about the outcome of Galileo's negotiations with the authorities of the Venetian republic on his remuneration; see Guidobaldo del Monte to Galileo, January 10, 1593, Galilei 1890-1909, X: 53f.

Sarpi, further evidence of Galileo's participation in performing the experiment is provided by entries immediately before and after Guidobaldo's report. The first entry in Guidobaldo's notebook that appears to be related to Galileo's visit at Guidobaldo's house in Monte Baroccio refers to Galileo's invention of a hydrostatic balance, the "Bilancetta." The problem of determining the specific weight of a substance following a procedure traditionally ascribed to Archimedes has been treated earlier in Guidobaldo's notebook, but without referring to Galileo's instrument (del Monte ca. 1587-1592, 119f). The fact, that such a reference to this instrument, and even a full treatment of it including a demonstration, is found in the last part of Guidobaldo's notebook together with other notes on Galileo's topics, is a strong indication that Guidobaldo received an explanation of this instrument from Galileo himself (del Monte ca. 1587-1592, 232-234).

On the facing page before Guidobaldo's report on the projectile trajectory experiment, another experiment concerning the resonance of strings is described which is also discussed in the *Discorsi*; we shall return to this experiment further on. The same coincidence of an entry in Guidobaldo's notebook and Galileo's writings is also found for the other entries on the same page, written immediately above and below Guidobaldo's report on the experiment concerning the projectile trajectory. In a short note above the description of the experiment Guidobaldo considers the flow of water along an inclined channel, serving to drive a mill. He writes:[1]

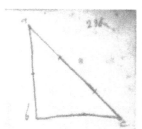

When a descent [caduta] will be of a height of ten, in order to give water to a mill the channel should be 15 [sic!], as the descent *ab* and the channel *ac*. But due to the general rule, *ac* should usually be elevated by ca. 45 degrees, according to the consideration of the quantity of water which one has. (del Monte ca. 1587-1592, 236)

The theme of water moving along an inclined plane is common in Galileo's writings, from his early treatise *De Motu*, via a letter to Guidobaldo del Monte in 1602 (Galileo to Guidobaldo del Monte, November 29, 1602, Galilei 1890-1909, X: 97-100), to his criticism in 1630 of a plan for straightening the river Bisenzio.[2] In the latter, Galileo explains more in detail that the motion of water along an inclined plane differs from that of a solid body since here one has to take into account not only the tilt of the plane but also the quantity of water flowing along it:

---

[1]   The entry has been discussed by Gamba 1995, 101. In this article the figure for the length of the channel has been transcribed as 25 as suggested by the appearance of the number in the text. Gamba has later (personal communication) convincingly argued that the figure has to be read as 15 consistent with the designation of length units in the figure.
[2]   See the discussion in Drake 1987, 320-329.

Now because in the acceleration of the course of the highest waters little part is played by greater slope and much by the great quantity of supervening water, consider that in the short channel although there is greater tilt than in the longer, the lower waters of the long [channel] are more charged by the great abundance of higher waters pushing and driving, by which impulse is more than compensated the benefit that could be derived from greater slope. (Galileo to Raffaello Staccoli, January 16, 1630, Galilei 1890-1909, VI: 639; translation quoted from Drake 1987, 328)

Both passages deal with the same physical effect, water flowing along a tilted channel or river. Guidobaldo apparently refers to a practitioners' rule according to which that tilt should usually be about 45 degrees in order to drive a mill and considers a situation in which a greater distance has to be bridged from the water source to the mill, resulting in a less steeply inclined channel (41.8 degrees). The question that must have motivated this consideration was surely that of the effect of this changed tilt on the flow of water; it may have well been triggered, as Gamba suggests in his paper, by a practical problem. Does the lesser inclination yield a smaller flow of water and is hence incapable of driving the mill? Guidobaldo's final remark refers to a consideration of the quantity of water that is required for the response to such a question and is hence in complete agreement with the essence of Galileo's argument in the passage quoted above. It is therefore plausible to assume that Galileo and Guidobaldo discussed how the laws of motion along inclined planes derived by Galileo in his treatise *De Motu* are changed if they are applied to channels or rivers, in which case, according to Galileo, the quantity of the liquid somehow compensates the effect of the tilt of the plane.

The short text written by Guidobaldo below his description of the experiment also summarizes an argument that is found in Galileo's writings. This text reads:

A cord which sustains a weight, sustains as much if it is short as it does when it is long; it is indeed true that in the long one [it breaks more easily], first, because of its gravity, second, because in the long one there can be many weak parts. Assume [può esser] that it [i.e. the long one] breaks more easily and by less weight. But if the cord would be sustained a little above from where it breaks because of its cracking and the weight would be a little underneath, without doubt it would break in the same way because it would have cracked in the same way. (del Monte ca. 1587-1592, 236)

Galileo inserted exactly the same argument between the propositions and proofs concerning the rigidity of bodies at the *Second Day* of the *Discorsi*. There one finds the following dialogue between his spokesman Salviati and the Aristotelian philosopher Simplicio:

Simplicio. (...) we see a very long rope to be much less able to hold a great weight than if shorter; and I believe that a short wooden or iron rod can support much more weight than a very long one when loaded lengthwise (not [just] crosswise), and also taking into account its own weight, which is greater in the longer.

Salviati. I think that you, together with many other people, are mistaken on this point, Simplicio, at least if I have correctly grasped your idea. You mean that a rope, say forty braccia in length, cannot sustain as much weight as one or two braccia of the same rope.

Simplicio. That is what I meant, and at present it appears to me a highly probable statement.

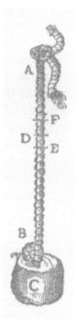

Salviati. And I take it to be not just improbable, but false; and I believe that I can easily remove the error. So let us assume this rope *AB*, fastened above at one end, *A*, and at the other end let there be the weight *C*, by the force of which this rope is to break. Now assign for me, Simplicio, the exact place at which the break occurs.

Simplicio. Let it break at point *D*.

Salviati. I ask you the cause of breaking at *D*.

Simplicio. The cause of this is that the rope at that point has not the strength to bear, for instance, one hundred pounds of weight, which is the weight of the part *DB* together with [that of] the stone *C*.

Salviati. Then whenever the rope is strained at point D by the same 100 pounds of weight, it will break there.

Simplicio. So I believe.

Salviati. But now tell me: if the same weight is attached not to the end of the rope, *B*, but close to Point *D*, say at *E*; or the rope being fastened not at *A*, but closer to and above the same point *D*, say at *F*; then tell me whether the point *D* will not feel the same weight of 100 pounds.

Simplicio. It will indeed, provided that the length of rope *EB* accompanies the stone *C*.

Salviati. If, then, the rope, pulled by the same hundred pounds of weight, will break at the point *D* by your own admission, and if *FE* is but a small part of the length *AB*, how can you say that the long rope is weaker than the short? Be pleased therefore to have been delivered from an error, in which you had plenty of company, even among men who are otherwise very well informed, and let us proceed. (Galilei 1974, 119-120)[1]

It is likely that by mentioning "men who are otherwise very well informed" who made the same error as Simplicio, Galileo had nobody else in mind but Guidobaldo himself. We shall see below that this is not the only case of Guidobaldo having entered an argument of somebody else into his notebook although he himself believed or had believed the contrary to be true. In any case, even if there were no independent evidence for Galileo having visited Guidobaldo at the very time when he performed the projectile experiment, the fact alone that the entries immediately before and after his protocol of the experiment reappear in Galileo's *Discorsi*, just as it is the case for the experiment itself, makes it difficult to imagine anything else than that Galileo was present when Guidobaldo entered these notes into his notebook.

But also the other parts of the notebook provide evidence for the dating of Guidobaldo del Monte's protocol of the projectile trajectory experiment into the year 1592.[2] A comparison of the entries in the notebook with the correspondence of Guidobaldo del Monte shows that most of the entries at the beginning of the notebook must have been written between the years 1588 and 1590. An unquestionable *terminus a quo* is given by entries related to publications of Fabricius Mordente (1585), Giovanni Battista Benedetti (1585), and Francesco Barozzi (1586); an equally unquestionable *terminus ad quem* by the fact that the last third of the notebook is mainly devoted to problems of perspective related to his work on a book which in great parts has been completed around 1593 and has finally been published in 1600. As far as a direct relation between entries of the notebook and issues mentioned in the correspondence can be established, they can all be dated into the years between 1588 and 1590; that is, work on a geometrical problem of Pappus (mentioned 1588)[3], the correspondence with Galileo on the center of gravity of paraboloids (beginning 1588)[4], and work on the cochlea (mentioned 1589 and 1590)[5]. Furthermore, a loose sheet of paper is inserted in

---

[1] It may be worth noting that in Sarpi's notebook shortly after the note on the outcome of the trajectory experiment one finds an entry about the resistance of bodies involving an argument similar to the one treated in Guidobaldo's notebook and Galileo's *Discorsi*, although no concrete experimental setting is described (see note number 543 in Sarpi 1996, 405). Sarpi merely limits himself to an application of Galileo's indirect proof to a general consideration of the resistance of continuous bodies against breakage. This note by Sarpi may represent a third case, in addition to those of the projectile trajectory and the catenary, in which an argument by Galileo is reported in both Guidobaldo's and Sarpi's notebooks.

[2] Guidobaldo's notebook has been widely neglected by historians of science. Certain passages of this notebook, in particular those on the hydrostatic balance and on motion in media (see below) have, however, been intensively discussed during a visit of Pierdaniele Napolitani and Pierre Souffrin in Berlin and during a workshop in Pisa organized by Pierdaniele Napolitani. The results of these discussions will appear in future publications.

[3] It is mentioned in Guidobaldo del Monte to Galileo, September 16, 1588, Galilei 1890-1909, X: 37 as a problem that Guidobaldo had already earlier communicated to Galileo.

[4] See Guidobaldo del Monte to Galileo, January 16, 1588, Galilei 1890-1909, X: 25f and the subsequent letters.

the notebook with astronomical data for a horoscope for a date in the year 1587.[1] Thus, the fact that the protocol of the projectile trajectory experiment performed in 1592 is written on one of the last pages of the notebook is in any respect in accordance with the dating of the other entries in the notebook.

Galileo was able to think and thus, as a competent mathematician, capable of drawing the obvious conclusion from the experiment that, as a consequence of the dynamical assumptions laid down in Guidobaldo's protocol, the trajectory cannot be hyperbolic but must be parabolic. In the following, we shall therefore simply assume that he must have been indeed, as he claimed, aware as early as 1592 of the parabolic shape of the projectile trajectory, possibly sharing this knowledge with his patron Guidobaldo. While this insight would qualify, by ordinary historiographical standards, as a "discovery," such a qualification leaves, however, completely open the question of what this "discovery" actually meant for both Galileo and Guidobaldo at that time. In order to understand the impact of this discovery one has to study the contexts in which it occurred – which may well be different for Guidobaldo and for Galileo.

Guidobaldo del Monte as an Engineer-Scientist

Guidobaldo del Monte represented a new type of engineer-scientists which emerged in the sixteenth and seventeenth centuries, in distinction from traditional academics.[2] The emergence of this new social group and its epistemological motives cannot be adequately understood without taking into account the technological development that had taken place at least since the Renaissance in certain European urban centers. The essence of this technological development is visible in the remarkable difference between large-scale projects in the early modern period and in ancient urban civilizations. Ancient large-scale projects, such as the construction of the Babylonian zikkurats and Egyptian pyramids, involved enormous challenges for labor organization, mastered by a class of high-rank officials with appropriate administrative knowledge about the acquisition, allocation, and maintenance of labor force. There is, however, no historical record of a comparative body of engineering knowledge adequately corresponding to the implicit technological complexity of these large-scale projects, nor of any social group representing technical in contrast to administrative knowledge. Even in the case of the most advanced construction projects of the Roman empire, there is hardly any trace of a technical intelligentsia which, beyond the organization of labor, developed a specific canon of technical knowledge—other than the type of com-

---

[5]   See Guidobaldo del Monte to Galileo, August 3, 1589, Galilei 1890-1909, X: 41 and Guidobaldo del Monte to Galileo, April 10, 1590, Galilei 1890-1909, X: 42f.

[1]   del Monte ca. 1587-1592, 212.

[2]   For historical discussions from which our account has benefited, see Bertoloni Meli 1992; Biagioli 1989; Micheli 1992; Gamba and Montebelli 1988; and Lefèvre 1978.

pilation of rules and standard models exemplarily represented by the work of Vitruv—that challenged ancient theories. The large-scale projects of the early modern period, such as the construction of the Florentine dome, on the other hand, are inconceivable without a group of specialized artisans, technicians, and engineers that combined administrative with technological competence. Due also to the limited availability of labor-force and other resources, these artisan-engineers were continuously confronted with technical and not only logistic challenges. In reaction to these challenges they were forced to explore the inherent potential of traditional technical knowledge in order to create new technical means, as for example the set of machines developed by Filippo Brunelleschi in order to build the Florentine cupola without an inaffordably more expensive scaffolding (see di Pasquale 1996). The engineers in early modern times were thus not only carriers of a traditional canon of knowledge, as it had been largely the case of the ancient administrators of large-scale projects, but were involved in a cumulative, self-accelerating process of innovation.

The technical knowledge of these engineers developed independently of the academic traditions and had itself, at first, little impact on the dominant scholastic Aristotelian interpretation of nature. While this knowledge was still largely transmitted in traditional social forms, that is by learning on the job within guild structures, it occasionally became the subject also of literary productions, as is illustrated by the writings of Leon Battista Alberti, Piero della Francesca, Leonardo da Vinci, and others. This knowledge thus became part of a new interpretation of nature and of man's place in it, entering an intellectual discourse in which alternatives were searched to the dominant scholastic interpretation of nature and society. Consequently the new technical knowledge, or rather its reflection in the new kind of technological literature, was brought, at least potentially, into conflict with Aristotelian interpretations of natural processes and technical devices. In the course of this entry of technical knowledge into an intellectual world it was brought into contact also with the heritage of antiquity, comprising not only alternatives to the Aristotelian theory of nature (e. g. Platonism or atomism) but also an unexploited richness of antique mechanical knowledge as represented by the *Mechanical Questions* of Pseudo-Aristotle and the writings of Archimedes.

In the sixteenth century this development led to the formation of a new category of intellectual (practical mathematicians and engineer-scientists such as Cardano, Tartaglia, Commandino, Oddi, Benedetti, Brahe, Kepler, Ricci, Stevin) who were no longer necessarily and, in any case, not completely involved in technical practice in the same way as the engineers themselves, but who rather specialized in the reflection of the new type of knowledge produced by this practice and, of course, in the attempt to make that reflection useful again for practical purposes. While their emergence as a technical elite gradually gained support from a new kind of institutionalized learning, exemplified by the Florentine Accademia del Disegno, their social status remained precarious throughout early modern times, making them dependent on the unreliable patronage of the courts and necessitating an equally unreliable overstatement of the practical relevance of their theoretical projects. It is exactly this group which formulated projects typical also for Galileo's research, for example Tartaglia's new science of ballistics or Benedetti's new science of motion in media. From what has been pointed out above concerning the inherent complexity of the new objects of knowledge it follows that in fact all of the engineer-scientists shared the problem of a considerable disproportion between their pretentious claims and their actual chances to attain success in their projects. Since they all searched for a new theoretical foundation of the practical knowledge in whose reflection they were engaged, they also necessarily shared an anti-Aristotelian attitude. Both their social status and their occupation make it understandable that they were, in addition, usually involved in competition and sometimes bitter controversies among themselves. Indeed, given the heterogeneous field of knowledge they were exploring, they could and did find reasons to search for alternative ways to create a "new science" of this or that subject. Nevertheless, for practically all of them the ancient works of mechanics, in particular the *Mechanical Questions* attributed at that time to Aristotle and the recently revived works of Archimedes, as well as the writings of Jordanus Nemorarius, provided a common core of mechanical knowledge which set the standard for any "new science" to be developed.

Guidobaldo del Monte precisely fits the characteristics of the engineer-scientist, apart from the fact that his social status as a feudal aristocrat made him independent of the unreliable patronage of the courts and of the search for a way to gain income from his passion.[1] His qualities as a practical man could not be better demonstrated than by the fact that, in 1588, he became inspector of Tuscan fortifications, but his competence was by no means restricted to the qualifications of a practitioner. At the beginning of his carreer he studied mathematics and philosophy, first at the university of Padua, later as a private disciple of Commandino at Urbino. Already shortly after he had finished his studies, he wrote one of the most influential books on mechanics of the century (published in Latin 1577, translated into Italian by Filippo Pigafetta and published 1581). His theoretical orientation may be represented by his commentary on Archimedes' work on the centers of gravity (published 1588). That he did, however, not conceive of theory as being separated from practical applicability is made clear by his credo as expressed in the preface to his mechanics:

> For mechanics, if it is abstracted and separated from the machines, cannot even be called mechanics.[2]

---

[1] For a succinct biography of Guidobaldo del Monte, see Drake 1987, 459, concerning his social status, see the discussions in Biagioli 1989 and in Allegretti 1992.
[2] Translation quoted from Drake and Drabkin 1969, 245.

This insistence of the relation between mechanics and machines was consequential also for the way in which Guidobaldo del Monte approached the problems of theoretical mechanics. In a letter from 1580 he wrote:

> Briefly speaking about these things you have to know that before I have written anything about mechanics I have never (in order to avoid errors) wanted to determine anything, be it as little as it may, if I have not first seen by an effect that the experience confronts itself precisely with the demonstration, and of any little thing I have made its experiment. (Guidobaldo del Monte to Giacomo Contarini, October 9, 1580; translated from Micheli 1992, 98)

Where this coincidence between theoretical conclusions and practical verification did not take place, such as it is, according to Guidobaldo, the case of mechanical processes involving motion, he tended to avoid the subject.[1]

Apart from his publications on mechanics, Guidobaldo del Monte wrote books on further topics which fit into the social pattern of an engineer-scientist with such an orientation. He published on geometry and perspective (*Planispheriorum universalium theorica* 1579; *Perspectiva* 1600). Further books have been published posthumously (*Problemata astronomica* 1609; *Cochlea* 1615). From a letter by his son Orazio written to Galileo after the death of his father[2] it is furthermore known that Guidobaldo had left several minor works unpublished (*In Quintum*; *De motu terrae*; *De horologiis*; *De Radiis in aqua refractis*; *In nono opere Scoti*; *De proportione composita*, and another booklet on instruments invented by him). In a letter by Muzio Oddi further minor works by Guidobaldo del Monte are listed, whose thematic range, as far as it is known, seems to fall within that represented by his other writings.[3]

Guidobaldo's intellectual profile which is represented by these publications and manuscripts is perfectly reflected by the contents of the notebook which contains also his protocol on the projectile trajectory experiment.[4] The notebook comprises 245 mostly numbered pages (together with some inserted sheets) and begins with extensive notes on sundials, continues with notes on a set of problems of plane geometry, which are followed by entries on mechanical problems, notes on spherical astronomy, and notes on geometrical problems of stereometry, and then, after a mixture of various entries comprising critiques of contemporary authors as well as further notes on sundials, the notebook contains extensive notes on perspective; at the end of the notebook one finds again a mixture of various entries, among them the protocol of the experiment on projectile motion. The notebook thus testifies both to Guidobaldo's practical and to his theoretical interests. It shows not only his familiarity with the tradition of antique mechanics

---

[1] See Gamba and Montebelli 1988, 76.
[2] Orazio del Monte to Galileo, June 16, 1610, Galilei 1890-1909, X: 371f.
[3] See Gamba and Montebelli 1988, 54.
[4] del Monte ca. 1587-1592, 236f.

but also his awareness of the works of his contemporaries, in particular of Commandino, Clavius, and Benedetti. Most remarkably, as mentioned already above, in his scattered entries on mechanical problems, Guidobaldo occasionally goes beyond his own book on mechanics, treating, e.g., the problem of the bent lever in close connection with the inclined plane. This improved treatment was apparently stimulated by a close reading and critique of Benedetti's book and most probably by a suggestion of Galileo. In addition to the more extensive sets of notes, one also finds scattered entries (in part discussed above) on such diverse topics as two methods of describing a hyperbola, problems of artillery, the motion of heavy bodies in media, the reflection of light by a mirror, the motion of the center of gravity of the earth, astrology, the sound of chords, and water supply for mills.

While the notes on the projectile trajectory experiment fit, as we have seen, perfectly into the chronological order of the entries in Guidobaldo's notebook, their content shows a certain contrast to the bulk of the other notes. In fact, with a few further exceptions, these notes correspond to Guidobaldo's intellectual profile as it is also known from his other writings and his correspondence. The protocol of the experiment, on the other hand, belongs to the small group of entries which have no counterpart in Guidobaldo del Monte's publications and seem not to belong to the areas of his main interests. The notes on projectile motion as well as other entries, some of which belong to this exceptional type, correspond, however, to topics known as having been subject of the work of Galileo during his time in Pisa. Besides the projectile trajectory, these subjects are the motion in media, Heron's crown problem, the inclined plane, and the sound of chords. From the correspondence between Guidobaldo and Galileo we know in fact that they exchanged since early 1588 not only letters but also copies of their work which unfortunately have not survived. It is therefore no surprise to find among Guidobaldo's notes some that appear to be related to Galileo's contemporary interests as they are represented by his early works *La Bilancetta* and his treatise *De Motu*. Therefore, whereas in general the young Galileo learned from his patron Guidobaldo del Monte, it may well be that, in this case, Galileo challenged his older colleague with subjects that did not belong to his familiar areas of competence. It may have been precisely because the study of motion was not among Guidobaldo del Monte's main concerns that he did, for all we know, not pursue the line of research suggested by the unexpected outcome of the experiment which would have led him immediately to discover the law of fall implicit in the result of this experiment. To take this last step remained, however, the privilege of Galileo, as far as we know.

Galileo in the Footsteps of Guidobaldo del Monte

Galileo's approach to the knowledge of his time followed along a path that brought him into closer contact with academic traditions than it was apparently the case for Guidobaldo del Monte. His intellectual development involved in fact two strands, a more technical one and a more philosophical one. Galileo's family background and education placed him among the engineer-scientists, while his early academic career introduced him into the scholastic philosophy of his time. In the following, we will see that it was primarily the contact with Guidobaldo del Monte which, in a decisive moment of Galileo's intellectual development, encouraged him to take up the life-perspective of the risky but rewarding career of an engineer-scientist.

Galileo's father Vincenzio was a musician and theoretician of music from which the young Galileo could learn much not only about the wide and curious field of acoustic phenomena produced by instruments but also about the possibility of applying mathematics to such phenomena. Thus it became an ongoing theme of interest to him, to which he dedicated a number of ingenious observations which are like gems interspersed with the wealth of his writings on diverse subjects, ranging from comets to mechanics. The experimental acuity of such observations may be illustrated by an episode of the *Discorsi* in which Galileo shows how to transform an artisanal operation, the scraping of a brass plate with a sharp iron chisel in order to remove spots from it, into an operation performed for the purpose of generating knowledge about the frequencies of sounds. In other observations of this type, he studied the dependence of the height of a tone from the size, material, and tension of a string, varying the latter by attaching different weights to it. Galileo also related the vibrations of the strings of an instrument to the swinging of a bell and to oscillations of a pendulum and derives from this comparison an explanation for consonance phenomena:

> The cord struck begins and continues its vibrations during the whole time that its sound is heard; these vibrations make the air near it vibrate and shake; the tremors and waves extend through a wide space and strike on all the strings of the same instrument as well as on those of any others nearby. A string tuned in unison with the one struck, being disposed to make its vibrations in the same times, commences at the first impulse to be moved a little; (...) it finally receives the same tremor as that originally struck, and its vibrations are seen to go widening until they are as spacious as those of the mover. This wave action that expands through the air moves and sets in vibration not only other strings, but any other body disposed to tremble and vibrate in the same time as the vibrating string. If you attach to the base of the instrument various bits of bristle or other flexible material, it will be seen that when the harpsichord is played, this little body or that one trembles according as that string shall be struck whose vibrations are made in time

with it. The others are not moved at the sound of the string, nor does the one in question tremble to the sound of a different string. (Galilei 1974, 99f)

Such observations and interpretations go probably back to experiments which Galileo's father may have performed together with him around 1588-89.[1] At least, when Galileo visited Guidobaldo del Monte in 1592 on his way to Padua, he must have been already so familiar with this explanation that it was made a topic in their discussions. In fact, as has been mentioned already, immediately before the protocol of the projectile trajectory experiment Guidobaldo del Monte made a note on experiments with different strings put in defined tensions by attaching weights to them. He describes the dependence of the height of a tone on the size, material, and tension of the strings, emphasizing the relation between the tone of a string and a characteristic motion ascribed to it. From this relation, an explanation of consonance is developed that is essentially identical with that given by Galileo in the *Discorsi*:

> From this one can also give the reason by which cause, if two instruments are close to each other which have many strings and if a straw is placed on the strings of one of them and if on the other one a string is touched, one then hears that that string of the other instrument which will be in unison with the one that is touched also sounds, and the others do not sound. And this could be produced by that [reason] that the air of the string that is struck because of its agitation moves all the other strings, but because those that are not in unison cannot receive the same motion of that which is struck, while that which is in unison can receive it, only this one sounds, and the others do not sound.[2]

The similarity between these passages confirms once more that some of the entries in Guidobaldo del Monte's notebook must have indeed been written under the influence of his discussions with Galileo. The dating to the year 1592 of this entry furthermore shows that it must have been indeed Galileo's family background that had brought him first into contact with the ways in which new knowledge was acquired on the basis of practical experience by the engineer-scientists.[3]

It seems, however, that Galileo's father was not only familiar with the fascination of searching for a theoretical formulation of practical knowledge but was also aware of the precarious social status of those who gave in to this fascination and made it the preoccupation of their professional careers. He decided, in any case, to save the young Galileo from the uncertain fate of an engineer-scientist and to rather secure him a more ordinary career by pressing him to enter a field with a low level of certain knowledge but a high level of guaranteed income, medicine.

---

[1] See Drake 1987, 17 and Settle 1996.
[2] del Monte ca. 1587-1592, 235, see Gamba and Montebelli 1988, 182 for a transcription.
[3] The role of Galileo's family background has been emphasized in particular in Settle 1996.

But even though Galileo initially ceded to his father's wish and began the study of medicine, he did not give up his interests and pursued, under the guidance of the engineer-scientist Ostilio Ricci, studies of mathematics and mechanics, comprising the works of Euclid and Archimedes.[1] Not long after these initial studies, Galileo's striking mastery of Euclidean geometry and Archimedian proof techniques testify to the success of Ricci's teaching. Both the fascination by and the competence in these matters eventually became so strong that the young Galileo even succeeded in convincing his father that mathematics and mechanics had to become his professional occupation. His first independent writing, the short treatise *La Bilancetta probably* written around 1586, accordingly dealt with the construction of an instrument based on Archimedian theory but designed for a practical purpose, that of determining specific gravities. Later he worked out highly specialized theoretical problems in the Archimedian tradition, such as the problem of the center of gravity of parabolas.

In the course of his university studies Galileo also encountered another tradition of antique knowledge, the Aristotelian philosophy dominating the intellectual world of his time. Whether he took up its intellectual challenge or simply because he wanted to increase his chances for gaining an income by teaching, he began to thoroughly familiarize himself with this philosophy, and in particular with Aristotelian physics. At that time he started to compose his treatise on motion which has already been mentioned above. It was originally written in dialogue form but later changed into a more systematic elaboration, treating the fundamental assertions of the Aristotelian theory of motion. Using Archimedian concepts, such as extrusion, in order to analyze traditional Aristotelian problems, such as motion in a medium, Galileo succeeded in giving his treatise an anti-Aristotelian twist that made it possible for him to pose as a natural philosopher developing a theoretical foundation of his own.

Galileo's technical elaborations of Archimedian problems brought him into contact with some of the leading mathematicians and engineer-scientists of his time, such as Clavius. Clearly, by far the most consequential contact was that to Guidobaldo del Monte. When Galileo began an exchange with him at the beginning of 1588, he happened to work on topics closely related to Guidobaldo del Monte's main occupation at that time. While initially Galileo and Guidobaldo just exchanged letters concerning the technicalities of a proof developed by Galileo, they soon came into close scientific cooperation. From the few surviving letters we can in fact conclude that they not only must have kept each other informed about their scientific interests by, at least over some periods, an almost day-by-day correspondence but that they also regularly exchanged their works for mutual criticism.

---

[1] For the contents of Ricci's teachings, see Settle 1971.

From the letters between Galileo and Guidobaldo del Monte we know, for instance, that the latter sent Galileo a copy of his commentary on Archimedes as soon as it was printed, asking for Galileo's criticism:

> I believe that in your modesty you say that you like my book which I sent you, but I pray you as much as I can, please warn me if there is anything with it, because I still have all the books at hand and it would be an easy thing to correct it where necessary. I would be very grateful if you would do me this favor. (Guidobaldo del Monte to Galileo, May 28, 1588, Galilei 1890-1909, X: 33)

It is therefore also plausible that an entry found in Guidobaldo del Monte's notebook containing a treatment of the problem of the inclined plane which contradicts the way he treated it in his *Mechanics* derives from an exchange between Guidobaldo and Galileo. In his entry, Guidobaldo treats the inclined plane in fact no longer following Pappus as it was done in his *Mechanics* and in earlier parts of the notebook, but rather in precisely the way found in Galileo's contemporary treatise *De Motu*, where it is derived from the bent lever. In his *De Motu*, Galileo actually criticizes the treatment of Pappus, thus implicitly criticizing also his patron Guidobaldo. In view of the close relationship to Guidobaldo precisely in these years, it is indeed unlikely that he should have hidden this criticism from his older colleague.

In spite of this intellectual closeness, there was, however, also a remarkable difference in their interests. It is, in particular, hardly imaginable that Guidobaldo with his emphasis on rigorous Archimedian proofs, on the one hand, and on practical applicability of mechanical knowledge, on the other, had much sympathy for the subtle problems of natural philosophy addressed by Galileo in his treatise *De Motu*. There are, in fact, a few hints pointing at this diversity of intellectual orientations. Guidobaldo was familiar with Benedetti's work which includes a theory of motion very similar to that of Galileo, that is, a theory also involving the Archimedian extrusion principle. Benedetti's major book, *Diversarum Speculationum*, is mentioned in Guidobaldo's notebook where its mechanical foundations are heavily criticized. In the notebook one also finds a short passage where a theory of motion in media is sketched which is similar to both Benedetti's and Galileo's approach in that it also makes use of extrusion but which diverges in the precise formulation of the basic laws of motion. This passage may well be a reaction by Guidobaldo to Benedetti's or Galileo's speculations but apparently remained without any consequence for his work.

This reluctance to accept theoretical considerations concentrated on problems of Aristotelian natural philosophy may also explain a short remark in a letter by Guidobaldo written at the time when it is generally assumed that Galileo had completed at least a first version of his *De Motu*. The remark expresses Guidobaldo's satisfaction with Galileo's return to problems of the center of gravity, the initial point of the common interest between them and the area in which he especially appreciated Galileo's competence:

> Because I did not have any letters from you for many days, your [letter] pleased me greatly (...)

> Moreover, I was very pleased to see that you have returned to the center of gravity; and you have done enough, having found what you wrote to me [et ha fatto assai haver trovato quanto mi ha scritto]; and I also have found some things but I cannot conclude my search for a certain tangent which drives me to despair, because it seems to me that I have found it by following a certain path, but I cannot demonstrate it and clarify it in my own mind with the demonstration: but your letter consoled me greatly, because I see that you search and do not conclude your search so quickly, whereas I am [usually] not surprised when I do not find [something]. But do not be surprised if I still do not send you what I promised to show you, due to the fact that I need to copy a lot of things; but as soon as I can, I will send them to you, because what I really appreciate above all else is having your opinion. (Guidobaldo del Monte to Galileo, December 8, 1590, Galilei 1890-1909, X: 45)

Considering the dramatic changes in Galileo's scientific activities after his move to Padua, the initial difference between the scientific interests of Guidobaldo del Monte and Galileo become even more evident. We do not know when Galileo first met Guidobaldo del Monte personally, but there is no evidence that this happened before he visited him on his way to Padua in 1592. It is true that Galileo already at that time was a multi-talented intellectual with a broad spectrum of interests and considerable competence also in the field of technology. However, compared with the scope of activities of an engineer-scientist like Guidobaldo del Monte as a supervisor of fortifications involved in large-scale projects, and running his own workshop which offered facilities for experimentation and production of instruments, Galileo's activities in such areas must have looked very modest. This, at least, would explain why Galileo in the following year drastically changed his fields of interests and in many respects copied the types of activities which were characteristic of engineer-scientists in general at that time, and in particular, the activities of Guidobaldo del Monte of running his own workshop, of inventing and producing instruments, and of reflecting and taking notes on a specific set of topics such as mechanics, military technology and architecture, practical geometry, and surveying. He thus returned to those activities which must have impressed him as a young man when he took private lessons from Ostilio Ricci.

In fact, one of the first works which Galileo produced after this "practical turn" of 1592 was his treatise on mechanics, very much in the style of Guidobaldo's book[1] in its combination of rigor and concentration on the simple machines. Galileo's mechanics was circulated only in manuscript, also because it was used for teaching purposes.[2] This was also the case for his introductions into military architecture and fortification dating from approximately the same time, which also have survived in manuscript copies. In 1596 he composed a treatise on measuring heights and distances by sighting and triangulation, probably also used for teaching purposes. Galileo's private disciples, among them young noblemen coming from various European countries, were in fact an important source of income for him, in addition to the modest salary he received from the university, and perhaps a remuneration he received for the horoscopes he prepared. His first real publication appeared only in 1606 and was dedicated to a military compass he had designed about 1597, following the example of Guidobaldo del Monte. This compass, as well as other instruments designed by Galileo, such as an instrument for gunners developed in 1595 or 1596 also following the example of Guidobaldo,[3] were produced by a Paduan artisan and eventually in a workshop of his own. The treatise on the military compass had also at first been used for private lessons and was finally published only because Galileo intended to protect his invention. His technological concerns are, in fact, perhaps more characteristic for the beginning of his Paduan period than his writings. As early as 1593, he received a Venetian privilege for a machine to raise water and was consulted by a Venetian official on matters of naval architecture. Around that time Galileo must have begun to frequent the recently expanded arsenal of Venice which made a lasting impression on him, as is well known from the opening to the *Discorsi*.

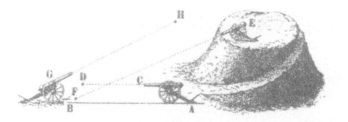

Figure 9. Artillery shots in Galileo's treatise on fortification[4]

---

[1] Historians of science usually try to point out the differences of Galileo's work on mechanics to all the numerous treatments of this subject by other authors, often following the *Mechanical Questions* of Pseudo-Aristotle. Even if one admits the alleged superiority of Galileo's treatment of the subject, it has to be emphasized that Galileo's work with regard to its canonical contents and the intention to improve the program of Pseudo-Aristotle's *Mechanical Questions* of explaining all mechanical devices by reducing them to the lever perfectly fits into this tradition.

[2] For this and the following, see Drake 1987, chap. III, 33- 49. See also Wohlwill 1993, 1: 141.

[3] See Schneider 1970.

[4] Figure reproduced from Galilei 1890-1909, II: 93

As a result of this practical turn, Galileo handled in his early Paduan period topics in a way quite different from how he dealt with them in his unpublished *De Motu* manuscript. Not the structure of the object but the purpose of the knowledge to be gained about it determined predominantly how an object was treated. A typical example is the treatment of different types of shots in his treatise on fortification. In contrast to the way he dealt with problems of artillery in the *De Motu* manuscript, here the trajectories are presented as if they were straight lines (see figure 9), because in the context of teaching the nomenclature of artillery any further differentiation would contribute nothing to the purpose of making his disciples learn the military vocabulary.

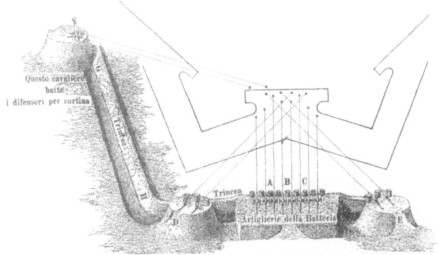

Figure 10. Artillery shots in Galileo's treatise on military architecture [1]

Also in his treatise on military architecture, Galileo depicted trajectories simply as straight lines (see figure 10), in this case, however, for a different reason. By the time of Galileo, the constructions of military architecture had achieved considerable improvements with regard to the resistance of this architecture against destruction by artillery. These improvements were based on a sophisticated application of geometry to the shapes of fortification buildings. Complex shapes of such buildings prevented the possible aggressor in the case of a military attack to bring his weapons into a position favorable for the intended destruction of these buildings. On the other hand, these improvements required also enhanced geometrical knowledge on the side of the aggressor in an attack. In the case of the example from Galileo's treatise on military architecture given here, Galileo argued for a suitable design and placement of platforms for the aggressor so that

---

[1] Figure reproduced from Galilei 1890-1909, II: 51

the defender is unable to use the geometry of the fortification building for an effective defense. In this case, the abstract representation of possible directions of the artillery shots by straight lines is even more adequate to the problem than a more realistic representation including the shapes of possible trajectories.

On the background of such a practical turn, it therefore comes as no surprise that Galileo was similar to Guidobaldo del Monte also in that one respect which is our central concern here: For a long time to come Galileo did not show any theoretical interest, just as it was the case for Guidobaldo del Monte, in the puzzling outcome of the projectile trajectory experiment and its implications, that is, the symmetry of the trajectory, the simultaneous effects of a violent and a natural component of the motion in every point of the trajectory, and the quadratic function determining the downward acceleration. What appears to the modern historian of science as a blunt contradiction to the theory developed in Galileo's unpublished *De Motu* manuscript was for Galileo himself, as we will see, merely an incentive for a slight modification which made this theory compatible with the outcome of the experiment. For the time being, however, he left his treatise on motion incomplete in favor of dealing with problems of motion only in the context of practical applications, such as problems of artillery.

A manuscript page probably from the Paduan period shows clearly what kind of use Galileo intended to make of the outcome of the experiment performed together with Guidobaldo del Monte. The page contains the outline of a treatise which shows that Galileo planned to make the curve of the trajectory the core topic of a specific treatise, which, however, can by no means be considered as substituting his discarded *De Motu*, but was rather designed as a work in the style of the treatises on artillery of the time. Galileo's treatise would thus have substituted Tartaglia's influential *Nova Scientia*. The treatise was planned to contain the treatment of the following topics:[1]

> Particular privileges of the artillery with respect to the other mechanical instruments.
> Of its force and from where it proceeds.
> If one operates with a greater force in a certain distance or from nearby.
> If the ball goes along a straight line if it is not [projected] along the vertical.
> Which line the ball describes in its [course].
> On the course and the time of charging the canon
> Impediments which render the canon defective and the shot uncertain
> On mounting [the canon] and dismounting it
> On the production of the caliber
> On the examination of the quality and the precision of the canon

---

[1] Galilei ca. 1602-1637, folio page 193r. In the dating we follow Drake, although no direct evidence for this dating is available. This dating is, however, strongly suggested by the text. The dominance of practical interests in the intended publication fits perfectly into the work of Galileo in Padua after his practical turn. The practical interests which the outline represents were pursued by him up to the composition of the *Discorsi*, but never appeared so pure again as here.

If, when the canon is longer, it shoots farther and why
In which elevation you shoot farthest and why
That the ball in turning downwards in the vertical returns with the same
forces and velocities as those with which it went up
Various balls [specially] prepared and lanterns and their use

This table of contents of Galileo's planned and never written treatise clearly
shows that he was well aware of the practical importance of the outcome of the
experiment recorded in Guidobaldo del Monte's notebook. In particular, he refers
to the symmetry of the trajectory, its continuously curved shape, and its dynamical composition exclusively in terms of improving the precision of artillery. Possibly, however, Galileo's planned treatise was intended only for purposes of
private teaching. Such a usage of the knowledge acquired by the experiment on
projectile motion certainly fits well with Galileo's efforts – extended over a
period of 40 years – not to make this knowledge publicly available.

In summary, after his move to Padua, Galileo not only came in close contact but
shared all essential characteristics with the engineer-scientists of his time. This
development indicates a radical reorientation of his life. Galileo had been earlier
engaged primarily in Aristotelian physics, which was not only the official dogma
of the church but also the main basis for interpreting nature at that time. He was
also familiar with and even fascinated by other ancient traditions, in particular the
mathematical methods represented by Euclid's Elements and even more by the
application of these methods by Archimedes.

Through Guidobaldo del Monte Galileo came into close contact with the experimental techniques and the research interests of a leading engineer-scientist. The
theoretical basis of these techniques was not so much Aristotelian physics but
rather the ancient tradition of mechanics as it was represented by the Pseudo-
Aristotelian *Mechanical Questions* and numerous contemporary treatises in this
tradition. As a consequence of this reorientation, Galileo's move to Padua not
only marks an advance in his academic career but also a practical turn, that
seemed to leave the theoretical interests of his Pisan years far behind. The discoveries of the parabolic shape of the trajectory and of the law of fall in 1592 therefore did simply not come at the right moment in his life for having any dramatic
consequences on his theory of motion.

Galileo and Paolo Sarpi—Towards a New Science of Motion

The fact that Galileo changed his activities so drastically towards contemporary
technology and its seemingly adequate theoretical foundation in mechanics does
not imply, however, that his previous activities were forgotten without any trace.
On the contrary, life in Padua had a dimension which was quite different from
what his patron Guidobaldo del Monte represented to him. Both in the house of

Giovanni Vincenzo Pinelli and at the university, Galileo encountered numerous intellectuals who were engaged in the great debates of the time, in theology, in Aristotelian philosophy, in astronomy, as well as in other fields. In these encounters Galileo had occasion to continue to pursue and even to develop the interests he had cultivated in the academic environment of Pisa.

Nobody within this group seems to have had a greater affinity with Galileo's own interests than Paolo Sarpi whom he met, for all we know, in 1592 in the house of Pinelli. A comparison of how discussions with Galileo were reflected in the notebooks of Guidobaldo and Sarpi shows that surely Paolo Sarpi appreciated the competence of the young Galileo for quite different reasons than Guidobaldo del Monte. In addition to the note on the result of the projectile trajectory experiment in the notebook of Sarpi, there are a number of topics which indubitably reflect discussions with Galileo Galilei, such as the latter's tidal theory summarized in notes that can be dated to 1595.

The affinity between Sarpi and Galileo and its distinction from the intellectual interests that had joined Guidobaldo and Galileo become clear from the range of topics covered by Sarpi's notes as well as from the theoretical focus in the treatment of these topics. An outstanding example is provided by their joint fascination with a theory of the type of Benedetti's theory of motion in media, a fascination that was, it seems, not shared by Guidobaldo. It is well known that Galileo's *De Motu* shows many similarities with Benedetti's writings published between 1583 and 1585. But even without these similarities, it is hardly believable that Galileo could have been unaware of these writings when he was composing his treatise, although Benedetti's name is not mentioned. It does appear, however, in the notebook of Guidobaldo del Monte, who heavily criticized Benedetti's mechanics on the very same pages which also contain his note on Galileo's derivation of the law of the inclined plane from the law of the bent lever. It is therefore plausible to assume that also the theories of motion advanced by Benedetti and Galileo were a subject of discussion between Guidobaldo and Galileo. In fact, the motion of bodies in media is treated in Guidobaldo del Monte's notebook, as we have mentioned. In his note on that subject Guidobaldo attempted, however, to formulate an alternative approach which remains closer to Aristotelian dynamics than that of Benedetti and Galileo.

This topic remained, in any case, only a passing interest of Guidobaldo's, whereas it was central to Sarpi's thinking on motion. In fact, the problem of natural motion and its explanation in contrast to Aristotelian physics was, as Sarpi's notes show, of central concern to him. In the 1580s, that is, long before he first met Galileo, he had intensively studied Benedetti's theory of motion in media, attempted to further develop it, performed experiments related to it, and dedicated extensive notes to this theory which show him as an ardent adept of Benedetti. In view of the close similarity between Benedetti's theory and Galileo's *De Motu*, it therefore comes as no surprise that Sarpi and Galileo must have immediately after their first encounter entered intense discussions on problems of motion.

Their common interests were, in fact, not limited to the laws of motion in media, but also comprised the problem of the motion of the earth, the relation between violent and natural motion, the explanation of violent motion by an impressed force, the relation between the effects of motion and of weight, the explanation of accelerated motion, and last but not least the shape of a projectile trajectory.[1] In spite of the many differences in detail between Sarpi's notes and Galileo's ideas as they are known from *De Motu*, their writings represent strikingly many common interests and questions, although they had not been acquainted with each other at the time these writings were produced. In particular, they both had views on projectile motion which suggested that, in general, the projectile trajectory cannot be symmetrical. In the tradition of Tartaglia's influential *Nova Scienza* Paolo Sarpi reconstructs, for instance, as Galileo did in his *De Motu*, projectile motion as resulting from the interplay of natural and impressed violent motion:

> The distinction between the violent and the natural is deduced from the principle that it is either outside or inside, the natural being the force that is inside and has formed the body and has provided it with its means [of motion]. The violent one which has been introduced by an external force soon diminishes, because it has no means to continue its action and finds a contrary and more powerful force from the inside so that the first [the natural] has the subject disposed in its way and fights and expulses the second [violent] one; the second is more powerful in the beginning and dominates but then it is forced to recede and loses. (Note number 100 in Sarpi 1996, 123f)

This reconstruction of projectile motion, according to the few dates which Sarpi inserted between the notes written at some time between 1578 and 1584, is in sharp contrast to the note corresponding to Guidobaldo del Monte's protocol on the projectile trajectory experiment, which Sarpi wrote in 1592 and, according to our interpretation, after the first encounter with Galileo. But Sarpi's note on the outcome of this experiment does not simply contradict his earlier reflection on that matter. At the same time, it shows how the puzzling symmetry of the trajectory can be made compatible with the conception of the interaction of violent and natural motion which was followed by both Sarpi and Galileo: In order to reconstruct the symmetrical trajectory it has to be merely assumed that the relation between violent and natural motion in ascending and descending is exactly the same, but with exchanged roles. This consequence has, indeed, explicitly been drawn in both Guidobaldo del Monte's protocol and Paolo Sarpi's note, and it is surely also the way Galileo brought the puzzling symmetry of the projectile tra-

---

[1] The editors of Sarpi's notebook conclude from these similarities that many of the ideas of Galileo from that time are taken from Paolo Sarpi; see Sarpi 1996, 400. This conclusion is unconvincing because it neither takes into account the common sources used by both of them, nor does it take into consideration the diverging developments of Galileo's and Sarpi's thought during the time of their close interaction.

jectory in accordance with his theory of motion as it was developed in his treatise *De Motu*. Thus, the discovery must have been conceived by all of them at the beginning as being much less dramatic than it must have looked from a later perspective. The reason for Galileo's negligence of the consequences of the discovery of the parabolic shape of the projectile trajectory was, therefore, not only the shift of his interests towards applicable technological knowledge which was associated with his move from Pisa to Padua but also the simplicity with which the new insight could be made compatible with his former beliefs.

That Sarpi's notes after 1592 indeed also reflect, albeit in an indirect way, discussions about views held by Galileo becomes clear from the context of Paolo Sarpi's entry on the symmetry of the projectile trajectory (comprising the notes with the numbers 537 and 538) among his other entries. He must have had extensive discussions on Galileo's theory developed in *De Motu* which are reflected in a series of entries immediately surrounding those on the projectile trajectory. These are essentially the notes from number 532 which is the first note on motion for the year 1592, to note number 542, which according to Sarpi's dates was written still in the same year. Furthermore a number of later notes show that Paolo Sarpi and Galileo kept in contact still discussing problems which are essentially based on Galileo's theory in *De Motu*.

This is not the place to discuss these reflections on Galileo's theory in detail. Nevertheless, a short overview has to be given in order to gain indications on Galileo's thoughts in this intermediate period of latent development of his natural philosophy, on which otherwise the historical documentation is so scarce. The notes start with several arguments focussing on the impression of force into a projectile and the force of percussion which a projected body exerts. The last entry preceding the notes 537 and 538, which deal with the projectile trajectory, essentially represents a summary of Galileo's explanation of acceleration in *De Motu*. The entries immediately following these notes concern the refutation of a physical counter-argument against Copernicanism involving the dissipation of objects from the rotating earth, an argument that is well known from Galileo's later *Dialogo*.[1] The subsequent two notes of the year 1592 are about bodies and their weights and motions in media (notes 541 and 542), the central topic of *De Motu*.

A note somewhat separated from the bulk of the "Galilean" notes described so far, but still dating from 1592, deals with the motion of a pendulum and the motion of a spinning top as examples of ongoing motions (note 547), a topic also discussed in Galileo's *De Motu* (Galilei 1890-1909 I: 335). The first entry made in 1593 (note 558) concerns the refutation of Aristotle's proportional relation between the weight and the speed of a falling body, again an argument exten-

---

[1]   One of the oldest indications of Galileo's adherence to Copernicanism, overlooked by the editor. In a later year, 1595, there are entries on Galileo's tidal theory; notes 569, 570 and 571 in Sarpi 1996, 424-427. Only concerning these latter entries the editor discusses their possible relation to Galileo's Copernicanism. See also Drake 1987, 37.

sively discussed in *De Motu*. When Galileo visited Venice once more in 1595, he apparently returned to his intensive discussions with Paolo Sarpi. Sarpi's notes of 1595 in fact contain a discussion of Galileo's tidal theory based on his Copernican views (notes 569-571).

This short overview of the notes which probably reflect discussions of Paolo Sarpi with Galileo shows that the results of the experiment on the projectile trajectory was not *the* central issue that bothered them really. The emendation of the Aristotelian explanation of the trajectory by the interplay of violent and natural motion required for explaining its symmetry had seemingly settled any dispute on this matter. On the other hand, Sarpi's notes nevertheless show that serious difficulties remained, and it even seems that Paolo Sarpi himself was somewhat hesitant to fully accept the symmetry of the trajectory. Note number 535 at least contains what was probably his strongest argument against the explanation of the symmetry of the trajectory by assuming a corresponding symmetry of violent and natural motion in the ascending and descending part of the trajectory and nothing indicates that Galileo had a sufficient answer to the problem. Paolo Sarpi compares in this note the force impressed into the projected body (for the sake of a drastic demonstration, his argument involves the arrow of an arquebus[1] shot vertically into the air) with the force which it is able to expend when it comes down, coming to the conclusion that these forces cannot be equal. It even cannot be excluded that he checked this conclusion by an experiment. In any case, the outcome of such an experiment would have been clear. Sarpi wrote:

> Reason would have it that when a heavy body has all the force to go up that it can receive, it weighs as much in going down as in going up in the same places distant from the starting points. But it is clear that an arquebus strikes through a table by way of the bullet that passes it, whereas, who would charge it even more, so that [the bullet] would go up much higher, [the bullet] in coming down would nevertheless hardly leave any mark on the table. (Note number 535 in Sarpi 1996, 391-393)

Even more than a decade later Paolo Sarpi confronted Galileo with this counterargument against the symmetry of upward and downward motion. They were still in touch around that time, had meanwhile exchanged their views on many other subjects, such as magnetism and its treatment by Gilbert,[2] but had apparently left their discussions of motion essentially at the point they had reached in the 1590s. In 1604 Paolo Sarpi returned to the subject, bringing up in a letter he wrote to Galileo once again questions on which he had made notes at the time when they first met, twelve years before.[3] On this occasion, he reminded Galileo of what they had agreed upon and on what the problems were that their discussions had left open. In his letter Sarpi thus confirms that Galileo was actually the source of

---

[1] An arquebus is an early type of a portable gun supported on a tripod or on a forked rest.
[2] See Paolo Sarpi to Galileo, September 2, 1602, Galilei 1890-1909, X: 91f.
[3] Paolo Sarpi to Galileo, October 9, 1604, Galilei 1890-1909, X: 114.

some of the notes he made in 1592 by unambiguously ascribing the views recorded in these notes to him. But the letter also shows that Sarpi himself continued to be puzzled by questions earlier recorded in these notes, in spite of his discussions with Galileo at the time. These questions concern, firstly, the quantities of impressed force which bodies of different kinds can receive and, secondly, the very counter-argument that Sarpi had earlier raised against Galileo's claim of the symmetry of the trajectory of projectile motion:

> We have already concluded that a body cannot be thrown up to the same point [termine] if not by a force, and, accordingly, by a velocity. We have recapitulated – so Your Lordship lately argued and originally found out [inventò ella] – that [the body] will return downwards through the same points through which it went up. There was, I do not remember precisely [non so che], an objection concerning the ball of the arquebus; in this case, the presence of the fire troubles the strength of the argument. Yet, we say: a strong arm which shoots an arrow with a Turkish bow completely pierces through a table; and when the arrow descends from that height to which the arm with the bow can take it, it will pierce [the table] only slightly. I think that the argument is maybe slight, but I do not know what to say about it. (Paolo Sarpi to Galileo, October 9, 1604, Galilei 1890-1909, X: 114)

Galileo responded to this question in his famous letter written a week later, on October 16, 1604:

> Concerning the experiment with the arrow, I believe that it does acquire during its fall a force that is equal to that with which it was thrown up, as we will discuss together with other examples orally, since I have to be there in any case before All Saints. Meanwhile, I ask you to think a little bit about the above mentioned principle. (Galileo to Paolo Sarpi, October 16, 1604, Galilei 1890-1909, X: 116)

Just as Sarpi had done in his letter, so also Galileo alluded to what both men had discussed before, in particular the law of fall and the symmetry of projectile motion. However, as is well known (and is also evident from Galileo's brief rebuttal in the above passage), the emphasis of this letter is not on the experimental aspect of the problem raised by Sarpi,[1] but a problem of a quite different kind, condensed in the invitation to think about a new "principle."

> Thinking again about the matters of motion, in which, to demonstrate the phenomena [accidenti] observed by me, I lacked a completely indubitable

---

[1]   The counter-argument of Sarpi was a serious problem for Galileo which he probably solved only late and eventually mentioned in the *Discorsi*, see Galilei 1974, 95f and 227-229 . That he solved this problem only late is indicated by a manuscript note in the hand of his son Vincenzio (born in 1606), where this problem is still mentioned as an open one, see Galilei 1890-1909, VIII: 446.

principle which I could pose as an axiom, I am reduced to a proposition which has much of the natural and the evident: and with this assumed, I then demonstrate the rest; i.e., that the spaces passed by natural motion are in double proportion to the times, and consequently the spaces passed in equal times are as the odd numbers from one, and the other things. And the principle is this: that the natural moveable goes increasing in velocity with that proportion with which it departs from the beginning of its motion; (...)

Without doubt, Galileo's letter to Sarpi documents a fundamental change in his thinking on motion. Whereas Galileo had earlier used his knowledge about the parabolic shape of the trajectory only in the context of practical mechanics, he now evidently aimed at constructing a deductive theory of motion based on a principle which seemingly follows directly from his experiences with the force of percussion and is apparently confirmed by the symmetry between vertical projection and free fall.[1] This principle postulates the increase of the degrees of velocity in proportion to the distance traversed. What is really new in his letter to Sarpi (and the reason why historians of science have paid the utmost attention to it) is the fact that Galileo at that time obviously had discovered a set of properties of accelerated motion, the most prominent one of them being the law of fall, which he now tried to arrange into a deductive system. Unfortunately, the letter is not very explicit in the enumeration of these discoveries. In his response to Sarpi he furthermore mentioned a proof for the law of fall but did not even give this proof but mainly served himself of the new principle in order to argue for the symmetry between vertical projection and free fall, that is, in order to treat precisely the issue that was controversial between him and Sarpi.

Is there any information about the developments that led Galileo from his occupation with practical mechanics to this challenging new theoretical program? Indeed, there has been preserved a letter to his patron Guidobaldo del Monte written two years earlier which makes perfect sense when interpreted in the context of a reorientation from practical mechanics to the goal of creating a new deductive theory of motion.

Galileo's letter to Guidobaldo del Monte from the 29th of November, 1602, is only one of a series of letters exchanged between Galileo and Guidobaldo del Monte at that time which are all lost with the exception of this one. From its content it can be concluded that Galileo must have informed Guidobaldo del Monte about a set of new discoveries, among them the isochronism of the pendulum and the isochronisms of motion on inclined planes which can be represented as chords of a circle. He furthermore had informed Guidobaldo del Monte about experiments to verify the discoveries.

---

[1] See Galilei ca. 1602-1637, folio 128 for Galileo's mention of percussion; for the argument concerning the symmetry between vertical projection and free fall, see the discussion in Damerow, Freudenthal, McLaughlin, and Renn 1992, 169f.

Galileo's earlier and now lost letter must have been interpreted by Guidobaldo del Monte as a deviation from what he thought to be standards for working on mechanics. Guidobaldo del Monte had meanwhile checked with his methods Galileo's discoveries and must have come to the result that they cannot be maintained. The letter by Galileo which has been preserved is a reaction to Guidobaldo del Monte's critique. It is friendly and humble in its tone, as was customary between them, but shows clearly that Galileo had, in Guidobaldo's eyes, departed from their common ground in mechanics as Galileo had learned it from his patron.

In particular, Guidobaldo del Monte had probably taken up one proposal of Galileo, that is, to check the isochronism of the pendulum by what supposedly was an identical arrangement, a ball rolling within a semi-sphere. The result, however, turned out to be negative; Guidobaldo del Monte tried directly to compare the times of two different balls descending from different heights to the bottom of a spherical "box" and did not find that these times of descent were the same. He possibly tried the same experiment also with inclined planes in a spherical container with an equally negative result. Guidobaldo del Monte must have informed Galileo about his negative result and must have added a critique stating that Galileo violated the principles of mechanics with the result that his discoveries were obviously absurd.

Galileo's answer, which fortunately has been preserved, starts with a humble, but insistent reiteration of his claims:

> Your Lordship, please excuse my importunity if I persist in wanting to persuade you of the truth of the proposition that motions within the same quarter-circle are made in equal times, because having always seemed to me to be admirable, it seems to me [to be] all the more so, now that your Most Illustrious Lordship considers it to be impossible. Hence I would consider it a great error and a lack on my part if I should allow it to be rejected by your speculation as being false, for it does not deserve this mark, and neither [does it deserve] being banished from your Lordship's understanding who, better than anybody else, will quickly be able to retract it [the proposition] from the exile of our minds. And because the experiment, through which this truth principally became clear to me, is so much more certain, as it was explicated by me in a confused way in my other [letter], I will repeat it here more clearly, so that you, by performing it, would also be able to ascertain this truth. (Galileo to Guidobaldo del Monte, November 29, 1602, Galilei 1890-1909, X: 97-100)

Then follows a detailed technical description of an experimental setting for comparing the oscillations of two equal pendulums, each consisting of a string of "two or three braccia"[1] length and a ball made of lead, together with a report of

---

[1]   The braccia measured in Padua at that time about 0.6 m, see Zupko 1981, 47.

Galileo's own results. This report is remarkable because, as often when Galileo speaks about his experiments, he obviously exaggerated the outcome; it seems technically impossible that he might have reproduced with two different pendulums 1000 oscillations in precisely the same time.[1] In any case, Galileo's claim suggests that he used quite heavy lead balls which, with all likelihood, where the ones he had ordered and received after July 1599 from a foundry.[2]

Next, Galileo criticizes Guidobaldo del Monte's own attempts to check the isochronism by means of a ball descending along a spherical path, attributing the negative results as being due to insufficient precision of the time measurement as well as to the deviations of the path and the smoothness of the surface from ideal conditions.

The second half of the letter deals with Guidobaldo del Monte's theoretical arguments against Galileo's discoveries. It starts with a critique of arguments against the two theorems of Galileo concerning the isochronism of descent along spherical paths and paths along inclined planes which Guidobaldo obviously must have based on plausibility instead of on a mathematical proof:

> With regard now to the unreasonable opinion that, given a quadrant 100 miles in length, two equal mobiles might pass along it, one the whole length, and the other only a span, in equal times, I say it is true that there is something wondrous about it; but [less so] if we consider that a plane can be at a very slight incline, like that of the surface of a slow-moving river, so that a mobile will not have traversed naturally on it more than a span in the time that another [mobile] will have moved one hundred miles over a steeply inclined plane (namely being equipped with a very great received impetus, even if over a small inclination). And this proposition does not involve by any adventure more unlikeliness than that in which triangles within the same parallels, and with the same bases, are always equal [in area], while one can make one of them very short and the other a thousand miles long. But staying with the subject, I believe I have demonstrated this conclusion to be no less unthinkable than the other.

---

[1] A closely related description is later given in the *Discorsi*, Galilei 1974, 226f. For a discussion of Galileo's exaggeration see Appendix B.

[2] See his bookkeeping accounts, Galilei 1890-1909, XIX: 132. The price for the two balls was 4 Lire. According to Galileo's notes in the years 1599 and 1606 on different kinds of brass, its price in these years varied between $\frac{1}{2}$ and 3 Lire for one Libra of brass, that is at that time in Padua about 0.4 kg; see Zupko 1981, 47, 134. Similarly a lot of $3\frac{1}{2}$ Libre of spoons, probably purchased for using the metal, was bought at a price of little less than $\frac{1}{2}$ Lira per Libra. This seems to be about the price for "cheap" metals. Assuming this price for the balls he bought and assuming also that they were indeed the ones made of lead he used in his experiment, they must have had a weight of about 1.6 kg and a diameter of about 6.5 cm. This is in accordance with the assumption that Galileo used quite heavy balls for his experiment. If the price would be considerably cheaper as one might expect in view of the fact that lead was used for artillery ammunition, this weight may have been correspondingly higher. A doubling of the diameter of the balls corresponds to an eight times higher weight or an eight times cheaper price.

Galileo continues his letter with a summary of his findings which he considered to be related to the isochronism of the pendulum. He gives a precise formulation of the isochronism of inclined planes inscribed into a circle which he claims to have been able to prove. He mentions the theorem that the descent along a broken chord (that is the sequence of two subsequent chords SI and IA) needs less time than the descent along the lower of the two chords (IA) alone, again maintaining to have found a proof. We do not have any indication that this claim was incorrect, all the more as there are several closely related folio pages in the manuscript Ms. Gal. 72 containing notes related to this theorem and even a complete proof.[1] But the only way that can be imagined for a proof under the conditions of Galileo's knowledge and which is attested to by manuscripts and the later publication in the *Discorsi* is based on the law of fall. In other words, in 1602 the law of fall as implication of the symmetrical parabolic shape of the projectile trajectory must have been so familiar to Galileo that he not even made a point of mentioning it.

Finally, Galileo explains in his letter the strategy of how he was trying by means of the two propositions on motion along inclined planes to provide a proof for the isochronism of the pendulum which he obviously considered at that time as an important discovery he had made:

> Until now I have demonstrated without transgressing the terms of mechanics; but I cannot manage to demonstrate how the arcs *SIA* and *IA* have been passed through in equal times and it is this that I am looking for.

Clearly, Galileo is trying here to reassure Guidobaldo del Monte that, in spite of the novelty of the subject for traditional mechanics, he is still adhering to the principles of this mechanics which had been the starting point of their exchange. For Guidobaldo these principles did not only comprise the theory as it is exposed in the ancient texts but also a strict correspondence between theory and practical experience. This attitude led him to be skeptical with regard to studies involving motion and is exemplified, in particular, also by his attempt to check Galileo's claims by experiments. In view of the failure of these experiments, Galileo was thus moving here on slippery territory. In the concluding passage of his letter, he attempted to appease Guidobaldo del Monte, in spite of his insistence that he had remained within the limits of mechanics, by explicitly agreeing with the latter's view that the practical verification of theorems involving matter in motion is indeed more problematic than in other parts of mechanics and in pure geometry:

> Regarding your question, I consider that what your Most Illustrious Lordship said about it was very well put, and that when we begin to deal with matter, because of its contingency the propositions abstractly considered by the geometrician begin to change: since one cannot assign certain science to

---

[1] See for instance Galilei ca. 1602-1637, folio 186v.

the [propositions] thus perturbed, the mathematician is hence freed from speculating about them.

I have been too long and tedious for your Most Illustrious Lordship: please excuse me, with grace, and love me as your most devoted servant. And I most reverently kiss your hands.

In spite of Galileo's friendly attempt to come to an agreement with his patron, it cannot be overlooked that he was on the way towards a radical break with the principles of Guidobaldo del Monte's mechanics, in particular with his strict adherence to the outcome of experiments. The achievements accumulated after his practical turn brought him finally back to questions of a theory of motion from which he had once started. The tension with Guidobaldo del Monte which becomes visible in his letter shows that his intellectual exchange with Paolo Sarpi begins to become influential for his Paduan activities.

Summing up, we have argued that the outcome of the projectile trajectory experiment performed in 1592 implied the law of free fall but that, as a result of Galileo's "practical turn," for a time period of about ten years there was no reason for him to elaborate the theoretical consequences of this experiment. Moreover, his exchange with Paolo Sarpi, which took up the issues of his treatise *De Motu*, led to a modification of his former theory in order to explain the symmetry of the trajectory; this modification seemingly made any drastic revision unnecessary. However, as early as in the year 1602, Galileo had derived already a set of propositions on motion that transgress not only the theory of *De Motu* and the reflections about this theory as they are represented by Paolo Sarpi's notes, but also the range of mechanics as it was conceived by Guidobaldo del Monte whom Galileo with his practical turn after his move to Padua tried to emulate so faithfully. Galileo's discoveries in the sequel of the projectile trajectory experiment comprise, as far as can be inferred from the few sources preserved, at least the law of fall, the isochronism of inclined planes in a circle, the broken chord theorem, and finally the isochronism of the pendulum. It comes therefore not as a surprise that only two years later we see Galileo on a new track, attempting no longer to integrate his discoveries into mechanics but trying to construct a completely new deductive theory which could substitute for the theory developed in *De Motu*.

Figure 11. Ms. Gal. 72, folio page 175 verso, containing a drawing with an erroneous construction of the projectile trajectory

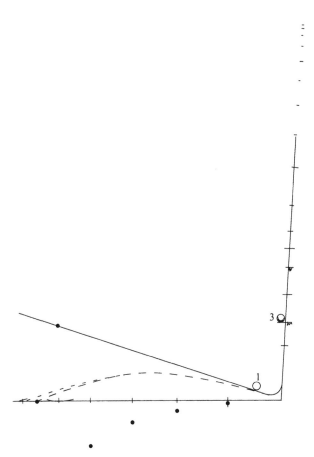

Figure 12. Based on construction lines and marks that are not drawn with ink Galileo's erroneous argument can be reconstructed

Did Galileo Trust the Dynamical Argument?—Galileo the Experimenter

Here is not the place to discuss the intricate development from the discovery of the law of fall to the final theory of motion as it was published in the *Discorsi*.[1] It is well known that it was still a long way to go for Galileo from the proof of the law of fall from the erroneous principle mentioned in his letter to Paolo Sarpi in 1604 to the comprehensive deductive theory of accelerated motion in his final publication

Moreover, the detailed analysis of this final achievement shows that he neither proved the law of fall in a way which could survive the restructuring of medieval natural philosophy into classical physics, nor did he find a complete proof of the parabolic shape of the trajectory which could at least stand up to his own criteria of theoretical consistency and rigor. Although coming close, he never in his life achieved an insight into the two principles that formed the basis of the scientific productivity of classical mechanics, which are, the principle of inertial motion and the principle of superposition of motions. His conceptual background remained that of *preclassical mechanics* and only his disciples were the first to reformulate his achievements in these categories of classical mechanics.

Here we shall concentrate instead on the question of how the Guidobaldo experiment which was at the outset of the discovery of the law of fall became finally incorporated into the new deductive theory of motion based on the law of fall. The way in which the law of fall was developed from the parabolic shape of the projectile trajectory immediately suggested how, vice versa, the parabolic shape of the projectile trajectory can be derived from the law of fall. Could that really work? Well, it worked at least for one particular case. The discovery of the law of fall resulted from a decomposition of the parabolic trajectory into two components which from a modern point of view have to be conceived of as a horizontal, uniform inertial motion and the vertically accelerated motion of fall. However, for Galileo there was no universal inertia. Only in the case of horizontal motion had he found already on the basis of his early treatise *De Motu* that this motion would persist without acceleration or deceleration as long as no forces or resistances intervene, a phenomenon which he designated as "neutral motion."

In the case of oblique projection, however, this interpretation seemed not to be applicable. Instead, in view of his discovery of the law of fall and the use he made of it as a basis for a new science exclusively constructed with the help of this powerful means of deduction, it must have appeared unavoidable to solve also

---

[1]   This has been treated elsewhere, see Damerow, Freudenthal, McLaughlin, and Renn 1992, chap. 3. This book also includes virtually exhaustive references to the literature up to 1991 on the development of Galileo's science of motion and its context, including works by Caverni, Dijksterhuis, Koyré, Wohlwill, as well as other essential contributions of the older literature in Italian, Dutch, French, and German which have been neglected by recent Galileo scholarship. According to our opinion, among the most significant contributions on this topic published after Damerow, Freudenthal, McLaughlin, and Renn 1992, chap. 3 are Hooper 1992, Takahashi 1993a, Takahashi 1993b, Porz 1994, Abattouy 1996, and Remmert 1998.

this question by a straightforward application of this law: Why on earth should the projectile trajectory be anything else than the composition of a decelerated motion in the direction of the projection, such as the motion on an inclined plane, and of the vertically accelerated motion of fall? In fact, a folio page dating also into Galileo's Paduan period[1] has been preserved which displays a construction aimed at determining the shape of the trajectory of oblique projection by applying exactly this idea (see figures 11 and 12).

Unfortunately, this transfer of Galileo's interpretation of horizontal projection to the case of oblique projection does not yield a vertically oriented parabola. Galileo must have realized that the curve resulting from this construction was not symmetrical, and he definitely gave up this attempt to compose oblique projection from two non-uniform motions, both governed by the law of fall.[2]

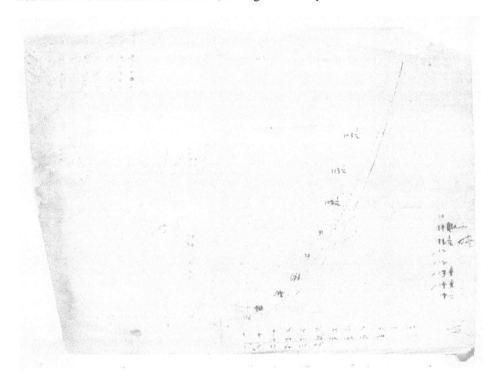

Figure 13. Ms. Gal. 72, folio page 107 recto

---

[1] Galilei ca. 1602-1637, folio 175v, see the discussion of this folio page in Damerow, Freudenthal, McLaughlin, and Renn 1992, 205-209.
[2] The correct superposition of an oblique inertial motion and free fall is found as a correction of a figure and an addition to the text in Galileo's annotated copy of the *Discorsi*, see folio pages 157 and 158 with annotations to the pages 262 and 263 of Galilei after 1638.

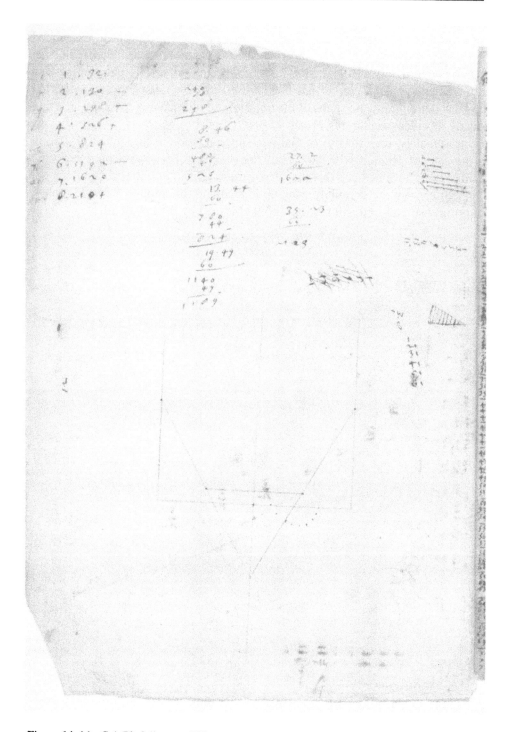

Figure 14. Ms. Gal. 72, folio page 107 verso

What he tried instead was to obtain a better understanding of projectile motion from experiments, supplementing the Guidobaldo experiment which had originally initiated his departure from traditional mechanics. According to the surviving documentation of these experiments, they seem to have failed altogether in providing unambiguous evidence of the symmetrical, let alone parabolic shape of the trajectory.[1] Thus, Galileo was forced to return to the source of his original insight, that is the common dynamical interpretation of the projectile trajectory and the curve of a hanging chain.

Given this return under the conditions of his richer theoretical knowledge as well as the facilities he had now for systematically performing sophisticated experiments,[2] did he really never realize that the catenary is not a parabola? In view of the evidence concerning his treatment of the subject in his last publication, the *Discorsi*, the answer seems clear: He seems either never to have realized this error or he must have consciously kept such an insight disguised.

Nevertheless, there is striking evidence that neither of these alternatives is true. Galileo obviously tried to compare the catenary and the parabola empirically, and he arrived at a definite, correct result. Among Galileo's notes on motion a folio has been preserved, folio 107, documenting this experiment (see figures 13 and 14).[3]

The folio is datable by its watermark to belong to the Paduan period.[4] It contains on the obverse two curves with a common upper endpoint and a common

---

[1] The few manuscripts bearing evidence to Galileo's experimental study of projectile motion are the folio pages 116v, 114v and 81r of his manuscripts on motion, Galilei ca. 1602-1637. They have been the subject of controversial discussions in the literature. Drake expressed his final account of the success of the experiments on oblique projection in the following statement with which we agree: "It is precisely because they [the experiments] did not succeed that they are of great interest, for they show how Galileo went about attacking a physical problem when it lay beyond his powers of solution." Drake 1990, 124

[2] For new findings on Galileo's experimental practise see Settle 1996.

[3] The reverse side of this folio has been interpreted as such a comparison in various papers by Naylor; for his views on Galileo's theory of projectile motion see Naylor 1974; 1975; 1976a; 1976b; 1977; 1980a; 1980b; Naylor and Drake 1983.

[4] Identification of watermarks has been introduced by Stillman Drake as a means of ordering and dating Galileo's folios, see Drake 1979. In the context of preparing an electronic representation of the manuscript, we have begun to systematically check Drake's assignments. Drake's identification of watermarks could be made only when the manuscript was still bound. Identifications of watermarks used here depend on a preliminary inspection of the unbound manuscript and differ therefore partly from Drake's identification.

The special form of the watermark of folio 107, a thin crossbow (Drake's type 6), occurs usually together with another watermark, a crown (Drake's "Mountains", type C12) on a double sheet of paper. Galileo mostly cut these sheets into two pieces so that only one of these watermarks can be found on the page. This makes it difficult to decide whether these watermarks occur also alone. Furthermore, another form of the watermark, a thick crown (Drake's type 15) occurs also together with the watermark depicting a crown. It is unclear if this variant indicates another type of paper. According to Drake, the earliest dates that these watermarks occur on dated letters is August 1607 (crown C12) and May 1608 (crossbow 15).

According to paper type, paper size and watermark, the following folios seem to be closely related to folio 107: f089b, f115, double page f116/117, double page f126/127, f129, f130, f131, f132, double page f134/235, double page f136/137, f138, double page f140/141, double page f142/143, f144, f145, f147, f148, f153, f155, double page f156/157, double page f158/159, f161, f165, f166, double page f168/169, f174, f176, f179, f185, f186, f187, double page f190/191, f192.

zero point at the bottom. One of the two curves continues symmetrically up to the top, while the second curve reaches only to the zero point at the bottom; it is similar to the first one but apparently deviates from it by a different curvature.

A set of partly corrected figures is written along the curves, a second set of partly corrected figures is contained in a small table at the right side of the curves. The figures in the table obviously represent the differences between the figures written along the curves. Three sequences of increasing integers are written along the base line of the curves. The first sequence represents square numbers, the third one cubic numbers. The intermediate sequence is constructed by the rule that the difference between the figures is each time increased by four.

The reverse side of folio 107 contains in its center a geometrical drawing. In one corner a table of figures is written representing a series of empirical data which are close to square numbers, the deviations being marked by plus and minus signs behind some of the figures.

Near this table there are several calculations giving figures which can also be identified, partly corrected, in the table. These calculations can easily be interpreted as conversions of measured figures into the sixty times smaller unit used in the table. Furthermore, the page seems to have been used as scratch paper. Some small drawings and columns of figures are written in different orientations near the upper right edge of the page

Although there is no text on the folio which could explain the function of the curves, drawings, and the figures, a meticulous investigation of the folio[1] made it possible to accomplish a detailed reconstruction of its contents and purpose. Precise measurements have established beyond any doubt that the symmetrically completed curve is a parabola whereas the deviating curve is a catenary, that is the curve of a hanging chain (see figure 15).

Close inspection of the original folio has furthermore revealed a considerable number of construction lines drawn without ink by means of a stylus or by compasses (see figure 16). These construction lines are invisible as long as the page is not illuminated with light under an extremely small angle. Moreover a great number of tiny impressions and wholes in the surface of the paper have been detected which result from transferring distances by means of compasses or from indicating points by pressing the sharp end of a stylus into the paper. The function of the figures along the curves could be identified as representing horizontal distances of the catenary from the middle axis. Thus, the figures in the table representing the differences between these measures provide a check of the extent to which the catenary deviates from a parabola. The corrections of the figures could be interpreted as resulting from manipulations performed in order to fit them to some rule for the sequence of the differences.

---

[1]   Our detailed reconstruction of folio 107 will be published elsewhere.

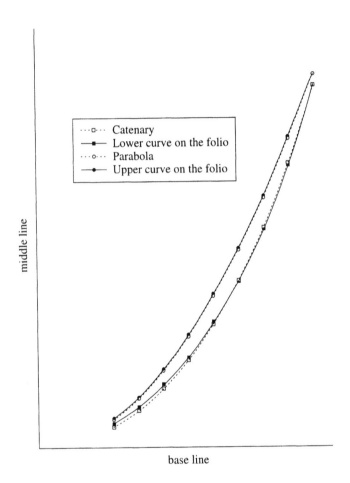

Figure 15. Comparison of the two curves on folio page 107 recto with the parabola and the catenary

The investigation of the reverse side of the folio has unquestionably revealed, as will be shown in the next sections, that the drawing in the middle is part of an unfinished attempt to construct a proof for the alleged parabolic shape of the catenary. The empirical figures in the table on the reverse side have been analyzed by fitting mathematical curves to the sequence.

It turned out that the figures fit perfectly to a catenary which is stretched to such an extent that the width is about twenty times the height which the chain is hanging down, whereas there are slight systematic differences of the data to a parabola.[1] In the case of such a stretched hanging chain, the difference of the catenary to a parabola is, however, so small that it cannot be excluded that the better fit of the catenary is just accidental.

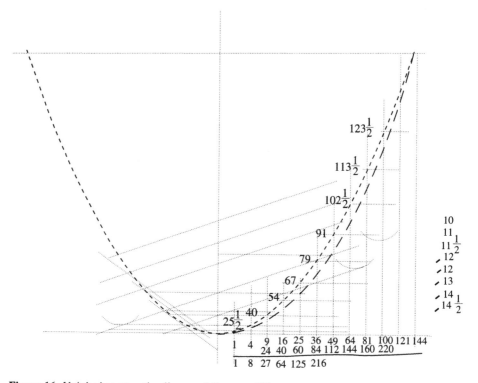

Figure 16. Uninked construction lines on folio page 107 recto

---

[1]   In contrast to the parabola, the catenary does not scale along a single axis. Therefore, the fit of the catenary implies a definite relation between length and width of the hanging chain. Assuming that the highest value of 2123 in the calculations for Galileo's table on folio page 107v is the maximal vertical measure of the hanging chain, the best approximation is achieved for a width of about 24000. This corresponds to a hanging chain with an inclination of 20 degrees with respect to the horizontal at the suspension point, which is in perfect agreement with the angles used on the template folio 41/42 discussed above. The measure used by Galileo was probably the "point," so that the experiment must have been performed with a hanging chain of about 24 meters width and 2 meters height. This estimation of the measures must, however, have been taken with caution. The determination of the width by means of the best fit of a catenary to the data involves a very high error variance. Furthermore, any real chain will systematically differ from the catenary with the result of a systematic error in the estimation of the width. Even a much smaller width still gives a very good approximation. In any case, it has to be taken into account that a catenary with the relation of width to height as it is reconstructed here differs so little from a parabola that unfortunately the better fit of the catenary cannot be taken as a proof for the given interpretation of the data as resulting from an experiment with a hanging chain.

In addition to its careful investigation, folio 107 has been included in an ongoing project in which the composition of inks in Galileo's manuscripts is analyzed using methods developed in nuclear physics.[1] The aim of these investigations is to determine ink differences on the folio pages which may provide clues for dating the entries. For folio 107 these investigations showed no substantial ink differences between the main entries on both sides of the folio.[2] In particular, the ink points of the parabola, the uncorrected figures along the curves and in the table beneath the curves, the quadratic and the cubic scale at the base line of the curves, the calculations and the table of measured figures on the reverse side, and the drawing in the middle of the reverse side show no differences in the ink composition (see figure 17). Although this result does not prove that the entries were made at the same time, this result can be nevertheless interpreted as a strong indication for a close connection of the entries.

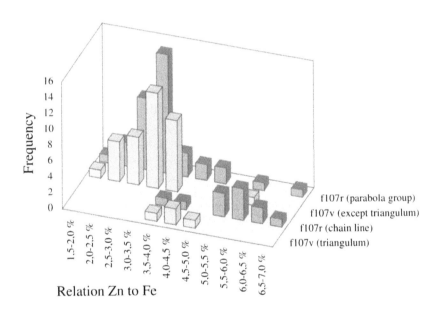

Figure 17. Result of the ink analysis of folio 107 recto. The diagram contains (from back to front): frequency distribution of the relation of zinc to iron at most points measured on the recto page (parabola, quadratic scale, cubic scale, uncorrected figures), on most points measured on the verso page (all points except the small triangle), at the chain line on the recto page, and at the small triangle on the verso page. Three slightly differing inks can be distinguished. The distributions show that all entries on both sides of the folio with the exception of the chain line and the triangle have been written with the same ink.

---

[1] The method used is called "Particle induced X-ray emission analysis (PIXE)," see e.g. Giuntini, Lucarelli, Mandò, Hooper, and Barker 1995. The project is a joint endeavor of the Biblioteca Nazionale Centrale in Florence, the Istituto e Museo di Storia della Scienza, the Istituto Nazionale di Fisica Nucleare in Florence, and the Max Planck Institute for the History of Science in Berlin. Its results will be published in the near future. See the first project report, Working-Group 1996.
[2] Relation to Fe: Zn 3%, Cu 1%, Pb 1% (on reverse slightly lower), Ni 0.5%.

It is remarkable that some entities on the two pages show a slightly different amount of zinc in its ink composition. Time-line measurements on another manuscript with dated entries have shown that during the use of the same ink there are sometimes characteristic changes in the ink composition which result in such slight differences of individual components. Two such ink differences could be identified on folio 107. First, the dashed line of the catenary, the intermediate scale at the bottom with constantly increasing differences, and partly the ink of the corrections show a higher amount of zinc than the main entries on both sides, which is an indication that they were added on a somewhat later occasion.[1] Similarly, a small triangular drawing on the reverse side shows a much less but still detectably higher zinc component.[2]

Taking all these results of the investigation of folio 107 together, the following scenario for its origin and purpose seems to be the most plausible one.[3] In order to check whether the curve of a hanging chain is in fact, as he believed, a parabola, Galileo at some point in his work at Padua constructed a parabola and superposed it with a catenary produced, as later described in the *Discorsi*, by means of a fine chain. The right side of the parabola was constructed by drawing a horizontal base line with ink and erecting on it without ink a set of perpendicular lines in equal distances from each other.[4] In order to draw the left side, the constructed curve was mirrored at the vertical middle line by means of a ruler.[5] Horizontal lines were drawn without ink with distances to the base line representing a sequence of squares.[6] A corresponding sequence of square integers was written as

---

[1] Relation to Fe: Zn 5.5%.

[2] Relation to Fe: Zn 4.5% (based on four measurements only)

[3] The results presented here are incompatible with the interpretation of folio 107 given by Drake. In his presentation of the manuscript he assumes that the entries on this folio were written at quite different times; see Drake 1979, 23, 40, 82, and 124, as well as his commentaries on pages XXXf, XXXV, and XXXVIII. According to Drake, the earliest entries are the figures in the table on the reverse, written in 1604 and representing empirical data of the discovery of the law of fall by means of the inclined plane experiment, see Drake 1990. Next, still in 1604, Galileo supposedly drew the curves and figures on the obverse, interpreted by Drake as a comparison of parabola and catenary, with the exception of the scales at the bottom which Drake considered as being added in 1609. Finally, according to Drake, Galileo added, also in 1609, the drawing on the reverse. He considered this drawing first as the depiction of an apparatus for measuring percussion, later as representing a water tube used by Galileo for the inclined plane experiment as a device for precise time measurements, see Drake 1990, 9-12.

[4] This distance may correspond to 1 uncia, but the measures reported in Zupko 1981, 174-179 are generally higher. The figures on the folio page show, in any case, that distance between the vertical lines was divided into 12 smaller units as it was the rule for the uncia throughout Italy in early modern times. It is remarkable that the measures on the page differ from that on folio 41/42 and the related folios, indicating that these folios, although also dealing with the catenary (see below), were probably written at a different place and time.

[5] The points which Galileo used to mirror the curve can easily be identified. These points were first indicated by impressions of compasses or stylus and then marked with ink in a particular way different from the way the rest of the curve is drawn.

[6] The scaling of the parabola was obviously chosen so that the parabola had about the size of the hanging chain. At the eight vertical the distance between the base line and the horizontal comes close to the horizontal distance to the middle axis, but does not fit perfectly. The difference is about 3%. It is therefore unclear if the scaling factor was consciously chosen or if it was implicitly determined by some procedure that was used to adjust the parabola to the hanging chain.

a scale at the bottom base line. The parabola was drawn with ink through the intersection points of the network of the inkless vertical and horizontal lines.

Next a hanging chain was matched to the parabola so that the curves coincide in the suspension points of the hanging chain and its lowest point.[1] The intersection points with the sequence of inkless horizontal lines with squared distances, which deviate from the intersection points through which the parabola was drawn, were marked by inkless stylus impressions, their distances to the middle axis were measured, and the results were written near the parabola, each of them at the level of the corresponding horizontal line. In order to check the parabolic shape the differences between the measurements were calculated and written down in a table to the right of the drawing. Since these differences would be equal if the curve were a parabola, Galileo must have realized immediately that the curve of the hanging chain did not only fail to match the constructed parabola, but any parabola that could be constructed (see table 1).

Distance to middle axis	Difference to next value	Distance of parabola to middle axis
$123^1/_2$	10	120
$113^1/_2$	$10^1/_2$	108
103	12	96
91	12	84
79	13	72
66	$12^1/_2$	60
$53^1/_2$	$14^1/_2$	48
39	$13^1/_2$	36
$25^1/_2$	---	24

Table 1. Table of the uncorrected distances of the catenary from the middle axis written along the curves and the uncorrected differences between the measurements, compared with the corresponding values of the parabola assuming a unit of one twelfth of the distance between the vertical lines

Facing this result, Galileo probably decided to repeat the attempt with a much longer chain in order to minimize possible errors caused by the limited flexibility of the material of the chain. He used a chain of several meters length which was stretched to such an extent that the curve was rather flat, thus necessarily produc-

---

[1]   The suspension points of the chain do not coincide with constructed points of the parabola. It is therefore more likely that the chain was empirically matched to the constructed parabola (as described in the *Discorsi*) and not the parabola scaled as to match the chain. The latter procedure would be much more difficult to execute with precision.

ing a better fit with a corresponding parabola, because the differences between the two curves increase with increasing relation of height to width as we know today from the correct formula of the catenary which was discovered much later.

The vertical height over the base line of the curve of the hanging chain was measured at eight points at regular distances from the lowest point of the curve. These measurements should increase in quadratic sequence. By calculations partly written down on the reverse side of the folio, the measurements were converted into the smallest unit, which was surely the same as the one used for measuring the catenary on the obverse side, and the results were written down in a small table left to the calculations in correspondence to a sequence of integers from 1 to 8 and their squares from 1 to 64 (see table 2).

1	1	32	
4	2	130	–
9	3	298	–
16	4	526	+
25	5	824	
36	6	1192	–
49	7	1620	
64	8	2104	

Table 2.   Galileo's table of measurements taken at a long chain

There are two remarkable differences between the measured values used in the calculations converting the units and the values Galileo actually entered into the table. While Galileo measured 1189 he wrote 1192, and while he measured 2123 he wrote 2104. The reason is obvious. Already when Galileo entered the data into the table he checked whether they correspond to his belief that they represent a parabola. A measurement at the double distance should yield a value which is four times greater. This rule applied to the first and the second measurement shows that the second value should be four times 32, that is 128; Galileo wrote this number down to the right of the table using tiny characters, obviously in order to keep this figure in mind. Applied to the third and the sixth measurement, the rule shows that the sixth value should be four times 298 which is 1192, and applied to the fourth and the eights value the rule gives four times 526 which is 2104. In both cases Galileo entered directly these theoretical instead of the slightly differing empirical values into the table.

Galileo finally attempted to check the deviations from the parabolic shape by calculating all values from the first value in the sequence. Since four times as well as nine times 32 fall short of the empirical values in the table, Galileo indicated these deviations by minus signs behind the values in the table. Realizing that all other values would also fall short, Galileo now probably changed the basic value from 32 to 33. Consequently, the next calculated figure of 16 times 33 is greater then the value measured; this is indicated by a plus behind the figure. Furthermore, the next calculated value nearly matches the measured value (824 instead of 825) so that this one remained unmarked. The next calculated value 36 times 33 again falls short, which is indicated by adding a minus sign. At this point Galileo stopped the procedure and left the last two figures without any indication of their deviation from the theoretical values. In spite of these small deviations, however, Galileo will, for good reasons, have considered the results to fit much better the expected parabolic shape than in the case of the small chain recorded on the other side of the folio.

Then Galileo started to work on a proof for the alleged parabolic shape of the catenary with a drawing below the table and the calculations, but left it unfinished. This proof attempt will be discussed in the next section.

According to the ink difference, Galileo must have returned some time later again to the comparison of the curves on the obverse side of the folio. In order to be better able to compare the curves he now inked the chain line, added the third scale with differences increasing each time by four, and tried to modify the figures along the chain line so that their differences would decrease with increasing distance from the middle axis monotonously, intending to find a simple rule for their decrease (see table 3).

Corrections of the distances to the middle axis	Differences to next values
$123^1/_2$ remained unchanged	10 remained unchanged
$113^1/_2$ remained unchanged	$10^1/_2$ changed in 11
103 changed in $102^1/_2$	12 changed in $11^1/_2$
91 remained unchanged	12 remained unchanged
79 remained unchanged	13 changed in 12
66 changed first in $66^1/_2$, then in 67*	$12^1/_2$ changed in 13
$53^1/_2$ changed in 54	$14^1/_2$ changed first in $13^1/_2$, then in 14
39 changed in 40	$13^1/_2$ changed in $14^1/_2$
25 $^1/_2$ remained unchanged	---

* If Galileo did not make a calculation error, the original entry cannot have been $66^1/_2$ because the unchanged previous value of 79 and the difference 13 which was changed in 12 make this impossible.

Table 3.   Table of corrections of the distances of the catenary to the middle axis and of the differences between the measurements

The result of these trial-and-error corrections[1] is a sequence of corrected numbers close to his original measurements. No value was changed more than one
unit of his scale, that is less than one millimeter. The differences between the values decrease now continuously from $14^1/_2$ to 10, but still failing to show a simple
rule of their decrease. At that point Galileo must have given up his attempts to
improve the data further.

This scenario of how Galileo wrote the entries on both sides of folio 107 is necessarily in part speculative, but it keeps close to the results of a careful investigation of the details of the page. It leaves unexplained several oblique construction
lines on the obverse side and the scratch notes and drawings on the reverse.[2] The
equally unexplained scratch notes and diagrams on the reverse page are too simple to allow for any substantial reconstruction of their purpose. However, they
may well belong to the context of the other entries on the folio representing
attempts to find an explanation for the puzzling difference between parabola and
catenary by ascribing them to modifications of the "moments" in Galileo's terminology by which, as will become clear soon, he tried to explain the shape of the
catenary.

Did Galileo Trust his Experiments?—Galileo's First Attempt of a Proof

The purpose of the drawing in the center of the reverse of folio 107 in the context
of an attempt to find a proof for the parabolic shape of the catenary would have
remained undiscovered if folio 132 had not been preserved. On this folio Galileo
worked more explicitly with two different geometrical constellations on a closely
related aspect of the same proof attempt, carrying it further than he did on folio
107. The most elaborate version is contained on the obverse of folio 132 (see figure 18), while another attempt which, however, was abandoned earlier is found

---

[1]   Probably starting from the top he changed first the 103 into a $102^1/_2$ and corrected correspondingly the differences $10^1/_2$ into 11 and 12 into $11^1/_2$ by writing first the new differences to the right of
the old ones, but then striking them out and writing the new figures directly over the old ones. The
result was a somewhat more regular sequence at the top. Then he started from the bottom where the
irregularities were greater. He changed now directly by overwriting the old figures, first the 39 into
a 40, correcting correspondingly the differences $13^1/_2$ into $14^1/_2$ and $14^1/_2$ into $13^1/_2$. He continued
by changing the 66 into $66^1/_2$ and started to correct the differences by changing the $12^1/_2$ into a 13.
Before he corrected the second difference, however, he must have realized that the differences
decreased too quickly so that they would not nicely match the upper part of the table. Thus, he left
the difference for the moment unchanged and started again from the figure below by changing now also
the $53^1/_2$ into 54, followed by changing back the already changed difference $13^1/_2$ into 14 which is in
the middle between the original value and the first change. He left the other difference which had
already been changed into 13 untouched and corrected instead again the next value from $66^1/_2$ into
67, correcting the corresponding next difference from 13 into 12.
[2]   No explanation has been found so far for the two sets of oblique construction lines on the
obverse side. Since these lines have only very few coincidences with intersection points of the
explained parts of the drawing, it may well be that these lines belong to an unrelated, unfinished
earlier drawing which was not yet inked so that the folio could be used again for a different purpose.

on the reverse of the same folio. As folio 107, this folio, too, can be dated by its watermark into the Paduan period of Galileo's work.

In contrast to the precise drawing in the center of the reverse of folio 107, the drawing in the left upper corner of folio page 132r is only a rough sketch, but essentially complete. It depicts two constellations of a hanging string fixed with its ends to two points *a* and *c* of a horizontal line, the middle of the line between these two suspension points being designated by the point *d*.

In the first position the hanging string is pulled down by a neatly drawn weight, fixed in its middle at point *b*, so that the string forms a triangle *abc*. In the second constellation two further weights have been fixed precisely in the middle of each half of the string at points *e* and *f* pulling the string outwards on circles around the suspension points *a* and *c* towards two points which are both designated by the letter *g*.

These additional weights raise the first weight from point *b* to point *o*. In addition to theses two constellations of the string, Galileo sketched a third constellation which results from pulling the points *e* and *f*, where the additional weights are fixed, symmetrically outwards to the extremes so that the string now goes vertically down from the points *a* and *c*.

A small table to the right of the drawing contains the intended measures of the drawing from which the actual dimensions of the drawing, however, deviate considerably. The table furthermore contains some calculated lengths which can easily be inferred from the given measures by applying simple geometry. In particular, the table gives for the distance of the suspension point *a* to the middle point *d* between the suspensions *a* and *c* the measure 30, so that the total distance between the suspension points is hence assumed to be 60.

Furthermore, the table gives for the distance *ab*, which is half of the total length of the string, the measure 90, and for the distances *ae* and *eb* to the point *e*, where one of the two additional weights had to be fixed, the measure 45 which is half of the length 90 of half of the total string.

The next entry in the table is an approximate measure of *db*, the height of the first constellation from the middle *d* of the horizontal connection of the suspension points *a* and *c* to the point *b* in the middle of the string. The value $84^7/_8$ given for this height has been calculated using the theorem of Pythagoras. The calculation of the square root of 7200 (that is, of the difference between 8100 and 900, the squares of the lengths 90 and 30 of *ab* and *ad*, respectively) can be found above the table, resulting in $84^{144}/_{169}$, rounded in the table to $84^7/_8$. From this value Galileo calculated its half $42^7/_{16}$, denoted as the measure for the distances *bi* and *di* from the end points *b* and *d* of the height to its middle point *i* which is also the middle point of the horizontal line *ef* connecting the points *e* and *f* where the additional weights have to be fixed.

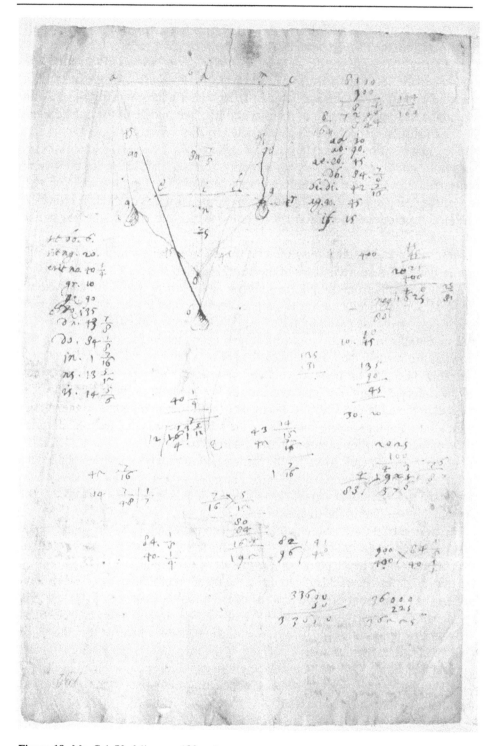

Figure 18. Ms. Gal. 72, folio page 132 recto

The table continues with the measure 45 for the parts *ag* and *go* of the string between the suspension point *a* and the points where the weights have been fixed in the second constellation – after they have been moved from *f* to *g* and from *b* to *o*. The next and final entry of the table gives the measure 15 for the distance *fi* between the end point *f* and the middle point *i* of the horizontal line *ef*, connecting the points *e* and *f* where the additional weights have to be fixed. The measure 15 for this length follows immediately from the similar triangles in the constellation to be half of the horizontal *dc*.

At this point the table stops, obviously because Galileo now encountered an obstacle which hindered him from calculating further the measures of the second constellation: How could he know how far the additional weights would move the string outwards from the points *e* and *f* to the points both designated as *g*, and how could he know how far, as a consequence, the first weight at point *b* would be raised to point *o*? In fact, the calculation of the measures of the second calculation resulting from the static equilibrium between the three weights fixed to the string requires knowledge about the compounding of static forces and algebraic techniques that were not available at Galileo's time. Galileo denoted the middle point of the line connecting the unknown positions of the additional weights in the resulting second constellation with the letter *n*. The only consequence that could be drawn was that the horizontal line *ng* had to be *longer* than the measure 15 of the corresponding line *if* in the original position, which he had recorded in the last line of the table, and that this line had to be *shorter* than the corresponding line in the extreme third constellation, that is, shorter than 30. In a similar way Galileo could have done all his calculations also for the other extreme constellation. This would have shown that the maximum distance by which the weight at point *b* could have been raised to point *o* was little more than 6.[1]

What Galileo really did in this situation is documented by a second table written to the left of the drawing near its bottom and by several calculations distributed all over the page. He bridged the unsolvable problem by a hypothesis written at the beginning of the second table. First he wrote "let *bo* be 6", but then cancelled this entry substituting it by the more realistic assumption "let *ng* be 20", that is, in between the original distance of the added weights of 15 and the extreme distance in the third constellation of 30. The entries on the folio page do not give any clue that might help to decide whether this assumption was based on an experiment or was simply guessed by Galileo. Fortunately, the drawing on the reverse of folio 107, to which we will turn below, shows that Galileo found an ingenious solution to this problem of determining the equilibrium position.

---

[1]   If the string goes vertically down 45 units from the top horizontal measuring 30 units from the suspension point to the middle line, the remaining string length of 45 units to the middle line results in an additional height which can be calculated as 15 times the square root of 5, that is, little more than $33^{1}/_{2}$ units. Added to the vertical distance of 45 units, this results in little more than $78^{1}/_{2}$ units, the value for the total height in the extreme position. Compared with the height of the first constellation calculated by Galileo as being $84^{7}/_{8}$ units, it follows that the maximum length by which the weight at the bottom can be raised is little more than 6 units.

The subsequent steps of Galileo's procedure on page 132r are well documented by the calculations on this page. Based on his hypothesis about the distance *ng*, which represents the distance of the added weights from the middle line after moving the string to the second constellation, he calculated the lengths of a number of lines belonging to this new constellation. The calculations were performed in the sequence of their arrangement on the page and the results were subsequently entered into the table. The general purpose of these calculations can be reconstructed from the last calculations on that page: Galileo checked whether the *squares* of the distances *ng* and *ad*, representing the horizontal distance of the assumed position of the additional weights from the middle line and the horizontal distance of the suspension points from that line, respectively, are proportional to the distances *no* and *do*, representing the corresponding vertical distances from the lowest point of the string. In other words, Galileo performed a "parabola check," i.e. he checked whether the suspension points and the weights lie on a parabola. This check is realized by comparing $ng^2 \times do$ and $ad^2 \times no$.

The calculations performed in order to make this comparison possible are straightforward and need not be reported here in detail, all the more as all calculations are explicitly given on the page and all rounding operations can be inferred from the values Galileo actually entered into the table of results. Having assumed *ng* to be 20 and having given at the beginning the length *ad* to be 30, Galileo needed to calculate only the values *no* and *do* using elementary geometry. That is what he actually did on the page via some intermediate steps.[1] He arrived at the rounded values $40\frac{1}{4}$ and $84\frac{1}{8}$ for the distances *no* and *do* respectively. Multiplying them with the squares 900 and 400 he achieved at the bottom of the pages as a result of his parabola check the values 36225 and 33650 which should have been equal if the points he checked would really be on a parabola.

After having finished the calculation of the values needed for the parabola check, Galileo calculated the distance *in*, that is, the vertical distance by which the additional weights moved downwards. Next Galileo added a point *s* below point *n* at $\frac{1}{3}$ of the distance *no*, that is, at the center of gravity of the second constellation, and calculated the distances *ns* and *is*. Finally, Galileo calculated the length of $\frac{1}{3}$ of the distance *bi* without entering, however, this last value of $14\frac{7}{48}$, which he rounded to $14\frac{1}{7}$, into the table. Obviously, he calculated this value in order to compare it with the length $14\frac{5}{6}$ of *is*, that is, he checked, whether the center of gravity of the two constellations would be at the same place, with the result that this is only approximately true.

This last operation provides an answer to the problem of how Galileo could know how far the additional weights would pull the string outwards. Guidobaldo del Monte's as well as Galileo's treatises on mechanics show that they both were

---

[1]   Using the length *cg* (given) and *ng* (assumed), Galileo calculated subsequently the lengths *no*, *gr*, *gq*, *dq* (cancelled), *cq*, and *dn* to arrive at *do*, the value he needed in addition to the value *no* for the parabola check. The calculated values *gq*, *dq* (cancelled), and *cq*, which are distances to the intersection *q* of the upper part of the string with the middle line through *d* and *b*, are not necessary for the calculation of *do*. The purpose of the calculation of these lines remains unclear.

well aware of the fact that such a constellation cannot be in equilibrium as long as its center of gravity does not reach the lowest possible point. In his notebook, Guidobaldo even considered the case that a change of the center of gravity of the earth, achieved by putting an additional weight on its surface, causes the earth to move as a whole

> because the center of gravity tends by its nature to the center of the world (...). (del Monte ca. 1587-1592, 54)

A similar consideration might have suggested to Galileo that he had to study how the center of gravity moves when the weights along the string are swinging into their equilibrium position.

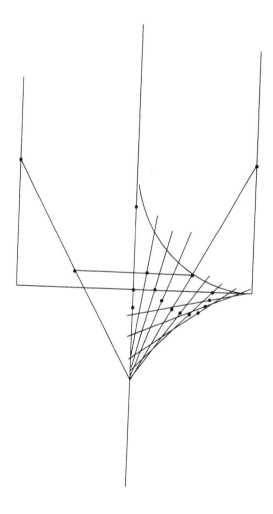

Figure 19. Drawing on folio page 107 verso with uninked construction lines

That this is, in fact, how Galileo solved the problem is documented by the precise drawing in the center of the reverse of folio 107 (see figures 14 and 19). In this drawing, particularly simple measures are used: the length of the string is exactly twice the horizontal distance of the suspension points[1] so that the extreme position of the two additional weights vertically below the suspension points can be reached only if the string between them is stretched out horizontally with all three weights hanging on the same level, the center of gravity being in the middle. Thus the drawing represents a constellation involving the shortest string that still makes it possible to pull the additional weights vertically below the suspension points. The centers of gravity of both constellations are marked by bold points, the original one where the added weights did not yet change the configuration and the extreme one where the weights are pulled vertically below the suspension points.

A great number of construction lines which are not drawn with ink can be identified on the folio page as being part of the drawing. They show that on the right hand side of the symmetrically hanging weights, Galileo systematically constructed a set of constellations. These constellations reach from the one extreme, in which the weights are pulled to the middle line, to the other, in which the weights are pulled vertically below the suspension points. From the circle along which the additional weight on the right-hand-side is moving, lines with the fixed length of the cord between the weights are drawn towards the middle line where the lowest weight is hanging. On each of these lines Galileo marked with ink by a bold point the distance of one-third of the total length. Since these points are necessarily on the same height as the center of gravity of the whole constellation, Galileo reached in an ingenious way a solution to the question of which constellation has the lowest possible center of gravity thus representing the equilibrium position.

What may have been the purpose of this arrangement of weights in equal distances on a string and of the check whether these weights lie on a parabola after reaching an equilibrium? An answer to this question is provided by a comparison with the strategy discussed above which Galileo used when he attempted to find a proof for the isochronism of the pendulum. Following the example of the Archimedian approximation of the circle by polygons, he tried to derive the isochronism of motions along the circle from the isochronism of motions along inclined planes which are chords in that circle by progressively substituting the chord by a sequence of smaller ones, thus treating the circle as polygon consisting of an infinite number of chords.[2] The comparison with this strategy suggests that Galileo tried to prove the alleged parabolic shape of the hanging chain by considering first a string with just one weight attached to its middle and then by

---

[1]   Nine units of twelve points each, if measured by the same units as the drawing on the other side.
[2]   Galileo's letter to Guidobaldo del Monte from November 29, 1602, Galilei 1890-1909, X: 97-100, shows that Galileo already at that time considered a proof of the isochronism of the pendulum along this line.

subsequently adding weights at intermediate points, thus treating the string as a weightless cord with an infinite number of equal weights attached to it in equal distances. If Galileo would have been able to prove that, for any arrangement with a finite number of equal weights attached in equal distances to a string, the weights would lie on a parabola, it would have been reasonable for him to conclude that the links of a hanging chain also lie on a parabola. Galileo's attempt to decide whether a constellation of three weights attached in equal distances to a string fulfills this condition finds a reasonable explanation if it is interpreted as the first step of realizing such a strategy for finding a proof for the parabolic shape of the catenary.

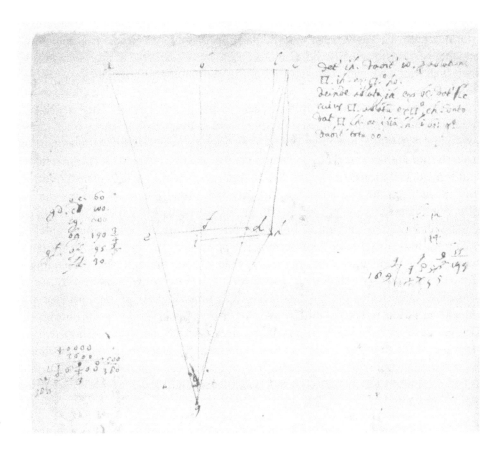

Figure 20. Ms. Gal 72, upper part of folio page 132 verso

This strategy, however, had no chance of being successfully pursued. Galileo could in fact not conclude anything from the fact that his calculation provided him with the result that his parabola check was nearly, but only nearly, fulfilled, i.e. the check whether the squares of the vertical distances of the weights from the bottom are proportional to the horizontal distances of the weights from the middle line. His attempt to prove the parabolic shape of the catenary ended up in a similar way as did his attempt to check the shape empirically. The least which Galileo would have needed to derive from the definition of the constellation representing the equilibrium was a proof that in this case the weights lie indeed on a parabola; only then could he have reasonably tried to generalize this result to more than three weights. However, the structure of the problem gave him no chance. Although his definition of the equilibrium position reduced the problem to a purely geometrical one, the relation between the motion of the added weights on their circles and the center of gravity of the entire constellation is far too complex as to be investigated by the mathematical tools available to Galileo; in modern terms, the relation is adequately represented by an irreducible equation of fourth degree. The only possibility he had was to solve the problem graphically, and that is what he actually did. But this solution does not lead to any better result than that which he had achieved by comparing empirically the catenary with a parabola.

Galileo was well aware of this obstacle to his proof attempt. This is evident from a second attempt of exactly the same kind as the one on folio page 132r, but involving different numerical values. This attempt, which can be found on the reverse side of the same folio, is documented again by a sketchy drawing, a table and some calculations (see figure 20). In this case, a width of 120 instead of 60 is chosen between the two suspension points and a total length of 400 instead of 180. The table contains only values of the first constellation which can be calculated without any assumption about the effect of additionally attached weights.

The second constellation representing the position of the string after additional weights have been attached is only roughly sketched at the right side of the drawing. A short cursory calculation documents that Galileo again started to use a somehow empirically or geometrically determined value for the distance by which the string is horizontally moved by the weights. At this point, however, Galileo gave up calculating any further values. Instead, he wrote a short text to the right of the drawing. In this text he expressed the dependency of the distance by which the weight in the middle of the string is raised – after attaching the additional weights – on the distance by which these additional weights move the string outwards.

> Let *ih* be given. *io* will be given by the subtraction of the square of *ih* from the square of *ho*. Hence by subtracting *ih* from *bc*, *lc* will be given, whose square, subtracted from the given square of *ch*, gives the square of *lh* and *lh* itself, that is *bi*: Therefore, the entire *bo* will be given.

Apparently, this text and the procedure of folio page 107 verso represent Galileo's *ultima ratio* in view of the poor outcome of his attempt to reduce the catenary to a sequence of strings with a finite number of weights attached. It turned out that, even in the simplest case, he was unable to theoretically determine the equilibrium position of the string needed for any further elaboration of the proof of the parabolic shape of the catenary which he probably intended to develop. As we will see in the next section, this failure to attain a proof of the alleged parabolic shape of the catenary did, however, not prevent Galileo from taking up the basic idea of this attempted proof in order to demonstrate another property of the chain which he intended to connect with his analysis of projectile motion, the impossibility to stretch a chain to a perfectly horizontal position.

Returning to the Dynamical Argument—The Final Proof

The failing early attempts of Galileo in Padua to empirically validate and theoretically prove the parabolic shape of the catenary are, however, not the last ones documented by his manuscripts and other contemporary sources. There is overwhelming evidence that the futile search for a satisfactory proof of the symmetry of the projectile trajectory directed his attention again to the alleged close relation between the projectile trajectory and the curve of a hanging chain due to the assumed common dynamical constellation.

Near the end of his life, Galileo composed the *Discorsi*, the final publication of the results of his life-long work on mechanics. It is known from Galileo's letters concerning this publication that the book was not really completed. As it was printed in the first edition, it ends with the *Fourth Day* which actually is the last part he managed to bring into a satisfying form to be published, apart from an appendix essentially reproducing a treatise on centers of gravity which Galileo had composed in his youth. In the following, we will argue that Galileo planned to complete the *Discorsi* with a *Fifth Day* which, among other topics, was intended to comprise a proof of the alleged parabolic shape of the catenary and an explanation of the practical utility of chains for determining projectile trajectories in artillery. It will also be shown that this *Fifth Day* eventually remained incomplete, not because Galileo had doubts about this demonstration but because of his failing health which hindered him from writing this final part of the *Discorsi* until the practical utility of chains for artillery was superseded by the introduction of another instrument, designed by Galileo's disciple Evangelista Torricelli.

As in the case of the *First Day*, the *Second Day*, and the *Third Day*, at the end of the *Fourth Day* the discussants postponed a topic for their meeting on the next day which, due to the rambling around which was characteristic of their extensive elaboration of the various topics, they were not able to complete on that day. At the end of the *First Day*, it was the main question of the resistance which bodies have to fraction that, due to the numerous digressions, had to be postponed to the next day. The dialogue at the end of the *First Day* is typical for such announcements in the *Discorsi*:

> *Salviati.* (...) But gentlemen, where have we allowed ourselves to be carried through so many hours by various problems and unforeseen discussions? It is evening, and we have said little or nothing about the matters proposed; rather, we have gone astray in such a way that I can hardly remember the original introduction and that small start that we made by way of hypothesis and principle for future demonstrations.

> *Sagredo.* It will be best, then, to put an end for today to our discussions, giving time for our minds to compose themselves tranquilly at night, so that we may return tomorrow (if you are pleased to favor us) to the discussions desired and in the main agreed upon.

> *Salviati.* I shall not fail to be here at the same hour as today, to serve and please you. (Galilei 1974, 108)

In the case of the *Second Day*, it was the reading of the book of the "Academician" that Galileo's spokesman announced[1] and that, indeed, became the issue of the following two days. And again, at the end of the *Third Day*, they postponed the treatment of projectile motion to the *Fourth Day* .[2] Hence, in all these cases, the postponed topics indeed became the center of the discussions on the next day.

Similarly, at the end of the *Fourth Day* a topic was brought up by Sagredo and Simplicio insistently, but kept dangling by Salviati. This time it is the very topic which had inspired Galileo's work on a new mechanics at its beginning: the alleged common parabolic shape of the projectile trajectory and the curve of a hanging chain, its dynamical foundation, and the resulting utility of the chain for drawing parabolic lines.

Sagredo introduces the topic into the discussion of the projectile trajectory referring to Galileo's interpretation of the trajectory as resulting from the composition of horizontal and vertical motion:

> *Sagredo.* I observe that with regard to the two impetuses, horizontal and vertical, as the projectile is made higher, less is required of the horizontal, but much of the vertical. On the other hand, in shots of low elevation there is need of great force in the horizontal impetus, since the projectile is shot to so small a height. (Galilei 1974, 255)

In the course of the discussion of the consequences of this composition Sagredo returns to the utility of the chain for drawing parabolic lines as it was raised

---

[1]   Galilei 1974, 142.
[2]   Galilei 1974, 215f.

already at the end of the *Second Day*, and Salviati announces further explanations:

> *Sagredo*. Then with a chain wrought very fine, one might speedily mark out many parabolic lines on a plane surface.
>
> *Salviati*. That can be done, and with no little utility, as I am about to tell you. (Galilei 1974, 257)

However, the discussion then turns away from this topic and returns instead to an argument that was used earlier in order to show that a projectile can never travel along a straight line along the horizontal, no matter how strong the impressed force driving it is (the "straightness question" which has been discussed also by other historians of science[1]). This argument is also based on the dynamical similarity between projectile motion and hanging chain. Galileo relates this property of projectile motion to the fact that it is similarly impossible to stretch a chain horizontally by whatever immense force may be applied; this claim remained to be proven and hindered Salviati for the moment from giving the announced further explanation on the relation of the projectile trajectory and the curve of a hanging chain.

The following proof deserves attention because it shows that Galileo's failed first attempt did not at all impel him to give up the underlying idea. It furthermore provides an experimental setting which makes Galileo's idea about the composition of a horizontal and a vertical force explicit which was conceived by him as to be the common dynamical basis of the projectile trajectory and the curve of the hanging chain. His proof uses a small weight hanging from the middle of a chain which is supported by two nails and stretched by two immense weights hanging from the loose ends of the chain; the proof essentially follows a line of reasoning similar to that discussed in the previous section. After the completion of this proof, the discussants return to the question of the utility of chains, which is now, however, definitely deferred to the next day, the *Fifth Day* of the *Discorsi*:

> *Simplicio*. (...) And now Salviati, in agreement with his promise, shall explain to us the utility that may be drawn from the little chain, and afterwards give us those speculations made by our author about the force of impact.
>
> *Salviati*. Sufficient to this day is our having occupied ourselves in the contemplations now finished. The time is rather late, and will not, by a large margin, allow us to explain the matters you mention; so let us defer that meeting to another and more suitable time. (Galilei 1974, 259)

---

[1] See Drake and Drabkin 1969, 103f.

A first clue for answering the question of what Galileo had in mind when he announced further explanations of the raised topic is provided by later comments of Vincenzio Viviani.[1] When he became Galileo's assistant in the second half of 1638, he began to study Galileo's works and must have carefully read the *Discorsi* as soon as they became available. When Viviani came across the first of the above quoted passages on the utility of the chain, he made the following marginal note in a copy of Galileo's book:[2]

> By means of this small chain perhaps Galileo found the elevations to hit a given target.

Viviani's note hits the nail on the head. It points to a problem that remained essentially unsolved in the *Fourth Day* of the *Discorsi*, that is, the lack of a satisfying theory of oblique projection.[3] Viviani assumed that the utility of the chain for Galileo must have been related to projectile motion and, in particular, to practical purposes of artillery. In the *Fourth Day* of the *Discorsi* Galileo presented, precisely in view of such practical purposes, tables relating the angles of shots to the amplitudes, altitudes, and sublimities of the resulting parabolic trajectories. These tables were, however, of limited use for artillery – even leaving aside problems such as air resistance etc. – since they do not relate the position of a given target to the properties of a shot. Indeed, the tables do not give the full trajectory but only certain key parameters. It was therefore plausible to supplement them with a way of constructing the trajectory that would be of more direct use to gunners. Viviani's remark suggests that Galileo still intended to use the alleged relation between parabola and catenary and the technique suggested by the projectile experiment reported in Guidobaldo des Monte's notebook, which has been discussed already,[4] in order to precisely fill this gap.

In another manuscript note of Viviani, he even considered the possibility that Galileo made use of chains also for more theoretical purposes, again in the con-

---

[1]   Viviani's comments discussed in the following were the basis of an earlier analysis of the role of the chain in the planned *Fifth Day* of the *Discorsi* by Raffaello Caverni, see Caverni 1972, V: 137-154. Caverni even claims to have found a substantial part of a dialogue on the chain supposedly written either by Galileo or by Viviani following Galileo. The authenticity of this alleged text has, however, been questioned, see Favaro 1919-1920. But in spite of the dubious character of this document, it has nevertheless also been taken seriously by modern historians of science, see Galilei 1958, 834-837. The text given by Caverni appears authentic in particular because it refers to a number of actually existing manuscripts in Galilei ca. 1602-1637, including folio 41/42 discussed above. The dialogue published by Caverni describes even how the curves on this folio page were drawn by Galileo with the help of carbon powder. Our analysis of the inks used on this folio page has shown, however, that these curves have actually been drawn by ink so that Caverni's dialogue on the utility of the chain is now definitely identified as a forgery, see Working-Group 1996.
[2]   This statement was written by Viviani on the margin of page 284 of the first edition of the *Discorsi* next to the passage in the first edition of the *Discorsi*: "potersi et ancora con qualche utilità non piccola come appresso vi dirò." Viviani's copy with this remark is kept in the Biblioteca Nazionale Centrale in Florence as Ms. Gal. 79.
[3]   See the analysis of Galileo's failed attempt in Damerow, Freudenthal, McLaughlin, and Renn 1992, chap. 3.
[4]   See the discussion of the folios 113r and 41/42 in note 2 above.

text of projectile motion. Referring to the same passage of the *Discorsi*, Viviani wrote:[1]

> See at page 284 [sic! not 384 as written by Caverni], the last phrase, which benefit Galileo meant, whether of measuring the parabolic line, or whether of the way of finding the propositions concerning projectile motion.

Viviani thus took into account the possibility that Galileo might still have in mind the theoretical program suggested by the projectile trajectory experiment of exploiting the alleged common dynamical foundation of the trajectory and the curve of a hanging chain as a heuristic device in order to find propositions on projectile motion.

In contrast to the modern historian, Viviani was in the unique position to simply ask Galileo what these obscure references to the utility of the chain in the *Discorsi* really meant – which is what he actually did when he was living with Galileo from late 1638 until the latter's death in 1642.[2] When he later included recollections of this period in a book which he published in 1674, Viviani explained what he had learned from Galileo himself about the utility of the chain and its relation to projectile motion:

> Now all I have left to say is how much I know about the use of chains, promised by Galileo at the end of the *Fourth Day*, referring to it as he intimated when, he being present, I was studying his science of projectiles. It seemed to me then that he intended to make use of some kind of very thin chains hanging from their extremities over a plane surface, to deduce from their diverse tensions the law and the practice of shooting with artillery to a given objective. But of this Torricelli wrote adequately and ingeniously at the end of his treatise on projectiles, so that this loss is compensated.

> That the natural sac of such chains always adapts to the curvature of parabolic lines, he deduced, if I remember well, from a reasoning similar [to this]: Heavy bodies must naturally fall according to the proportion of the momentum they have from the places from which they hang, and these momentums of equal weights, attached to points of a balance [which is] supported by its extremities, have the same proportion as the rectangles of the parts of that balance, as Galileo himself demonstrated in the treatise on resistances. And this proportion is the same as the one between straight lines, which from the points of that same balance [taken] as the base of a parabola, can be drawn in parallel to the diameter of this parabola, according to the theory of conic sections. All the links of the chain, that are like so many equal weights hanging from points on that straight line which con-

---

[1]   See Ms. Gal. 74, folio page 33r, kept in the Biblioteca Nazionale Centrale in Florence. For a transcription (erroneously referring to folio page 23r) see Caverni 1972, V: 153.
[2]   See Drake 1987, 394.

nects the extremities where this chain is attached and serves as base of the parabola, have in the end to fall as much as permitted by their momentums and there [they have to] stop, and [they] must stop at those points where their descents are proportional to their momentums from the places where those links hang, in that last instant of motion. These then are those points which adapt to a parabolic curve as long as the chain and whose diameter, which raises from the middle of the said base, is perpendicular to the horizon. (Viviani, Vincenzo 1674, 105f)

The first part of Viviani's text confirms what we have discussed above, that Galileo intended to introduce the chain as an instrument by which gunners could determine how to shoot in order to hit a given target. The main part of Viviani's text represents the sketch of a proof based on the determination of the moments of weights hanging from a beam supported at its two ends and on the assumption that the links of the chain descend according to these moments. If Viviani's report on that proof is reliable, it implies that Galileo had planned to crown his life-long reflections about the relation between projectile trajectory and chain with a proof of the alleged parabolic shape of the catenary, a proof that would have become a key subject of the never-finished *Fifth Day* of the *Discorsi*.

The reliability of Viviani's report is confirmed by a folio page in Galileo's own hand, folio page 43r, in which one finds a chain line (marked by little rings) representing a projectile trajectory, short texts, and some calculations (see figure 21). The texts perfectly confirm Viviani's report on Galileo's intentions to continue the ideas developed at the end of the *Fourth Day*. They deal with three issues related to the end of the *Discorsi* and contain:

- a brief explanation of the utility of chains for purposes of artillery
- a sketch of the resolution of the "straightness question" based on the dynamical justification of the alleged relation between projectile trajectory and hanging chain
- a note on the main idea of the proof sketched by Viviani.

The first text concerning the practical utility of the chain reads:

Let the little chain pass through the points *f* and *c*, and, given the target *z*, stretch the chain so much that it passes through *z*, and you will find the distance *sc* and the angle of elevation etc. (Galilei ca. 1602-1637, folio page 43r)

This text fits the interpretation given above that, for Galileo, the practical utility of the chain consisted in determining the shooting angle necessary in order to hit a given target if other parameters, in this case the amplitude of the parabola, are given. Immediately underneath Galileo sketched how he planned to resolve the "straightness question":

It is to be demonstrated that, just as it is impossible to stretch a chain into a straight line, it is likewise impossible that the projectile ever travels along a straight line, if not along the perpendicular upwards, just as also the chain as a plumb-line stretches itself along a straight line.

The text sketches the line of argument followed by Galileo at the end of the *Fourth Day*. The third text on folio page 43r is a short note written next to the chain line and evidently referring to magnitudes in Galileo's diagram:

The heavy body in *g* presses with less force than in *s* according to the proportion of the rectangle *fgc* to the rectangle *fsc*.

Just as in the longer explanation by Viviani, this short text by Galileo also focuses on the "rectangle" which corresponds to the product of the two parts of the base-line of the catenary, whose division is obtained by vertically projecting a given point of the catenary onto this base-line. Clearly, Galileo's note refers to the same proof-idea as Viviani's sketch. We can therefore indeed rely on this more explicit text in order to reconstruct the proof of the parabolic shape of the catenary which Galileo intended to incorporate into the *Fifth Day* of the *Discorsi*.

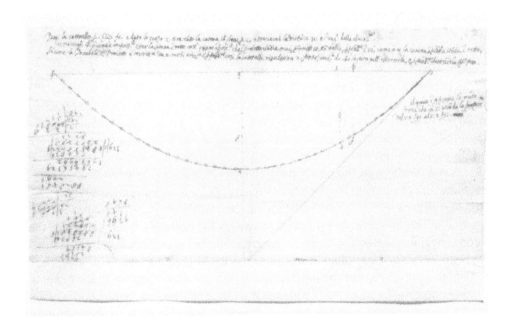

Figure 21. Ms. Gal. 72, folio page 43 recto

Viviani's text implies that the basis for the proof is a theorem on the resistance of a beam supported at both ends, proven by Galileo in his theory on the strength of materials. The theorem determines the proportion between limit resistances to breaking of a cylinder supported at both ends by the inverse proportion of the rectangles whose sides are the distances of the breakage points from the two ends:

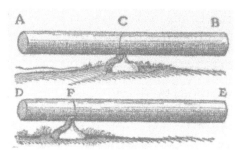

> If two places are taken in the length of a cylinder at which the cylinder is to be broken, then the resistances at those two places have to each other the inverse ratio [of areas] of rectangles whose sides are the distances of those two places [from the two ends.][1]

In the proof of this proposition the forces, represented by weights hanging down from the beam, are determined which are necessary at any particular place to break the beam. In a manuscript version (the "Pieroni manuscript") of this part of the *Discorsi*, the proof of Galileo's theorem is directly expressed in terms of moments of weights, just as it is done in Viviani's text.[2] In the formulation of the Pieroni manuscript, it then follows immediately that the moments of equal weights hung from a beam supported at its extremities are to each other as the rectangles whose sides are the distances of the points at which they are attached from the two ends of the beam. From the geometrical properties of the parabola it then follows further that these moments, taken at given points of the beam, are proportional to the vertical distances of these points from the corresponding points of a parabola whose base-line is given by the beam. The geometrical representation of the moments by a parabola leads to a figure quite similar to the drawing on folio page 43r with the beam of the proposition corresponding to the horizontal line between the two points at which the chain is suspended. Thus, the geometrical representation of the moments at the beam makes clear what Galileo had in mind when he applied this proposition to the hanging chain.

Galileo's application of the proposition on the stability of beams to the hanging chain is essentially based on the idea that a chain can be conceived of as a beam which is cut in small pieces linked in a way that makes them move down in pro-

---

[1] Galilei 1974, 133, corresponding figure taken from Galilei 1890-1909, VIII: 175.
[2] Galilei 1890-1909, VII: 176; for a discussion of the concept of moment see Galluzzi 1979.

portion to the moments or forces acting on them as if they were weights suspended from the corresponding points of the beam. Viviani indeed argues that the links of a chain can be considered as so many little weights suspended from the beam and that the descent of these little weights from the horizontal is proportional to their moments. The latter proportionality follows from that between effect and cause if the moments of the links of the chain are considered to be the causes of their descents, a conclusion familiar also from other parts of Galileo's mechanics.

What is wrong with this proof of an evidently fallacious statement? The basic idea of the proof is correct also from the viewpoint of classical mechanics, only that Galileo did not take into account that actually the number of links of a hanging chain is not equally distributed over the horizontal. Given a fixed distance of the suspension points at the ends of the chain, Galileo's basic assumption deviates the more from the real situation the greater the length of the chain is, so that it is more steeply hanging down at its ends.

From the viewpoint of classical mechanics, Galileo's assumption merely amounts to the determination of an approximation. His construction can fairly well be justified by the fact that he, of course, did not have available the mathematical means necessary to take the length of the hanging chain within a given horizontal interval into account, instead of using the length of the horizontal beam as a model.

In principle, this type of approximation is characteristic of all physical laws, which indeed at some time in a revised theory of the future appear, at best, as approximations. There is no greater difference between Galileo's chain that stands for real chains as that between mass points that stand for extended rigid bodies, ideal gases that stand for real gases, and Newtonian masses that stand for rest masses as conceived in relativity theory.

From the viewpoint of Galileo, however, there was no reason for considering his argument as dealing only with an approximation, or even more, to consider it as fallacious. He was simply arguing, as any physicist does, in a framework of a given mechanical theory. His proof involves basic concepts of his mechanics, such as the concept of momentum, and does not contain any obvious "errors" if taken within his conceptual system. His conceptualization of the chain as consisting of links that are able to move independently from each other along the vertical according to his mechanical model of the hanging chain seems problematical from a modern point of view but must have appeared quite natural to Galileo in view of the dynamical justification he could give for the comparison with the case of projectile motion.

In view of the striking success that this proof of the parabolic shape of the catenary must have represented for the closure of Galileo's theory of projectile motion, it is all the more surprising that it was neither incorporated into the version of the *Discorsi* published in 1638 nor into the drafts for the *Fifth Day* that have survived. In order to reconstruct the fate of this proof and the role Galileo intended for it in this last day of the *Discorsi*, we have to briefly recapitulate the history of the final composition of this book. The reconstruction of this final composition will also allow us to determine at least approximately in which period this proof was first formulated by Galileo.

After Galileo's condemnation in 1633 he returned to the scientific work that had been central to his concerns before he made his telescopic discoveries and engaged in his struggle for Copernicanism, that is, the theory of motion and the strength of materials. Even shortly after his condemnation, when he was still in Siena as a guest of the archbishop Ascanio Piccolomini, he began to write on the strength of materials, composing most of the treatise later contained in the *Second Day* of the *Discorsi*.[1] While a substantial part of the insights to be incorporated into this treatise had been attained already during Galileo's Paduan time, it was only now that he derived the propositions making up the final part of that treatise, including the crucial proposition quoted above for the derivation of the parabolic shape of the catenary.[2]

By mid-1635 much of the material later to be incorporated into the first two Days of the *Discorsi* had been completed.[3] Meanwhile, Galileo had formed the idea of composing four dialogues dealing with both the strength of materials and the theory of motion.[4] In the same year 1635, he probably also began to rework and edit his material on motion, dealing with motion along inclined planes, projectile motion, and the force of percussion for inclusion into this larger publication. After various failed efforts to find a publisher for the planned book, Galileo finally reached an agreement with Elzevir in May 1636.[5] In mid-1636, he managed to complete the *Third Day* (dealing with motion along inclined planes) and sent it to Elzevir.[6] Next Galileo turned to working on projectile motion, to be treated in the *Fourth Day*; by the end of 1636 he also had decided to amplify the original project of the book by an appendix containing his early work on centers of gravity.[7] In March 1637 Galileo sent an incomplete version of the *Fourth Day* to Elzevir, comprising the announcement of a further *Day* treating the force of percussion.[8] By that time, Galileo must have been working already on a dialogue on this topic.[9] An extensive draft of this part of the *Fifth Day* has survived and

---

[1]   See e.g. Galileo to Andrea Arrighetti, September 27, 1633, Galilei 1890-1909, XV: 283f and for historical discussion Drake 1987, 356.

[2]   How far Galileo's work on the strength of materials was progressing can be inferred from his contemporary correspondence; see e.g. Niccolò Aggiunti to Galileo, September 10, 1633, Galilei 1890-1909, XV: 257f.

[3]   See Fulgenzio Micanzio to Galileo, April 7, 1635, Galilei 1890-1909, XV: 254, and Galileo to Elia Diodati, June 9, 1635, Galilei 1890-1909, XV: 272f.

[4]   Galileo to Elia Diodati, June 9, 1635, Galilei 1890-1909, XV: 272f.

[5]   See e.g. Galileo to Fulgenzio Micanzio, June 21, 1636, Galilei 1890-1909, XVI: 441f, and for historical discussion Drake 1987, 374.

[6]   Galileo to Fulgenzio Micanzio, June 28, 1636, Galilei 1890-1909, XVI: 445, and Galileo to Fulgenzio Micanzio, August 16, 1636, Galilei 1890-1909, XVI: 475.

[7]   Galileo to Elia Diodati, December 6, 1636, Galilei 1890-1909, XVI: 524.

[8]   Galileo to Elia Diodati, March 7, 1637, Galilei 1890-1909, XVII: 41f, and Fulgenzio Micanzio to Galileo, March 7, 1637, Galilei 1890-1909, XVII: 42.

[9]   See Drake 1987, 383.

was first printed in a later edition of Galileo's collected works; when exactly this draft was written remains unknown.[1] This draft dialogue on percussion was clearly written without the intention to include the topic of the chain. It thus corresponds to the original announcement of a further Day early in the *Fourth Day*. The incomplete version of the *Fourth Day* which Galileo sent in March 1637 ended with the tables on projectile motion and still lacked the treatment of the chain at the end of the printed version; it also still lacked the reiteration of the announcement of a further Day at the end of the printed version, now amplified by the theme of the catenary. In other words, by March 1637 Galileo did not yet dispose of the proof of the parabolic shape of the catenary.

In May and June of 1637, Galileo sent further material to Elzevir, probably comprising the Appendix on centers of gravity and also material for the *Fourth Day*.[2] In September 1637, the Dutch printers complained that they still had not received the manuscript of the *Fifth Day* and sent Galileo a memorandum to that effect.[3] In November Elzevir acknowledged having received from Galileo another folio related to the *Fourth Day*.[4] By the latest at this point, but possibly already in June, the *Fourth Day* was concluded in the way it later appeared in print, that is, comprising the treatment of the chain. Since we know from the analysis of folio page 43r that the main argument of this concluding section, dealing with the "straightness" question, was sketched at a time when Galileo possessed the proof idea for his demonstration of the parabolic shape of the catenary, it seems plausible to assume that this proof was conceived at some point between March and November 1637.

Since February 1637 Galileo suffered from problems with his sight which delayed his work on the *Fifth Day*.[5] In November Elzevir informed Galileo that he would go on with the printing but, if possible, await the completion of the *Fifth Day*.[6] In the beginning of January 1638 Elzevir offered to complete the *Discorsi* according to Galileo's orders if he should be unable to do it himself:

> Concerning the treatise on percussion and on the use of the chain, if you cannot bring it to perfection, I will complete it according to your order. (Louis Elzevir to Galileo, January 4, 1638, Galilei 1890-1909, XVII: 251)

By the end of January, however, the fate of the *Fifth Day* was sealed, at least for the first edition of the *Discorsi*. At that point in time, Elzevir requested from Galileo everything that was still needed in order to finalize the book and suggested to

---

[1] See Galilei 1974, 281-306 and for a discussion of the chronology of Galileo's work on the *Fifth Day* see Galilei 1890-1909, VIII: 26-33.

[2] Fulgenzio Micanzio to Galileo, May 2, 1637, Galilei 1890-1909, XVII: 71f, Fulgenzio Micanzio to Galileo, May 9, 1637, Galilei 1890-1909, XVII: 76f, and Fulgenzio Micanzio to Galileo, June 20, 1637, Galilei 1890-1909, XVII: 114f.

[3] See Justus Wiffeldich to Galileo, September 26, 1637, Galilei 1890-1909, XVII: 187f.

[4] Fulgenzio Micanzio to Galileo, October 17, 1637, Galilei 1890-1909, XVII: 199f and Louis Elzevir to Galileo, November 1, 1637, Galilei 1890-1909, XVII: 211.

[5] See Drake 1987, 384.

[6] Louis Elzevir to Galileo, November 1, 1637, Galilei 1890-1909, XVII: 211.

him to add an explanation concerning the absence of material on the force of percussion, if none was to be included.[1]

What was the fate of the *Fifth Day* after the *Discorsi* had been finally published
in 1638? Galileo continued to work on his theory of motion, clearly also because
he was dissatisfied with the exposition of this theory in his book. Following a
suggestion of his disciple Viviani, who became his assistant in the second half of
1638, Galileo first turned to a problem in the logical foundation of his theory of
motion, whose solution he intended to insert into the second edition. He focussed
his attention to problems of the deductive structure of his theory also because he
found it difficult to elaborate new theorems given his problems of sight, as he
wrote in a letter to Baliani of 1639.[2] But in the same letter, he also wrote that he
planned to enrich a future edition of the *Discorsi* with material on other scientific
subjects, including the force of percussion, had he ever a chance to do so. This
plan must have comprised also the promised treatment of the catenary.

This plan remained, however, unrealized. Not only was it difficult for Galileo to
bring his numerous hitherto unpublished scientific results into a publishable form
in view of his failing health, but he must have definitely abandoned his original
plan to deal with the utility of the chain for artillery in a separate day of the *Discorsi* when he discovered that his theory of projectile motion had meanwhile
been substantially elaborated by somebody else, Evangelista Torricelli.

When Galileo, in March 1641, saw Torricelli's treatise on motion, he not only
found there that Torricelli had succeeded in solving some of the key problems of
Galileo's theory of projectile motion, such as the derivation of the parabolic
shape of the trajectory in the case of oblique projection, but also that Torricelli
had himself designed an instrument that would make these insights useful for
gunners, thus effectively replacing the chain in its presumed utility for artillery.[3]
Instead of elaborating another treatise dealing with novel physical problems such
as percussion, Galileo thus settled for securing what he had already achieved and
pursued his approach to polish the deductive foundation of his theory of motion.
He therefore began, in October 1641, to compose a dialogue on the theory of proportions which he dictated to Torricelli.[4] From its beginning, it is clear that this
dialogue was intended to replace the earlier plan of a *Fifth Day* on the catenary
and on percussion and was supposed to directly follow the Appendix and the
Fourth Day of the *Discorsi* as they were published in 1638.

The unfortunate fate of the *Fifth Day* of the *Discorsi* apparently definitely
sealed also the fate of the chain as a key subject of the theory of motion inaugu-
rated by Galileo. Definitely? Well, not quite. It saw a striking revival in a context
we had to neglect here, that of Galileo's study of the motion of the pendulum.

[1]   Louis Elzevir to Galileo, January 25, 1638, Galilei 1890-1909, XVII: 265.
[2]   Galileo to Giovanni Battista Baliani, August 1, 1639, Galilei 1890-1909, XVIII: 78.
[3]   Torricelli 1919 and Benedetto Castelli to Galileo, March 2, 1641, Galilei 1890-1909, XVIII:
303.
[4]   See Drake 1987, 421f and Giusti 1993.

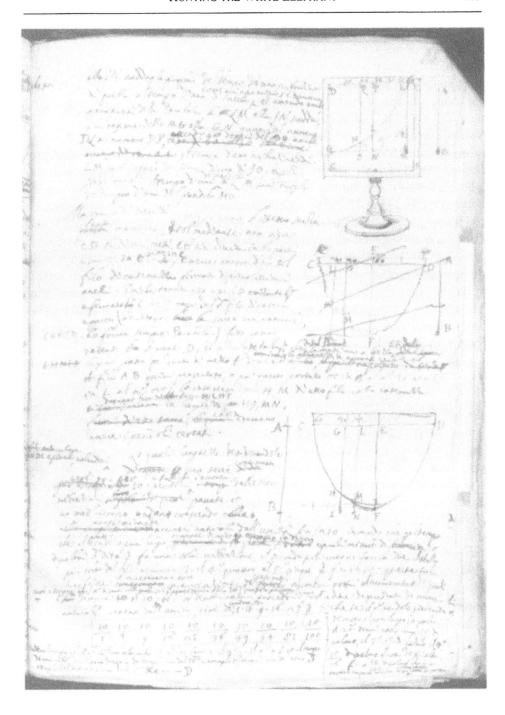

Figure 22. Vincenzio Viviani's Application of a hanging chain for determining the length of a pendulum with a given time period

We have seen above that Galileo's faithful disciple Viviani made sure that Galileo's proof of the parabolic shape of the catenary was preserved for posterity by including it in a book reporting on Galileo's unpublished achievements. But Viviani also thought of giving practical significance to Galileo's discovery, even after the chain had lost, as we have also pointed out above, its practical utility for gunners which Galileo had in mind due to a new instrument introduced by Torricelli. Viviani considered instead the practical utility of the supposedly parabolic shape of the catenary for determining the lengths of pendulums swinging with a certain desired period of time.[1] He designed an instrument consisting of a horizontal rod whose one half is divided into 60 equal parts with a chain hanging underneath the entire rod (see figure 22). If the distance between the vertex of the chain and the rod is chosen to be such that a pendulum of that length would swing with a period of one second, then the length of a pendulum swinging with any given fraction of a second can be found by first selecting the corresponding value on the scale along the horizontal rod. The distance between the rod at that point and the corresponding point of the chain underneath gives the desired length of the pendulum. The line of a hanging chain, supposedly of parabolic shape, is hence used by Viviani as a mechanical representation of the functional relation between times and lengths of the pendulum.

When and How did Galileo Discover the Law of Fall?

At the beginning of this investigation the problem has been raised whether the standard answer to the question of when and how Galileo made his celebrated discoveries of the law of fall and of the parabolic shape of the projectile trajectory is correct. But even after our extensive examination of the historical evidence, comprising some hitherto neglected or unknown documents, the difficulties of answering this question did not disappear. On the contrary, the apparent answer to this question achieved here makes it evident that this question does not really hit the point. Whatever answer will be given, it necessarily reduces the origin of a new conceptual structure, which is the outcome of a complex human interaction determined by both tradition and innovation, to the activities of one individual subject, Galileo Galilei. If the reconstruction of the history of the discovery as it is given here is reliable, an answer to this question necessarily detaches his activities from the context which made them meaningful in the social process of emerging knowledge.

At least in the case of the discovery of the law of fall and the parabolic shape of the projectile trajectory, the context of discovery is indistinguishably intermingled with the context of its justification. Neither the statement of the law of fall

---

[1]   See Viviani, Vincenzio after 1638, folio page 64r. For a transcription see Caverni 1972, IV: 428.

nor that of the parabolic shape of the trajectory are *per se* meaningful. Since they do not correspond to any immediate experience they have to be theoretically justified in order to attain their exceptional status within the body of mechanical knowledge. Without any theoretical context they can not be related to practical experiences with falling or projected bodies, a context on which the truth of these statements heavily depends. They were only the results of a collective process which created classical mechanics and thus provided Galileo's discoveries with that meaning which is implicitly presupposed in the question of when and how Galileo made them. It is this context that gives to statements such as the law of fall the appearance of being empirical facts which are relatively independent of their theoretical justification by means of proofs within a particular representation of mechanical knowledge.

However, this context of later developments was not the historical context of Galileo's discoveries. The context which made up the stage for Galileo was rather determined by a moderately anti-Aristotelian conception of motion which Galileo shared with contemporaries such as Benedetti and Sarpi and which formed the basis of his treatise *De Motu*. It is obvious that none of them had at the beginning any idea of the law of fall. Consequently, they could not draw any conclusion from such a law concerning the shape of the projectile trajectory. Moreover, the question of according to which law acceleration takes place would probably not even have been a meaningful question to them. Galileo, at least, explicitly discussed in *De Motu* the acceleration of a falling body as an accidental phenomenon. The issues which actually bothered Galileo and some of his contemporaries at that time were problems such as the question of what ratio exists between the speeds of different falling bodies. This question, for instance, was generally considered to make perfect sense, only the correct answer to this question being a controversially disputed, open problem. Another problem of this type was the question of what the true cause of speed and slowness of natural motion really was. A further question was, how exactly the process of impressing motive force into a body functions. How can natural and forced motion interact? In contrast to these questions which were considered to be questions that could unambiguously be settled, a variety of different explanations for the initial acceleration of the motion of a falling body were conceivable, non of them making the question of what the law is according to which this acceleration takes place into a meaningful question. In short, with regard to acceleration, there really seemed to be nothing to discover in this context!

Galileo's *De Motu* contains, on the other hand, a number of arguments which can well be considered as important discoveries, although they do only make sense in the historical context they were raised. We have shown above, for instance, that the concept of neutral motion which in *De Motu* complements the concepts of natural and forced motion and which to a certain extent contradicted assumptions on motion which, in spite of his anti-Aristotelian attitude, he shared with Aristotle, was an essential precondition that permitted Galileo (contrary to

Guidobaldo who did not share Galileo's preoccupation with a science of motion) to identify the parabolic shape of the trajectory and to infer from this shape the law of fall. In a similar way, the proposition that the ratio of speeds of bodies moving down inclined planes with the same vertical heights is equal to the inverse ratio of the corresponding lengths of the planes, became an important precondition for the development of the theory which Galileo finally published in his *Discorsi*. Given that neither Galileo's concept of motion nor his concept of speed corresponds to the notions of classical mechanics, both examples do not represent discoveries in the sense which is inherent in the question of when and how Galileo discovered the law of fall. Even if Galileo would have derived this law from observations as they are analyzed in *De Motu*,[1] or if he would have found the law experimentally he could not have recognized it as *the* important achievement he considered it later.

When and how has the law of fall been discovered? Is the discovery to be dated to the moment when somebody for the first time happened to stumble upon the idea that the spaces traversed by a falling body might increase in the same proportion as the squares of the times? We have argued that in the context of a theory of motion as the one Galileo and several of his contemporaries adhered to at the time of *De Motu*, a discovery of the law of fall in that sense could not have been recognized as a substantial challenge of this *De Motu* theory. According to our reconstruction, this situation occurred indeed a short time later when Galileo performed together with Guidobaldo del Monte the projectile trajectory experiment recorded in Guidobaldo's notebook. The outcome of this experiment appears with hindsight as a pathbreaking achievement which forced Galileo to give up his misconception of the projectile trajectory and to develop the idea of its parabolic shape. It seems therefore that the now available evidence for Galileo's role in performing this experiment, which is provided by the note about its outcome in Paolo Sarpi's notebook, demonstrates unambiguously that the discovery must be dated into the year 1592.

But does it really make sense to attribute a discovery to somebody who gives this discovery quite a different meaning? What Galileo reported to Sarpi must have been perceived by him as a discovery. But what he considered to be the discovery was probably the observation – amazing in the context of the *De Motu* theory – that the trajectory seemed to be symmetrical like a hyperbola or a parabola. The fact that in his protocol Guidobaldo left open whether the curves achieved by the experiment were parabolic or hyperbolic and also the fact that Sarpi's note does not say anything at all about the precise mathematical shape of the trajectory makes it, on the other hand, questionable that the precise shape of

---

[1] This would not have been absolutely impossible. Galileo came close to the discovery of the isochronism of chords in a circle which, in fact, he discovered only shortly afterwards. Together with his considerations on the speeds and forces on inclined planes, this discovery makes it nearly possible to infer the law of fall. He only needed the relation between speeds and times for motions along inclined planes in the form of his later "postulate" as he used it in the *Discorsi*; see Damerow, Freudenthal, McLaughlin, and Renn 1992, 156-158.

the trajectory was the point that mattered to them, even though Galileo could not have had any trouble in recognizing its close connection to the dynamical interpretation of the experiment. The discovery of the symmetry, however, was amazing because it seemed to indicate that natural and forced motion, although allegedly quite different in nature, were symmetrically exchanging in ascending and descending their roles. This was not to be expected, but could nevertheless easily be made compatible with the prevailing conception of natural and forced motion.

Another aspect of the "discovery" was surely considered by both Galileo and Guidobaldo to be even more important, namely, the (fallacious) conclusion that the same explanation can be given for the symmetrical shape of the catenary and that of the projectile trajectory; they assumed that both curves result in the same way from the interaction of a horizontal and a vertical force. As we have shown, Galileo believed that to be true till the end of his life when the law of fall for a long time already had become the cornerstone of his new science of mechanics and he regarded the discovery of the parabolic shape of the trajectory—as he claimed in the conflict with Cavalieri—as "the first fruit" of his studies and as a "flower (...) broken from the glory." This re-evaluation of the discovery of the parabolic shape of the projectile trajectory was not accompanied by a reinterpretation of its early dynamical justification. On the contrary, he even underlined once more the importance of the common dynamical explanation of catenary and trajectory when he planned to make the proof of the parabolic shape of the catenary a final highlight of the *Fifth Day* of the *Discorsi*.

Given that neither Guidobaldo nor Galileo initially and partly even later fully recognized the theoretical consequences of the outcome of their projectile trajectory experiment, does it then make more sense to date the discoveries of the parabolic trajectory and the law of fall not to the year when they performed it but rather to a later time when Galileo realized its implications? Should the discovery, for instance, be attributed to his first interpretation of the curve generated in the experiment as a parabola and not as a hyperbola or any other similar curve, thus accepting as a consequence the validity of the law of fall? Or does it make more sense to date the discoveries even later to the time when he was able to prove the law of fall and the parabolic shape of the trajectory? Should perhaps even stronger criteria be applied? Can he, for instance, be credited already with the discovery of the law of fall as long as he was still erroneously proving it from the assumption that the "degrees of velocity" increase proportional to the spaces traversed?[1] Or should the discovery be attributed to him only when he had found the proof which he finally published in the *Discorsi*? Is it relevant for this attribution that even this proof is still not a proof valid in classical mechanics? Or is the famous inclined plane experiment as an empirical demonstration of the validity of the law of fall a better indication of the discovery of the law of fall than any

---

[1] For this assumption see the letter by Galileo to Paolo Sarpi, October 16, 1604, Galilei 1890-1909, X: 115f and the discussion in Damerow, Freudenthal, McLaughlin, and Renn 1992, chap. 3.

theoretical speculation? Galileo later claimed that when he had repeated this experiment "a full hundred times, the spaces were always found to be to one another as the squares of the times." (Galilei 1974, 170) Does it depend on whether such a claim is true that Galileo can be credited to have discovered the law of fall? Any dating later than 1592 when the projectile trajectory experiment was performed for the first time has obviously to take an act of interpreting this experiment as an indication of the discovery and is thus liable to doubts whether such a discovery can really be established as an indubitable historical fact.

What date after the day in the year 1592 when Galileo and Guidobaldo performed the experiment might be considered as the day when Galileo discovered the parabolic shape of the trajectory or the law of free fall? It has been shown that the "practical turn" in Galileo's life after his move to Padua made him look at the "discovery" in a specific way. Knowledge such as the recognition of the parabolic shape of the trajectory seemed at that time to be relevant to him only insofar as it was applicable to the solution of technological problems. He made, for instance, the "discovery" the basis of his intended treatise on artillery which, however, was never written. But even if he had written this treatise, it would probably have added nothing to what he knew already before. Galileo's experiments and studies in that period made him an experienced engineer-scientist, but did not substantially change his interpretation of the projectile trajectory experiment in the theoretical framework of a revised *De Motu* theory.

Turning to the time when he returned to the study of motion, we are apparently better off. The letter to Guidobaldo of 1602 which attests to this reorientation may be regarded as a testimony for a changed relation to Guidobaldo del Monte and his way of drawing the boundaries of mechanics and of validating a discovery. At that time the law of fall was used already substantially as a means of proving other propositions in order to assure the validity of observations such as the isochronism of inclined planes in a circle or, without success, the isochronism of the pendulum. What else should be necessary to credit him with the "discovery" of the law? The available evidence does, however, in no way indicate a dramatically new interpretation of the projectile trajectory experiment that could justify taking this reorientation as representing the real "discovery" of the law of fall.

The following period when Galileo already intensively worked on a deductive theory of motion based on the law of fall as its core theorem provides good reasons to worry whether stronger criteria should not perhaps be used for accepting a scientific activity as attesting to the discovery of a central theorem of classical mechanics such as the law of fall. In fact, at that time Galileo still had no answer to the problem raised by Sarpi who objected against Galileo's dynamical interpretation of the symmetry of the trajectory that an arrow shot vertically upwards has a much greater force than an arrow falling down from the maximum height of the shot. The theoretical program Galileo offered instead, as it becomes visible in his letter to Sarpi written in October 1604, is far from being convincing from the viewpoint of classical mechanics. Galileo intended to built up a theory based essentially on one fallacious principle which he expected to be able to cover "the other things," which probably refers to such diverse topics as those of the former *De Motu* theory enriched by his experiences as a practitioner, the theoretical discussions in his early years in Padua, in particular those with Sarpi about natural motion, the force of percussion, the isochronism of the pendulum, the length-time relation of the pendulum, and, of course, the law of fall and the projectile trajectory.

By returning to the origin of the "discovery," that is the experiment performed together with Guidobaldo del Monte, Galileo topped his ambitious theoretical program with the challenge of still another issue, that is, the derivation of the catenary. The situation became worse when Galileo realized that his claim that the catenary is a parabolic curve just as the projectile trajectory could not be verified empirically except for very flat hanging chains. For understandable reasons Galileo did what he apparently always did in such a situations, he trusted a proof which he believed to be true within his theoretical framework more than the outcome of an experiment. He did so for good reasons. As an experienced practitioner he knew many reasons why an experiment could fail. It would have been silly to give up such beliefs as the truth of the law of fall, the parabolic shape of the projectile trajectory, and the parabolic shape of the catenary only because he could only approximately demonstrate their validity by some experimental arrangements. Given such circumstances, what can then be a reliable distinction between a discovery and an error? What meaning can be ascribed to a statement such as "Galileo discovered the law of fall," or "Galileo discovered the parabolic shape of the projectile trajectory" when he "discovered" in exactly the same way also the parabolic shape of the catenary? Galileo finally stuck to everything he thought to be able to prove; his deductive theory in the *Discorsi* is the final outcome of his discoveries, representing an integration that legitimately can be considered as the outset of the development of a new science of motion although the development of this theory was completed only long after his death. As we have shown extensively in our study of the origins of classical mechanics, it was not individual achievements of Galileo designated as "discoveries" but their participation in a collective process of constructing a new body of knowledge that made his activities meaningful in spite of the seemingly chaotic path of his stumbling from one error into the other.[1] We hope that the internal consistency of the activities of Galileo, which is rather denied than confirmed by a description of them as a series of "discoveries," has been made evident by the reconstruction presented here.

According to our opinion, the conclusion from this reconstruction can be generalized. Independent of the contemporary systematic contexts of a developing body of knowledge and the contexts of its practical application, single elements of a structured body of knowledge such as the mental representations of the

---

[1] Damerow, Freudenthal, McLaughlin, and Renn 1992.

accelerated motion in free fall or of the trajectory of projectile motion are just meaningless tokens that trigger the phantasy of those who are separated from these contexts by a historical distance. Outside of their own contexts which make them meaningful, the actions of discoverers such as Galileo appear erratic and confront the historian of science only with a series of unsolvable riddles. The activities of the heroes in the history of science as well as the activities of the numerous practitioners on whose shoulders they stand regain their meaning only from the reconstruction of the continuity and change of the processes that transform these contexts. Hence, trying to solve the riddles of the history of science by determining the exact points in time in which the great discoveries of human history were made and describing how precisely they took place is nothing else but hunting the white elephant.

Appendix A: Selected Letters

The appendix comprises four letters which are partially quoted in the main text, a letter by Guidobaldo del Monte to Galileo from 21 February 1592, a letter by Galileo to Guidobaldo del Monte from 29 November 1602, a letter from Paolo Sarpi to Galileo from 9 October 1604, and Galileo's response to Sarpi from 16 October 1604. Although most of these letters are well-known and have often been discussed in the literature, they are here presented in a new English translation prepared by Fiorenza Z. Renn and June Inderthal. Previous translations into English were either partial translations or have misrepresented key passages of these letters. According to the argument of our paper, these letters have to be read as being closely related to each other. They provide a glance at two crucial intellectual contexts for Galileo's research on the law of fall.

Letter of Guidobaldo del Monte to Galileo in Pisa, February 21, 1592.

My Most Magnificent and Honorable Lord.

Because I did not have news of your Lordship for many days, I got my son Horatio to ask you [about them]. From what I see, I realize that your Lordship has written [letters] to me on previous occasions and I did not receive them, just as I did not receive that one which your Lordship told me you have written to me concerning your father's death. Indeed, when I heard about it, I was very sorry, both for the love of him and for the love of your Lordship; he did not seem so old to me that he could not have lived many more years. I condole with your Lordship, but we must be content with these upsets which the world deals us.

It also saddens me to see that your Lordship is not treated according to your worth, and even more it saddens me that you are lacking good hope. And if you intend to go to Venice in this summer, I invite you to pass by here so that for my part I will not fail to make any effort I can in order to help and to serve you; because I certainly cannot see you in this state. My forces are weak but, whatever they may be, I will employ them all in serving you. And I kiss your hands, as well as those of S.r Mazzone if he happens to be in Pisa. May the Lord grant your wishes.

In Monte Baroccio, 21st February 1592.
From your Lordship's Servant

Guidobaldo dal Monte. (Galilei 1890-1909, X: 46f)

Letter of Galileo to Guidobaldo del Monte in Montebaroccio, November 29, 1602.

Your Lordship, please excuse my importunity if I persist in wanting to persuade you of the truth of the proposition that motions within the same quarter-circle are made in equal times, because having always seemed to me to be admirable, it seems to me [to be] all the more so, now that your Most Illustrious Lordship considers it to be impossible. Hence I would consider it a great error and a lack on my part if I should allow it to be rejected by your speculation as being false, for it does not deserve this mark, and neither [does it deserve] being banished from your Lordship's understanding who, better than anybody else, will quickly be able to retract it [the proposition] from the exile of our minds. And because the experiment, through which this truth principally became clear to me, is so much more certain, as it was explicated by me in a confused way in my other [letter], I will repeat it here more clearly, so that you, by performing it, would also be able to ascertain this truth.

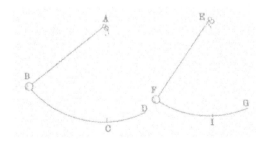

So now I take two thin threads of equal length, each being two or three braccia long, and let them be *AB, EF.* [I] hang them from two small nails, *A* and *E*, and at the other ends, *B* and *F*, I tie two equal lead balls (although it would not matter if they were unequal). Then, by removing each of the above-mentioned threads from its perpendicular, but one very much [so], as through the arc *CB* and the other very little, as through the arc *IF*; I let them go freely at the same moment of time. The one begins to describe large arcs, like *BCD*, and the other describes small ones, like *FIG*; but yet the mobile *B* does not consume more time moving along the whole arc *BCD* than the other mobile *F* in moving along the arc *FIG*. I make absolutely sure of this in the following way:

The mobile *B* moves along the large arc *BCD*, returns along the same *DCB*, and then comes back towards *D*, and it does this 500 and 1000 times, reiterating its oscillations. Likewise, the other one goes from *F* to *G*, and from here returns to *F*, and will likewise make many oscillations; and in the time that I count, let us say, the first hundred large oscillations *BCD*, *DCB* etc., another observer counts another hundred very small oscillations through *FIG*, and he does not count even a single one more: a most evident sign that each particular of these very large [oscillations] *BCD* consumes as much time as each particular of those minimal ones [through] *FIG*.

Now, if all *BCD* is passed [through] in as much time as *FIG*, then, in the same way, half of them, these being descents through the unequal arcs of the same quadrant, will be done in equal times. But even without staying on to enumerate other [oscillations], your Most Illustrious Lordship will see that the mobile *F* will not make its very small oscillations more frequently than the mobile *B* [will make] its larger ones, but rather, they will always go together.

The experiment which you tell me you have done in the box can be very uncertain, either because its surface has perhaps not been well cleaned or maybe because it is not perfectly circular, and because one cannot observe so well in a single passage the precise moment in which the motion begins. But if your Most Illustrious Lordship still wants to take this concave surface, let the ball *B* go freely from a great distance, such as from point *B*. It will pass to *D*, at the beginning producing its oscillations with large intervals, and at the end with small ones; but the latter, however, [will not be] more frequent in time than the former.

With regard now to the unreasonable opinion that, given a quadrant 100 miles in length, two equal mobiles might pass along it, one the whole length, and the other only a span, in equal times, I say it is true that there is something wondrous about it; but [less so] if we consider that a plane can be at a very slight incline, like that of the surface of a slow-moving river, so that a mobile will not have traversed naturally on it more than a span in the time that another [mobile] will have moved one hundred miles over a steeply inclined plane (namely being equipped with a very great received impetus, even if over a small inclination). And this proposition does not involve by any adventure more unlikeliness than that in which triangles within the same parallels, and with the same bases, are always equal [in area], while one can make one of them very short and the other a thousand miles long. But staying with the subject, I believe I have demonstrated this conclusion to be no less unthinkable than the other.

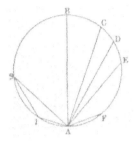

In the circle *BDA*, let the diameter *BA* be erected on the horizontal, and let us draw from the point A to the circumference any lines *AF, AE, AD, AC*: I demonstrate identical mobiles falling in equal times both along the perpendicular *BA*, and along the inclined planes of the lines *CA, DA, EA, FA*; so that, by starting at the same moment from the points *B,C,D,E,F*, they will reach the end point *A* at the same time, and let the line *FA* be as small as we want it to be.

And maybe even more unthinkable will appear the following, also demonstrated by me; that wherever the line *SA* being not greater than the chord of a quadrant, and [given] the lines *SI* and *IA*, the same mobile, starting from *S*, makes the journey *SIA* quicker than just the journey *IA*, starting from *I*.

Until now I have demonstrated without transgressing the terms of mechanics; but I cannot manage to demonstrate how the arcs *SIA* and *IA* have been passed through in equal times and it is this that I am looking for.

Please do me the favor of kissing the hand of Sig.r Francesco in return, telling him that when I have a little leisure, I will write to him about an experiment which has already entered my imagination, for measuring the moment of the percussion. Regarding your question, I consider that what your Most Illustrious Lordship said about it was very well put, and that when we begin to deal with matter, because of its contingency the propositions abstractly considered by the geometrician begin to change: since one cannot assign certain science to the [propositions] thus perturbed, the mathematician is hence freed from speculating about them.

I have been too long and tedious for your Most Illustrious Lordship: please excuse me, with grace, and love me as your most devoted servant. And I most reverently kiss your hands.

In Padua, 29th November 1602
From Your Illustrious Lordship's Most Obliged Servant

Galileo Galilei. (Galilei 1890-1909, X: 97-100)[1]

---

[1]   Since the original of this letter has not been preserved the diagrams may not be reliable.

Letter of Paolo Sarpi to Galileo in Padua, October 9, 1604.

My Most Excellent Lord and Most Respected Master.

With the occasion to send you this enclosure, I thought of proposing to you an argument to be resolved, and a problem which keeps me in doubt.

We have already concluded that a body cannot be thrown up to the same point [termine] if not by a force, and, accordingly, by a velocity. We have recapitulated – so Your Lordship lately argued and originally found out [inventò ella] – that [the body] will return downwards through the same points through which it went up. There was, I do not remember precisely [non so che], an objection concerning the ball of the arquebus; in this case, the presence of the fire troubles the strength of the argument. Yet, we say: a strong arm which shoots an arrow with a Turkish bow completely pierces through a table; and when the arrow descends from that height to which the arm with the bow can take it, it will pierce [the table] only slightly. I think that the argument is maybe slight, but I do not know what to say about it.

Here is the problem: if there are two bodies different in species, and any of them receives a force that is smaller than that of which it is capable; if now the force is communicated to both of them, will they receive the same amount of it? Thus, if gold were able to receive from a maximum force [an amount of] 20 and not more, and silver of 19 and not more; if they are now moved by a force of 12, will they both receive a force of 12? It seems that this is the case because the force is entirely communicated, the body is capable [of receiving it], hence the effect is the same. It seems [on the other hand] that this is not the case because, [if it were so], two bodies of different species, driven by the same force, would reach the same point with the same velocity.

If someone said: a force of 12 will move silver and gold to the same point but not with the same velocity. Why not [we may respond] if both of them are capable of receiving even more than that which [the force of] 12 can communicate to them?

I do not want to oblige Your Lordship to answer. Just in order not to send this paper blank, which had a peripatetic appetite of being filled with these characters, I wanted to satisfy it, acting like the agent does with the prime matter. And now, I stop here and kiss your hands.

In Venice, 9th October 1604
Your Most Excellent Lordship's Most Affectionate Servant

Brother Paulo from Venice (Galilei 1890-1909, X: 114)

Letter of Galileo to Paolo Sarpi in Venice, October 16, 1604.

Very Honorable Lord and Most Cultivated Father.

Thinking again about the matters of motion, in which, to demonstrate the phenomena [accidenti] observed by me, I lacked a completely indubitable principle which I could pose as an axiom, I am reduced to a proposition which has much of the natural and the evident: and with this assumed, I then demonstrate the rest; i.e., that the spaces passed by natural motion are in double proportion to the times, and consequently the spaces passed in equal times are as the odd numbers from one, and the other things. And the principle is this: that the natural moveable goes increasing in velocity with that proportion with which it departs from the beginning of its motion; as, for example, the heavy body falling from the point a along the line $abcd$, I assume that the degree of velocity that it has at c, to the degree it had at b, is as the distance ca to the distance ba, and thus consequently, at d it has a degree of velocity greater than at c according as the distance da is greater than ca.

I should like your Honorable Lordship to consider this a bit, and tell me your opinion. And if we accept this principle, we not only demonstrate, as I said, the other conclusions, but I believe we also have enough in hand in order to show that the naturally falling body and the violent projectile pass through the same proportions of velocity. For if the projectile is thrown from the point d to the point a, it is manifest that at the point d it has a degree of impetus sufficiently powerful to drive it to the point a, and not beyond; and if the same projectile is in c, it is clear that it is linked with a degree of impetus sufficiently powerful to drive it to the same point a, and, in the same way, the degree of impetus in b is sufficient to drive it to a, whence it is manifest that the impetus at points d, c, b goes decreasing in the proportions of the lines da, ca, ba; whence, if it goes acquiring degrees of velocity in the same (proportions) in natural fall, what I have said and believed up to now is true.

Concerning the experiment with the arrow, I believe that it does acquire during its fall a force that is equal to that with which it was thrown up, as we will discuss together with other examples orally, since I have to be there in any case before All Saints. Meanwhile, I ask you to think a little bit about the above mentioned principle.

Concerning the other problem posed by you, I believe that the same bodies receive both the same force, which, however, does not create the same effect in both; as for example the same person, when rowing, communicates his force to a gondola and to a larger boat, both being capable of assuming even more of it, but it does not result in one and in the other [boat] the same effect concerning the velocity or the distance-interval through which they move.

I am writing in the dark, this little may rather suffice to satisfy the obligation of answering than that of finding a solution of which to speak orally I reserve to a meeting in the near future.

And with all respect I kiss your hands.

In Padua, 16th October 1604
Your Very Honorable Lordship's Most Obliged Servant,

Galileo Galilei (Galilei 1890-1909, X: 115f)

Appendix B: Galileo's Claims from the Perspective of Modern Physics

BY DOMENICO GIULINI

Part 1. On Galileo's Exaggerations

That Galileo somewhat exaggerated the outcome of experiments described in his *Discorsi* is often suspected. Leaving alone the question as to why this might happen, it seems useful to also produce some precise quantitative estimations of such suspected exaggerations. This we shall do in the present part of this appendix for the famous case concerning the isochronism of the pendulum. Compare e.g. Drake 1990, chapters 1 and 14.[1]

---

[1] With respect to this example S. Drake states on p. 210-211 that "when the arc to the vertical for the pendulum having the wider swing is no more than 25°, the difference in times for it and the other pendulum is not very great and it keeps on diminishing." After all, the following quantitative estimation shows that such differences are observable after at most 20 swings.

Our estimations will be based on the *exact* formula for the period of a pendulum *without friction*.[1]

In a famous part towards the end of the first 'day' of the *Discorsi* (Galilei 1974, 87-88), Galileo (i.e., Salviati) gives the following account of an experiment:

> Ultimately, I took two balls, one of lead and one of cork, the former being at least a hundred times as heavy as the the latter, and I attached them to equal thin strings four or five braccia long, tied high above. Removed from the vertical, these were set going at the same moment, and falling along the circumferences of the circles described by the equal strings that were the radii, they passed the vertical and returned by the same path. Repeating their goings and comings a good hundred times by themselves, they sensibly showed that the heavy one kept time with the light one so well that not in a hundred oscillations, nor in a thousand, does it get ahead in time even by a moment, but the two travel with equal pace. The operation by the medium is also perceived; offering some impediment to the motion, it diminishes the oscillations of the cork much more than those of the lead. But is does not make them more frequent, or less so; indeed, when the arcs passed by the cork were not more than five or six degrees, and those of the lead were fifty or sixty, they were passed over in the same times.

Taking for the *braccio* 0.6 meters and hence the length of the pendulum between 2.4 and 3.0 meters, we see that we talk about periods certainly larger than 3 seconds.

The amplitude α is taken to be the angle between the thread of the pendulum and the vertical (direction of the gravitational field). The exact expression for the period T as function of α is an elliptic integral of first kind whose expansion in terms of $\sin(\alpha/2)$ begins as follows (Sommerfeld 1994 Reprint, Mechanik):

---

[1] Friction has two effects:1) It leads to an exponential damping of the amplitudes, 2) it enhances the period by an amount depending on the damping. The first affects our considerations insofar as we will calculate accumulated phase differences for pendulums of substantially different amplitudes. Hence we must check that the actual damping indeed allows to maintain such a difference in amplitudes for the considered periods of accumulation. Regarding 2) we need to estimate this effect since it threatens to level our calculated phase difference which is solely based on the enhancement of the period with amplitude.

Applied, as below, to a situation of two pendulums, one with large amplitude and small damping, the other with smaller amplitude because of stronger damping, we see that *both* pendulums will suffer an enhancement of their periods, albeit from different sources. However, the estimation of the enhancement due to damping is easily done and shows that a levelling of these two effects does not occur. To see this, let σ denote the number of full swings after which the amplitude has dropped by a factor of e^{-1}, the difference ΔT to the undamped period T is then given by $\Delta T/T = (8\pi^2 \sigma 2)^{-1}$ (plus higher powers in $(2\pi\sigma)^{-2}$, which we can safely neglect). Hence the corresponding number of swings after which a phase difference of $2\pi/n$ to the undamped pendulum has occurred is given by $N_{ph} = \sigma^2 8\pi^2/n$. Note in particular the *quadratic* dependence on σ and the relatively large prefactor $8\pi^2 \approx 79$. They imply that even for a considerable damping, like $\sigma = 5$, we would have to wait around 200 full swings to see a phase difference to the undamped pendulum of $2\pi/10$. This is a much smaller effect than the one discussed below.

$$T(\alpha) = 2\pi\sqrt{\frac{l}{g}}\left\{1 + \frac{1}{4} + \sin^2\frac{\alpha}{2} + \frac{9}{64}\sin^4\frac{\alpha}{2} + \dots\right\}. \qquad (A1)$$

Hence the period increases with the amplitude resulting in the lead-pendulum falling behind the cork-pendulum. We denote by $N_n(\alpha)$ the smallest integer number of full swings beyond which a pendulum of constant amplitude α will have fallen behind a time of at least T/n against a pendulum of period sufficiently close to $T := 2\pi\sqrt{l/g}$ (i.e. of sufficiently small amplitude, like 3°). After N_4 full swings the phase difference is at least $\pi/2$ and certainly detectable by be naked eye, since then the pendulums start to move in opposite directions. More careful but still unsophisticated observations should reveal deviations from synchrony by, say, one tenth of T, that is, after N_{10} swings.[1]

By definition of $N_n(\alpha)$ we have

$$N_n(\alpha) = \text{smallest integer} \geq \frac{T}{n \cdot (T(\alpha) - T)}. \qquad (A2)$$

Using (A1) we get for the various values of α and $n = 4$ or $n = 10$:

α	10°	15°	20°	25°	30°	35°	40°	70°	80°
N_4	132	59	33	21	15	11	9	3	2
N_{10}	53	24	14	9	6	5	4	2	1

From these values we infer that a situation with amplitudes $\alpha_{\text{lead}} = 25°$ and $\alpha_{\text{cork}} = 3°$ certainly cannot have appeared synchronous for longer than about 20 full swings.

The situations becomes even more drastic in a later description of a similar experiment, reported shortly after the beginning of the fourth day (Galilei 1974, 226). In this second experiment two balls of lead are suspended on equally long strings of 4-5 braccia and the periods compared for amplitudes $\alpha_1 = 5°$ and $\alpha_2 = 80°$ (!). Here again the assertion is that no deviations from synchrony could be detected, whereas our values for N_4 show that it must have been clearly apparent after 3 full swings the latest. After 4 full swings the two pendulums will even cross the origin approximately simultaneously with oppositely directed velocities.[2]

---

[1] For example, by letting two separate experimenters count and voice the passages of zero amplitude for the two pendulums respectively. Such a method is in fact suggested in the *Discorsi* (Galilei 1974, 227).

[2] In Galilei 1974, footnote 12 on page 227, S. Drake states that "a disagreement of about one beat in thirty should occur with pendulums of length and amplitudes described here." Unfortunately he did not state how he arrives at this result, which, seen from our analysis, seems to be an underestimation of the real effect by more than a factor of 3.

Part 2: Theory of the Hanging Chain and its Parabolic Approximation

In this part we first describe the usual theory, $\mathbf{T}_{ex}$, of the hanging chain in terms of Newtonian concepts, and then its approximation, $\mathbf{T}_{ap}$, for small slopes y', which leads to parabolic shapes as would be the case for constant mass distributions along the horizontal projection of the chain. On the level of physical quantities ("Observables") this approximation corresponds to expansions in terms of d/D to various degrees, depending on the observable, where

$$2D = \text{horizontal distance of the suspension points}$$

and $d = $ sag, i.e. the distance between the lower apex and the horizontal line joining the suspension points.

The Exact Theory $\mathbf{T}_{ex}$

We will think of the hanging chain as being given by a function $y(x)$ in a Cartesian xy-plane. A point in this plane is denoted by its coordinates (x,y), so that the curve is the set of points $(x,y(x))$, where x ranges over an interval which we take to be $[-D,D]$. y' *and* y'' denote the first and second derivatives of y with respect to x.

The fundamental equation for the theory of the chain is obtained from a simple and typical argument based on a local application of the principle of *balance of forces*. To do this, we imagine the chain being cut at $(x, y(x))$ and consider one end. We denote by $F(x)$ the strength of the force that one would have to apply to one end in order to keep the corresponding part of the chain in its place. This is also called the chain's tension. We can decompose $F(x)$ into a horizontal component, $F_h(x)$, and a vertical component, $F_v(x)$. Since by definition a chain can only support tangential forces, these components must satisfy

$$y'(x) = \frac{F_v(x)}{F_h(x)}. \tag{B1}$$

If the external force (gravitation) has no horizontal component, $F_h(x)$ must in fact be independent of x. Otherwise a piece of chain with different strengths of the outward pointing horizontal forces could not stay at rest; hence

$$F_h(x) = F_h = \text{const.}$$

Differentiating $(B1)$ once more then leads to

$$y''(x) = \frac{F'_v(x)}{F_h}. \tag{B2}$$

It is now easy to express the right hand side of (B2) as function of x and $y'(x)$, since $F_v(x + dx) - F_v(x)$ must clearly be equal to the weight of the piece of chain between $(x, y(x))$ and $(x + dx, y(x + dx))$. If we denote by μ the mass per unit length of the chain, which we take to be constant[1], then its weight is given by $\mu g\, ds(x)$, where $ds(x) = \sqrt{1 + [y'(x)]^2}\, dx$ is the length of the (infinitesimal) piece of chain that we consider. Hence (B2) results in

$$\frac{y''}{\sqrt{1 + [y']^2}} = \frac{1}{h} := \frac{\mu g}{F_h}, \qquad (B3)$$

which is our fundamental equation defining $\mathbf{T}_{ex}$ ['ex' for exact].

Upon integration with boundary data $y(x = \pm D) = 0$ one obtains the famous cosh-form[2]

$$y(x) = h\,[\cosh(x/h) - \cosh(D/h)]. \qquad (B4)$$

For an engineer, say, it would be more appropriate to eliminate the non geometric parameter h in favour of the *length* of the chain, given by

$$L := \int_{-D}^{D} dx \sqrt{1 + [y']^2} = 2h \sinh(D/h), \qquad (B5)$$

or its sag

$$d := -y(x = 0) = -h[1 - \cosh(D/h)] = 2h \sinh^2(D/2h), \qquad (B6)$$

i.e., to solve (B5) for $h(L, D)$ or (B6) for $h(d, D)$ respectively and insert this into (B4). But, unfortunately, this cannot be done in terms of elementary functions. Hence (B4) (B5) or (B4) (B6) should be thought of as *implicit* representation of the hanging chain as function of the parameters L, D or d, D respectively.

Finally, the total tension $F(x)$ of the chain is easily computed:

$$F(x) = \sqrt{F_h^2 + F_v^2(x)} = F_h \sqrt{1 + [y'(x)]^2} = \mu g h \cosh(x/h), \qquad (B7)$$

which, using (B4), can also be read as saying that F grows linearly in y.

The Approximating Theory $\mathbf{T}_{ap}$

Galileo's approximative modelling of the hanging chain by a parabola can be understood within the larger context of an *approximation of theories*. It is

---

[1] The following formula (B3) remains valid for variable μ. It then implies that the hanging chain can be made to assume *any* convex shape by letting $\mu > 0$ vary appropriately along the chain.
[2] An equivalent form, obtained by applying the addition laws for cosh-functions, is

$$y(x) = 2h \sinh((x + D)/2h) \sinh((x - D)/2h). \qquad (B4')$$

obtained as *first* approximation of the fundamental equation (*B3*) for small slopes y'. Such approximations clearly make sense only for $y' < 1$, which is just the regime for which the parabolic approximation of the hanging chain is claimed in the relevant part of the *Discorsi* (Galilei 1974, 256-257; see The Neglected Issue: Trajectory and Hanging Chain). Hence we expand the square-root in (*B3*) in terms of powers of y' and truncate the second and all higher powers. But since y' appears already in squared form under the square-root, this amounts to simply replacing this square-root by 1. *From the derivation of (B3) it is clear that this is equivalent to taking the mass-distribution as homogeneous along the horizontal projection (x-axis) rather than being homogeneous along the proper length. This, in turn, is precisely the [implicit] assumption that underlies the application of Galileo's results on the distribution of moments along a solid and homogeneous cylindrical beam which rests horizontally supported at both ends; see the main text.* In first approximation one simply expands (*B3*) in terms of powers of y' discarding the second and all higher powers.

The fundamental equation that defines the approximating theory $\mathbf{T}_{ap}$ is now simply given by:

$$y'' = \frac{1}{h} \tag{B3'}$$

and for the same boundary data as above one obtains

$$y(x) = \frac{1}{2h}(x^2 - D^2). \tag{B4'}$$

Formally this corresponds to a quadratic expansion of the cosh-function in (*B4*) in terms of the dimensionful parameter $1/h$, which should be understood as expansion in terms of a dimensionless parameter $(\frac{1}{h}) \times$ (intrinsic length) $\cong D/h$. The sag, d, is now given by the simple formula

$$d = \frac{D^2}{2h}, \tag{B6'}$$

which, in contrast to the exact theory, can now be easily solved for h. This allows us to explicitly parameterise the curve by the geometric quantities d and D. An expansion in terms of D/h is hence equivalent to an expansion in terms of d/D.

Note that in general it will not be the case that the exact expressions of an approximating theory correspond to certain approximations of the exact theory, but only that simultaneous expansions in both theories coincide up to some order. For example, the expression for the length L in $\mathbf{T}_{ap}$ has the complicated structure

$$L = \int_{-D}^{D} dx \sqrt{1 + (x/h)^2} = D[\sqrt{1 + (D/h)^2} + (h/D)\operatorname{asinh}(D/h)] \quad (B5')$$

but the quadratic expansions of (B5) and (B5') in terms of d/D (i.e. in terms of D/h and then h eliminated using (B6') both lead to

$$L = 2D\left[1 + \frac{2}{3}\left(\frac{d}{D}\right)^2\right]. \quad (B5'')$$

The same holds for the total tension, which in $\mathbf{T}_{ap}$ takes the form

$$F(x) = \mu g h \sqrt{1 + (x/h)^2}, \quad (B7')$$

whereas the quadratic expansions of (B7) and (B7') in terms of d/D coincide in the following "engineer-formula"

$$F(x) = \mu g \frac{D^2}{2d}\left[1 + 2\left(\frac{x}{D}\right)^2\left(\frac{d}{D}\right)^2\right]. \quad (B7'')$$

Finally we can raise the question of how to grade the quality of the approximation of $\mathbf{T}_{ex}$ by $\mathbf{T}_{ap}$. This can be done for each observable (here observables are e.g. $y(x)$, d, $F(x)$ and L) by looking at the orders of the first non-vanishing correction to $\mathbf{T}_{ap}$ by $\mathbf{T}_{ex}$. Let O_{ex} and O_{ap} be the values of an observable on "corresponding" [see below] solutions of the fundamental equation of $\mathbf{T}_{ex}$ and $\mathbf{T}_{ap}$ respectively. Then one considers

$$\Delta(O) := \frac{O_{ex} - O_{ap}}{O_{ex}} \quad (B8)$$

and defines as usual $o(\Delta(O))$ to be that integer which characterises the leading order in the expansion of $\Delta(O)$ with respect to the expansion parameter (here d/D). The grade $g(O)$ of the expansion can then be defined as

$$g(O) := o(\Delta(O)) - 1. \quad (B9)$$

In our case we obtain

$$g(y(x)) = g(d) = 1, \quad g(F(x)) = g(L) = 3. \quad (B10)$$

From the definition of $\mathbf{T}_{ap}$ together with $y' = \sinh(x/h) = x/h + \dots$ one could not have expected a grade of approximation better than 1 [linear approximation]. But, as we just saw, the approximation might come out to be much better. This mirrors a well known phenomena in physics: that some formulae "are better than their derivation". In our case this is for example true for the tension (formulae ($B7'$) ($B7''$)), which deviates from the exact expression only in *fourth order* in d/D, thereby slightly *underestimating* the real tension.

Finally we wish to comment on the notion of *corresponding solutions*. In order to define a correspondence one has to make a choice of preferred observables whose values uniquely fix solutions of the fundamental equations ($B3$) and ($B3'$). Solutions with coinciding values on these observables are then defined to correspond to each other. Such a definition should therefore always be thought of as *relative* to the choice of preferred observables. So, for example, for given horizontal distance $2D$ of the suspension points one may either take the horizontal tension F_h (as we did) or the length L or the sag d to fix the solution. A non-trivial consequence of this general observation is that the grade of an approximation of some observable will in general *depend* on the choice of preferred observables which are used to fix the solutions.

Acknowledgment

This paper makes use of the work of research projects of the Max Planck Institute for the History of Science (MPIWG) in Berlin, some pursued jointly with the Biblioteca Nazionale Centrale in Florence, the Istituto e Museo di Storia della Scienza (IMSS), and the Istituto Nazionale die Fisica Nucleare in Florence. In particular, we have made use of results achieved in the context of a project dedicated to the development of an electronic representation of Galileo's notes on motion (together with the Biblioteca Nazionale Centrale and the Istituto e Museo di Storia della Scienza, both in Florence; this electronic representation is freely accessible from the websites of the IMSS, www.imss.fi.it and the MPIWG, www.mpiwg-berlin.mpg.de, see also Damerow and Renn 1998), of results achieved in a study of the time-sequence of entries in Galileo's manuscripts by means of an analysis of differences in the composition of the ink (together with the Biblioteca Nazionale Centrale, the Istituto e Museo di Storia della Scienza, and the Istituto Nazionale die Fisica Nucleare, all in Florence), and finally of results achieved in the context of a central research project of the Max Planck Institute for the History of Science, dedicated to the study of the relation of practical experience and conceptual structures in the emergence of science. We would especially like to acknowledge the generous support of several individuals involved in these projects: Jochen Büttner, Paolo Galluzzi, Wallace Hooper, Franco Lucarelli, Pier Andrea Mandó, Fiorenza Zanoni-Renn, Urs Schoepflin, Isabella Truci, and Bernd Wischnewski. From several other individuals we received helpful suggestions acknowledged at appropriate places throughout the paper.

References

Abattouy, M. 1996. *Galileo's Manuscript 72: Genesis of the New Science of Motion (Padua ca. 1600–1609).* Preprint 48. Berlin: Max Planck Institute for the History of Science.

Allegretti, G. 1992. *Monte Baroccio: 1513-1799.* Mombaroccio: Commune di Mombaroccio.

Arend, G. 1998. *Die Mechanik des Niccolò Tartaglia im Kontext der zeitgenössischen Erkenntnis- und Wissenschaftstheorie.* München: Institut für Geschichte der Naturwissenschaften.

Bertoloni Meli, D. 1992. "Guidobaldo dal Monte and the Archimedean Revival". *Nuncius 7*:3-34.

Biagioli, M. 1989. "The Social Status of Italian Mathematicians 1450-1600". *History of Science 27*:41-95.

Camerota, M. 1992. *Gli Scritti De Motu Antiquiora di Galileo Galilei: Il Ms. Gal. 71. Un'analisi storico-critica.* Cagliari: Cooperativa Universitaria.

Caverni, R. 1972. *Storia del metodo sperimentale in Italia.* New York: Johnson Reprint.

Damerow, P., G. Freudenthal, P. McLaughlin, and J. Renn. 1992. *Exploring the Limits of Preclassical Mechanics.* New York: Springer.

Damerow, P., and J. Renn. 1998. "Galileo at Work: His Complete Notes on Motion in an Electronic Representation". *Nuncius 13*:781-789.

del Monte, G. ca. 1587-1592. *Meditantiunculae Guidi Ubaldi e marchionibus Montis Santae Mariae de rebus mathematicis.* Paris: Bibliothèque Nationale de Paris, Manuscript, Catalogue No., Lat. 10246.

di Pasquale, S. 1996. *L'arte del costruire.* Venice: Marsilio.

Drake, S. 1973. "Galileo's Discovery of the Law of Free Fall". *Scientific American 228*:84-92.

---. 1979. "Galileo's Notes on Motion Arranged in Probable Order of Composition and Presented in Reduced Facsimile". *Annali dell'Istituto e Museo di Storia della Scienza.*

---. 1987. *Galileo at Work: His Scientific Biography.* Chicago: University of Chicago Press.

---. 1990. *Galileo: Pioneer Scientist.* Toronto: University of Toronto Press.

Drake, S., and I. E. Drabkin. 1969. *Mechanics in Sixteenth-Century Italy.* Madison: University of Wisconsin Press.

Favaro, A. 1919-1920. "Scritture Galileiane apocrife". *Atti e memorie della R. Accademia di scienze lettere ed arti in Padova 36*:17-29.

---. 1966. *Galileo Galilei e lo studio di Padova.* Padova: Antenore.

Fredette, R. 1969. *Les De Motu "plus anciens" de Galileo Galilei: prolégomènes.* Ph.D.diss. Montreal: University of Montreal.

Galilei, G. 1890-1909. *Le opere di Galileo Galilei.* Edizione Nazionale, ed. A. Favaro. Florence.

---. 1958. *Discorsi e dimonstrazioni matematiche intorno a due nuove scienze attinenti alla mecanica ed i movimenti locali*, ed. A. Carugo and L. Geymonat. Torino: Boringhieri.

---. 1960. *On Motion and On Mechanics: Comprising De Motu (ca. 1590)*. Madison: University of Madison.

---. 1967. *Dialogue Concerning the Two Chief World Systems*, ed. S. Drake. Berkeley: University of California Press.

---. 1974. *Two New Sciences*, ed. S. Drake. Madison: University of Wisconsin Press.

---. 1989. *Two New Sciences* (2 ed.), ed. S. Drake. Toronto: Toronto Press.

---. after 1638. *Discorsi (annotated copy of Galileo)*. Florence: Biblioteca Nazionale Centrale, Florence, Manuscript, Ms. Gal. 79.

---. ca. 1602-1637. *Notes on Motion*. Florence: Biblioteca Nazionale Centrale, Florence, Manuscript, Ms. Gal. 72.

Galluzzi, P. 1979. *Momento*. Rome: Ateneo e Bizzarri.

Gamba, E. 1995. "Guidobaldo dal Monte tecnologo". *Pesaro città e contà: rivista della società pesarese di studi storici* 5:99-106.

Gamba, E., and V. Montebelli. 1988. *Le scienze a Urbino nel tardo rinascimento*. Urbino: QuattroVenti.

---. 1989. *Galileo Galilei e gli scienziati del ducato di Urbino*. Urbino: Quattroventi.

Gillispie, C. C. (ed.). 1981. *Dictionary of Scientific Biography*. New York: Charles Scribner's Sons.

Giuntini, L., F. Lucarelli, P. A. Mandò, W. Hooper, and P. H. Barker. 1995. "Galileo's writings: chronology by PIXE". *Nuclear Instruments and Methods in Physics Research B* 95:389-392.

Giusti, E. 1993. *Euclides reformatus: la teoria delle proporzioni nella scuola galileiana*. Torino: Bollati Boringhieri.

Hooper, W. E. 1992. *Galileo and the Problems of Motion*. Ph.D.diss. indiana: Indiana University.

Klemm, F. 1964. Der junge Galilei und seine Schriften "De motu" und "Le mecaniche". In E. Brüche (ed.), *Sonne steh still: 400 Jahre Galileo Galilei*, 68-81Mosbach: Physik Verlag.

Koyré, A. 1966. *Etudes Galiléennes*. Paris: Hermann.

Lefèvre, W. 1978. *Naturtheorie und Produktionsweise*. Darmstadt: Luchterhand.

Libri, G. 1838. *Histoire des Sciences Mathématiques en Italie*, vol. 4. Paris: Renouardi.

Micheli, G. 1992. Guidobaldo del Monte e la meccanica. In L. Conti (ed.), *La matematizzazione dell'universo*, 87-104S. Maria degli Angeli - Assisi: Porziuncola.

Miniati, M., V. Greco, G. Molesini, and F. Quercioli. 1994. "Examination of an Antique Telescope". *Nuncius* 9:677-682.

Naylor, R. H. 1974. "The Evolution of an Experiment: Guidobaldo Del Monte and Galileo's "Discorsi" Demonstration of the Parabolic Trajectory". *Physis 16*:323-346.

---. 1975. "An Aspect of Galileo's Study of the Parabolic Trajectory". *ISIS 66*:394-396.

---. 1976a. "Galileo: the Search for the Parabolic Trajectory". *Annales of Science 33*:153-172.

---. 1976b. "Galileo: Real Experiment and Didactic Demonstration". *ISIS 67*:398-419.

---. 1977. "Galileo's Theory of Motion: Processes of Conceptual Change in the Period 1604-1610". *Annales of Science 34*:365-392.

---. 1980a. "Galileo's Theory of Projectile Motion". *ISIS 71*:550-570.

---. 1980b. "The Role of Experiment in Galileo's Early Work on the Law of Fall". *Annales of Science 37*:363-378.

Naylor, R. H., and S. Drake. 1983. "Discussion on Galileo's Early Experiments on Projectile Trajectories". *Annales of Science 40*:391-396.

Porz, H. 1994. *Galilei und der heutige Mathematikunterricht. Ursprüngliche Festigkeitslehre und Ähnlichkeitsmechanik und ihre Bedeutung für die mathematische Bildung.* Mannheim: B.I. Wissenschaftsverlag.

Remmert, V. R. 1998. *Ariadnefäden im Wissenschaftslabyrinth: Studien zu Galilei: Historiographie - Mathematik - Wirkung.* Bern: Lang.

Sarpi, P. 1996. *Pensieri naturali, metafisici e matematici,* ed. L. Cozzi and L. Sosio. Milano: Ricciardi.

Schneider, I. 1970. *Der Proportionalzirkel, ein universelles Analogrecheninstrument der Vergangenheit.* München: Deutsches Museum.

Settle, T. B. 1961. "An Experiment in the History of Science". *Science 133*:19-23.

---. 1971. Ostilio Ricci, a Bridge between Alberti and Galileo. In *XII Congrès International d'Histoire des Sciences, Actes.,*vol. IIIB. Paris: Blanchard.

---. 1996. *Galileo's Experimental Research.* Preprint 54. Berlin: Max Planck Institute for the History of Science.

Sommerfeld, A. 1994 Reprint. *Mechanik* (8 ed.). Vorlesungen über theoretische Physik, ed. E. Fues, vol. 1. Frankfurt am Main: Deutsch.

Takahashi, K. i. 1993a. "Galileo's Labyrinth: His Struggle for Finding a Way out of His Erroneous Law of Natural Fall. Part 1". *Historia Scientarum 2-3*:169-202.

---. 1993b. "Galileo's Labyrinth: His Struggle for Finding a Way out of His Erroneous Law of Natural Fall. Part 2". *Historia Scientarum 3-1*:1-34.

Tampone, G. 1996. *Il restauro delle strutture di legno.* Milano: Ulrico Hoepli.

Tartaglia, N. 1984. *La Nova Scientia.* Bologna: Forni.

Torricelli, E. 1919. De motu gravium naturaliter descendentium, et projectorum. In G. Loria and G. Vassura (ed.), *Opera di Evangelista Torricelli.,*vol. 2. Faenza: Montanari.

Viviani, V. 1674. *Quinto libro degli Elementi di Euclide, ovvero scienza universale delle proporzioni, spiegata colla dottrina del Galileo*. Florence: Condotta.

---. after 1638. *Notes on Mechanical Problems*. Florence: Biblioteca Nazionale Centrale, Florence, Manuscript, Ms. Gal. 227.

Wohlwill, E. 1899. "Die Entdeckung der Parabelform der Wurflinie". *Abhandlungen zur Geschichte der Mathematik* 9:577-624.

---. 1993. *Galilei und sein Kampf für die Copernicanische Lehre*. Vaduz: Sändig.

Working-Group. 1996. *Pilot Study for a Systematic PIXE Analysis of the Ink Types in Galileo's Ms. 72: Project Report No. 1*. Preprint 54. Berlin: Max Planck Institute for the History of Science.

Zupko, R. E. 1981. *Italian Weights and Measures from the Middle Ages to the Nineteeth Century*. Philadelphia: American Philosophical Society.

2. The Context of the Artists: Astronomy and its New Representations

HORST BREDEKAMP

Gazing Hands and Blind Spots: Galileo as Draftsman*

The Argument

The article deals with the interrelation between Galileo and the visual arts. It presents a couple of drawings from the hand of Galileo and confronts them with Viviani's report that Galileo had not only wanted to become an artist in his youth but stayed close to the field of visual arts throughout his lifetime. In the ambiance of these drawings the famous moon watercolors are not in the dark. They represent a very acute and reasonable tool to convince the people who trusted images more than words. The article ends with Panofsky's argument that it was Galileo's anti-Mannerist notion of art that evoked a repulsion of Kepler's ellipses. It tries to show that it was again an aesthetical prejudice that hindered Einstein from accepting Panofsky's theory.

I. Cigoli, Galileo, Michelangelo

The interrelation between Galileo Galilei and the artists of his time is still a major problem to be solved.[1] The most spectacular event happened in Rome. In the year 1610, the Roman painter Ludovico Cigoli was entrusted with the task of painting the chapel of Pope Paul V in Santa Maria Maggiore. The complicated project included a depiction of the Queen of Heaven standing on the moon in the central axis of the cupola. Surprisingly, in 1612 the painter frescoed the moon under the feet of the Mary with an uneven surface (fig. 1). The lunar sphere, only half lit, shows shadows and streaks of light in its illuminated part, suggesting heights and

* This paper was presented in different forms at the Hamburger Kunstakademie in June 1993, as an inaugural lecture at the Humboldt University in December 1994, and at the Zürich Galilei-Symposium from January 1996. Parts of it were published under the title, "Galileo Galilei als Künstler" (Bredekamp 1996a); parts were published under the title, "Zwei frühe Skizzenblätter Galileo Galileis" (Bredekamp 1996b) and parts appeared as Bredekamp 1995, 366ss. I am grateful to Jürgen Renn for his critique and help.
[1] Recently, attempts have been made to parallel Adam Elsheimer's moon on his copper "Flight to Egypt" which he worked on 1609 in Rome, with Galileo's results of his observations of the moon from the same year (Cavina 1976), or to find a reflection of the surface of the moon which Galileo had detected in the work of Artemisia Gentileschi, a close friend of Galileo (Garrard 1989, 334s). A quite recent study tries to illuminate the whole horizon of interrelation between artists and scientists of the early Seicento, mainly Galileo (Reeves 1997).

Figure 1. Lodovico Cigoli, Moon Madonna, cupola fresco in the Capella Paolina of Santa Maria Maggiore, Rome.

depths, as if the flawless moon had suddenly contracted a case of the measles or smallpox. Thus Cigoli confirmed the illustrations of the moon in Galileo's "Sidereus Nuncius" which had been published two years before.[2]

That Cigoli painted the irregular surface of the moon at a time when Galileo's lunar observations were still controversial is one of the Roman contradictions of the time. Members of the progressive wing of the Church which strove for its reconciliation with modern science were able to support Galileo's view of the moon all the more as Apocalypse 12 allowed to connect the spotted moon with evil and heresy over which the *Immacolata* celebrated her victory.[3]

The orthodox view, instead, resisted massively. Up to this point the moon, like all celestial bodies, had been thought to be a perfectly round, platonic form, which unlike the earth — "the filth and dregs of the universe" bore not the faintest irregularity.[4] In its smooth, fully illumined form it served as a symbol of the Mariological Church, still appearing as such, for example, in Bartolomé Esteban Murillo's *Immacolata* of ca. 1660.[5] In 1616, only four years after completion of Cigoli's fresco, a cross section of the chapel for Paolo de Angelis' book on Santa Maria Maggiore showed Mary standing on a smooth, unmarred sphere in accord with traditional iconography.[6] Not until the restoration work of 1931 was it discovered that the engraving had falsified the moon of the fresco by perfectly smoothing its surface.

In 1612 Cigoli honored Galileo a second time, and more successfully. He compared him to Michelangelo, the breaker of the Vitruvian rules and rebel against the order of tradition: "And I think that the same happens to you as to Michelangelo when he started to build outside the order of the others till to his times, where all of them unanimously, turning against him, said that Michelangelo had ruined the architecture with all of his misuse fare from Vitruv."[7]

Cigoli's bridging between Michelangelo and Galileo went beyond a metaphorical connection when Vincenzio Viviani, assistant and biographer of Galileo, dated Galileo's birthday in his biography one day after Michelangelo's death, February 19, 1564, and later, 1692, fixed it precisely on February 18.[8] Thus, in a remarkable

---

[2] Artists in Rome, and among them Cigoli, had used the telescope by themselves. Elsheimer might have used one of the earliest pieces in Rome already by 1610 (Cavina 1976, 142f.). The painter Domenico Passignano had made observations of his own before February, 1612 (Lodovico Cigoli to Galileo, February 3, 1612, in Galilei 1890–1909, 268). By March Cigoli was in possession of another telescope (Cigoli to Galileo, March 3, 1612, ibid., 287). On Cigoli, see Matteoli 1980; 1982; on Cigoli and Galileo: Faranda 1986, 95–96; Contini 1991, 110–112; Puppi 1992, 244ss.; Magani 1992, 145f.

[3] The events have been known since Erwin Panofsky's pioneering study of Galileo's relationship to art: (Panofsky 1954, 5; cf. more specific: Wolf 1991/92, 313; Ostrow 1996b, 233f.; Ostrow 1996a, 244ss.). The general context is given by Rivka Feldhay (Feldhay 1995).

[4] Blumenberg 1980, 23.

[5] Edgerton 1991, 231f; cf. Ostrow 1996b, 222–229; Reeves 1997, 144ss.

[6] de Angelis 1621, 194.

[7] "et mi credo avengha lo istesso come quando Micelagniolo cominciò a architetturare fuori del'ordine degli altri fino ai suoi tempi, dove tutti unitamente, facendo testa, dicevano che Micelagniolo avea rovinato la architettura con tante sue licenze fuori di Vitruvio" (Ludovico Cigoli to Galileo, July 7, 1612, in Galilei 1890–1909, XI:361).

[8] The components of the complicated story of the investigation and manipulation of Galileo's

effect of providence, the relay baton of the immortals seemed to have passed directly from Michelangelo to Galileo.

At first glance one could take the bridging between Galileo and Michelangelo as a merely rhetorical fantasy, given in order to connect Galileo with the artistic quality of the "uomo universale." But Viviani's manipulation was more than just an act of piety. Michelangelo's reincarnation in Galileo was meant to transfer the artist's fame as a supporter of the Counter-Reformation to the reputation of Galileo, whose condemnation in 1633 continued to be a dark spot of his life.[9] On the other side Galileo, who had been protected by the Medici, could transform Michelangelo, the defender of the Florentine republic against the imperial army that brought the Medici to power, a partisan of the ruling family. Thus both "profited" crosswise from each other in the name of Florence and the Medici.

The connection between Michaelangelo and Galileo became a *topos*. Even still Kant spoke of the "metempsychosis of three geniuses: Michaelangelo, Galilei, and Newton,"[10] and in 1793 the historian Giovanni Battista Clemente de' Nelli confirmed the overlapping of Michaelangelo's death and Galileo's birth.[11] Not until 1887 did the editor of the collected works of Galileo, Antonio Favaro, find out that this coincidence had occurred not through a portentous play of nature, but rather through careful manipulation.[12]

II. Artistic Tools

Viviani, using Giorgio Vasari's Lives of the Most Famous Painters, Sculptors and Architects as a model, had also spiced Galileo's biography with motifs from the youth of Giotto.[13] There were reasons to give Galileo an artist's aura as he was close to being an artist in social terms and in practice. From the beginning, Galileo turned the duties usually reserved for artists to his advantage in order to become an employee of the Medici court, for after centuries of tenacious social advancement, it was the court artist that embodied Galileo's dream of a fulfilled life of research: an independent position free of public teaching responsibilities and without specific work obligations.[14]

It is well known that, following the maxim of his friend and supporter Giovanni

---

birthday are reported by Michael Segre (Segre 1989, 221–225; 1991, 116–122). Most important was the marble inscription at the facade of Viviani's house in the Via dell'Amore (Via San Antonio), nr. 11, prepared in 1692, claiming the day of Michelangelo's death as the day of Galileo's birth (Favaro 1880, 41).

[9] Galluzzi 1994.

[10] "Metempsychosis dreyer genies: Michelangelo, Galilaei und Newton" (Kant 1923, 826; cf. Maio 1978, 3 , 11 n.2).

[11] de' Nelli 1793, 21–22.

[12] Favaro 1887, 703–711.

[13] Segre 1989, 219ss., 225f.; 1991, 112–122.

[14] Westfall 1984, 191. The rise of the court artist is studied by (Warnke 1985).

Ciampoli, "blessed is he who is able to strengthen his fortune through gifts,"[15] Galileo proffered his research results and inventions as presents to the powerful and highly-placed. Among them one finds his instruments such as the compass and the telescope as well as his books and letters and even his astronomical discoveries.[16] No less significant is the fact that Galileo followed one of the traditional preroga- tives of the court artist, designing emblems and coats of arms for the potentates, in which the glory of their rule could be manifested to the senses. Thus Jacques Callot, a court artist to the Medici in Florence at that time, sketched a series of heraldic designs.[17]

With the same intent and comparable iconography, Galileo designed an emblem for a Medici in the year 1608. First he purchased a round lodestone for the crown prince Cosimo de' Medici in order to win his favour — an investment in the future, as it were. The spherical form was, of course, a play on the *palle*, the six balls that adorned the Medici coat of arms. On the occasion of Cosimo's wedding, Galileo suggested as an impresa a round lodestone to which a series of small iron balls would adhere:

> That the magnetic sphere ... is admirably suited to the person of the most illustrious prince is apparent: first, because the balls are an ancient insignia of the Medici house, and furthermore, because it has been extensively written by the greatest philosopher and confirmed by clear demonstration that our earthly world in its primary and universal substance is nothing but a great magnetic sphere; and since the name Cosmo [cosmos] signifies the same thing as mondo [world], one could understand our great Cosimo under the most noble metaphor of the magnetic sphere.[18]

After the lodestone and its emblematic elaboration, Galileo handed over four additional spheres in the year 1610. With the help of his telescope, Galileo had discovered four planets in the field of Jupiter, which he bequeathed to the grand duke of Florence as a divine gift:

> The creator of the stars himself now seemed to direct me through clear signs to ascribe these new planets to the glorious name of your Highness before all others .... Because I have discovered these stars, unknown to all previous

---

[15] "Beato chi col donare può accellerare la sua fortuna!"(Ciampoli 1978, 232; cf. Biagioli 1993, 39).

[16] Biagioli 1993, 48ss.

[17] Callot 1971, 1432.

[18] "Che poi per la palla di calamita acconciamente si additi la persona del Ser.mo Principe, è manifesto: prima, per esser le palle antica insegna della Casa; in oltre, essendosi da grandissimo filosofo diffusamente scritto, et con evidenti dimostrazioni confermato, altro non essere questo nostro mondo inferiore, in sua primaria et universal sustanza, che un gran globo di calamita, et importando il nome Cosmo il medesimo che mondo, potrassi sotto la nobilissima metafora del globo di calamita intendere il nostro gran Cosimo." (Galileo Galilei to Christina of Lorraine, September 1608, in Galilei 1890–1909, 10:222). For more on the entire context cf. Biagioli 1993, 120ss.; Bredekamp 1993, 59ss.

astronomers, under your protection, most serene Cosimo, I am entirely
justified in my decision to call them by the august name of your family.[19]

Herewith the Medici were immortalized: released from the contingencies of earthly
mutability, their glory now hovered in the astral plane, inseparably bound to the
stars.

Of Galileo's iconographical and emblematical enterprises there is also visual
evidence. It must have been around 1610 that Galileo recorded two designs for
Medici coats of arms on the reverse of a page covered with numerous tables (fig.
2).[20] Galileo first jotted down a rough sketch of the cartouche of the coat of arms in
which the crown in the middle shows a fleur-de-lis; the uppermost of the six
individual spheres arranged below likewise shows the petals of the lily. The lower
design confirms this arrangement, but chooses a more oblique perspective to
better emphasize the depth of the cartouche and more concretely suggest the
embedding of the spheres.

Designs of this sort for emblems and coats of arms also served as a kind of
calling card for the artist himself. When Leonardo da Vinci was in the service of the
Milanese court, he designed an emblem with a compass whose needle pointed to a
sun with the three lilies of the French king. It was to receive the epigraph,
"Whoever is fixed on a star does not waver." Presumably it was meant not only as a
declaration of loyalty to the Milanese duke, but also as a hint by Leonardo to be
called by the French king.[21] That Galileo considered the possibility of bringing the
moon, as well, into association with Cosimo ("cosmos") Medici is suggested by the
line at the left edge of the page bearing the coat of arms, at the lower end of which
hangs the crescent of a half moon. Galileo may also have intended to include the
sphere of the moon among the six Medici balls.

Galileo attained the fulfillment of all his wishes in 1610 when he was given the
post of court philosopher with a salary among the ten highest in Tuscany. It was
more than half again as great as that of the best-equipped court sculptor, Giovanni
da Bologna.[22] At once, he was freed from teaching. And so he assumed that
privileged status which Michelangelo introduced into the history of labor law after
he took over absolute control of the construction of St. Peter under no obligation,
responsible to God alone.[23] With this status, as has been argued, Galileo had
reached his goal of becoming the "Michelangelo of mathematicians."[24]

---

[19] "Ut autem inclito Celsitudinis tuae nomini prae ceteris novos hosce Planetas destinarem,
ipsemet Siderum Opifex perspicuis argumentis me admonere visus est...Quae cum ita sint, cum, te
Auspice, COSME Serenissime, has Stellas superioribus Astronomis omnibus incognitas exploraverim,
optimo iure eas Augustissimo Prosapiae tuae nomine insignire decrevi." (Galilei 1610, 56f.). On this
act of bequeathal, cf. (Biagioli 1993, 52f.).

[20] In Mss. Galileiani, 50:32ʳ.

[21] (Warnke 1985, 75; cf. Reti 1959, 40), who alone surmises a political significance.

[22] Biagioli 1993, 104.

[23] Vasari 1962, 84.

[24] Biagioli 1993, 86f.

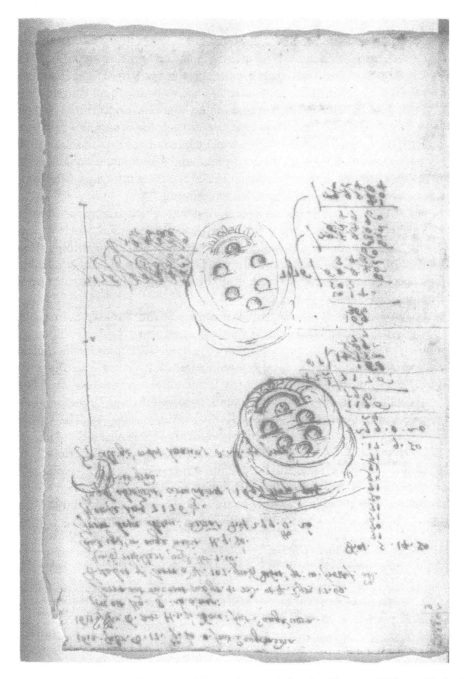

Figure 2. Galileo Galilei, design for Medici coat of arms. Ink drawing, Florence, Biblioteca Nazionale Centrale, MS. Gal. 50, fol. 32r.

III. Pen Drawings

The affinity should not be carried too far, but there do indeed exist intimations of a connection beyond mere similarity of artistic tools and working conditions. The painter Cigoli and Galileo had been friends since around 1585, when they had studied with Ostilio Ricci, the court mathematician, who later taught mathematics at the "Academia del disegno" in Florence, in the house of the artist and engineer Bernardo Buontalenti.[25] Viviani was surely reflecting this situation when he reported that Galileo busied himself "with great delight and marvelous success in the art of drawing, in which he had such great genius and talent that he would later tell his friends that if he had possessed the power of choosing his own profession at that age, ... he would absolutely have chosen painting."[26]

Viviani's remark is confirmed by several drawings which are preserved in the manuscripts of the Biblioteca Nazionale Centrale of Florence, but which were not included in the national edition, since they seemed to contribute nothing to the image of Galileo. Until the scanned version of all the manuscripts is made available we may expect other surprises as well, upon inspection of the precious original manuscripts.[27]

Among the earliest evidence are two of three pages from Galileo's examination essay of 1584 on Aristotle's *De coelo*. These two pages form a protective binding at the front and back of the second part of the treatise.[28] It is conceivable that the pages served as a covering from the beginning and that Galileo used them as a surface for drawing and writing when the treatise was returned to him. This, however, is unlikely in view of the fact that older scribblings are found on one of the two pages. It is more probable that Galileo, having received back his manuscript, found a sheet of paper lying in his room that was suitable in size and thickness to serve as a protective covering for each of the two parts. At first glance the appearance of these pages might give the impression of a non-explainable complexity, but it can be, at least hypothetically, put into a differentiated order.

The obverse, which presents a rather confused impression, shows numbers and figures written in pen (fig. 3), stemming without a doubt from Galileo's hand.[29] In

---

[25] (Olschki 1965; Settle 1971; Masotti 1970–1980, XI:405f.; Wazbinski 1987, I:283; Reeves 1997, 6). On the Academia, see also (Reynolds 1974).

[26] "Trattenevasi ancora con gran diletto e con mirabil profitto nel disegnare; in che ebbe così gran genio e talento, ch'egli medesimo poi dir soleva agl'amici, che se in quell'età fosse stato in poter suo l'eleggersi professione, averebbe assolutamente fatto elezione della pittura" (Viviani 1890, 602).

[27] The project is presently undertaken by the Biblioteca Nazionale Centrale and the Istituto e Museo della Scienza in Florence. Galileo's manuscripts on mechanics have been made available on the Internet as a result of a joint project of the Max Planck Institute for the History of Science, the Biblioteca Nazionale Centrale and the Istituto e Museo della Scienza in Florence under: http://galileo.imss.firenze.it/ms72/index.html
http://www.mpiwg-berlin.mpg.de/Galileo—Prototype/MAIN.HTM. On first scale the manuscripts on fortifications in the Ambrosiana, Milan, could evoke new insights: (Tabarroni 1984).

[28] (Mss. Galileiani, 46:56–102). The first quire extends from fol. 1–55. I am grateful to Michele Camerota for calling my attention to these pages.

[29] Mss. Galileiani, 46:56ʳ.

Figure 3. Galileo Galilei, sketch, ink on paper, ca. 1584 and earlier. Florence, Biblioteca Nazionale Centrale, MS Gal. 46, fol. 56r.

the upper half of the page a woman is shown from behind, her curls partly pinned up and partly falling onto her shoulders. Her left arm is raised, the hand swelling as if, perspectively enlarged, it were swinging back toward the viewer. This extreme and obviously unsuccessful experiment with the possibilities of perspectival fore-shortening is marked through with every sign of displeasure. Draped over the woman's left shoulder is a robe that evidently covers her thighs as well. Her legs are turned to the right and bent so that the more vigorously rendered right arm can reach back toward the feet.

No direct model is discernible, but possible comparisons can be made with the female figures of the Neptune fountain of the Piazza della Signoria in Florence, created before 1575 in the workshop of Bartolomeo Ammannati. Even closer are individual figures from Ammannati's Fountain of the Elements for the Palazzo Vecchio, which in the 1580s was located in the garden of the Villa Pratolino. Giovanni Guerra's drawing of the figures after their removal to the grounds of the Palazzo Pitti,[30] presumably drawn from memory around 1598 and arranged into an ensemble, enables a comparison in the same graphic medium (fig. 4). The "Fiorenza" bending back on the right in the middle of the arch comes relatively close to the head and shoulder of Galileo's female figure, while the Hippocrene shows a similar, if more relaxed, seated posture. These somewhat vague references are not intended to suggest Galileo's dependence on Ammannati, but rather serve to elucidate the formal climate in which Galileo sought to orient himself.

The other motifs appear too unspecific to necessitate a search for models. Below and to the left of the female figure, a horse's head is sketched, flanked by the head of a bearded man and writing exercises for the letter "g." The lower area contains red chalk drawings that appear to stem from the hand of a child: columns of squares and geometric building blocks, like those used in the game of hopscotch.

In its subdivision into three areas, this page is no anomaly. Paper was available in any form,[31] but it was not cheap, and as a rule was reused further times. As mentioned before, the relation between Galileo and Michelangelo suggested by Viviani should not be exaggerated, but chance has it that an early example from the hand of Michelangelo bears witness to a similar use of the page (fig. 5).[32]

While the obverse bears an entry from the year 1501, the reverse is covered with written characters and three sketches: a right hand, a left leg seen from the front, and a nude seen obliquely from behind. Nevertheless, both sides are executed in the same ink and with the same pen, so we may assume that the approximately 12-year-old Michelangelo undertook writing and poetry exercises in different styles of handwriting, as may have been prescribed by his grammar teacher: at the top, the script of official business, and below, that of personal letters and notes.

---

[30] Heikamp 1978, 146.

[31] (Kemp 1979, 59). On the significance of drawing, see (ibid., 57ss.; cf. also Westfehling 1993, 98ss.).

[32] (Florence, Archivio Buonarroti, II/III:fol. 3ᵛ), in (Dussler 1959, 56, No. 27, Fig. 35); the interpretation followed here stems from (Perrig 1991, 68ss.).

Figure 4. Giovanni Guerra, the Ammannati fountain, drawing, ca. 1598. Vienna, Graphische Sammlung Albertina.

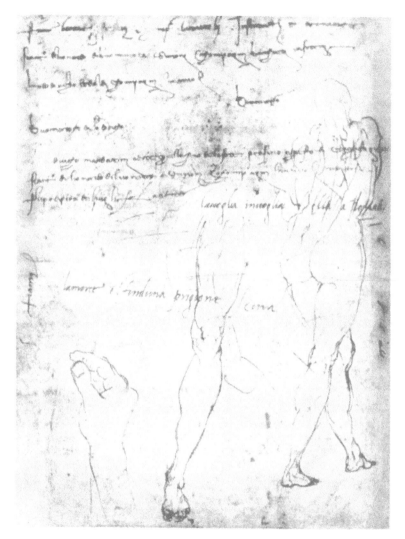

Figure 5. Michelangelo, Sketches, Archivio Buonarr, Florence, II/III fol. 3v.

The lines in the older style of script are nonsensical and can be explained as mere writing practice. In addition to the names of his uncle and brother, Michelangelo here wrote the designations for different vocations and his surname in several variations: "Buonarotto" in the fourth line, but "Buonnarootto" in the fifth.

Along with the evident deficiencies of a beginner, the drawings immediately manifest the self-willed development of the autodidact as well. In the sketched hand, the enlarged knuckle areas and the swollen ball of the thumb, as well as the alteration in the two first joints of the index finger, bear witness to the groping character of the attempt, while the perspective foreshortening already bespeaks a willingness to tackle problems characteristic of both Michelangelo here and Galileo later.

In its mixture of writing and drawing exercises, the page from Michelangelo evinces a structure similar to that of Galileo's sheet from the Aristotle treatise. The latter was presumably filled in three stages: first, the red chalk drawings of the perhaps 5 to 10-year-old, then the writing exercises of the 12 to 15-year-old, and finally the calculations and drawings of the student, presumably in his early twenties.

The reverse side of the second quire of Galileo's examination treatise, as well, shows a similar pattern of use (fig. 6).[33] Here, however, figural representations

Figure 6. Galileo Galilei, sketch, ink on paper, ca. 1584. Florence, Biblioteca Nazionale Centrale, MS. Gal. 46, fol. 101v.

---

[33] Mss. Galileiani, 46:fol. 101ᵛ.

predominate. In the upper half of the page, a man pulls a figure upward toward himself; the two rounded breasts presumably indicate the figure is a woman. In a second layer, turned 90 degrees, the outline of a horse's head appears amid writing exercises, while an awkward hand with widely spread thumb and index finger opens up toward the right, a subject Michelangelo had also sketched; clearly it belonged to the usual material of the autodidact.

Below and to the left are the legs of a figure clad in tight-fitting breeches, while at the bottom of the page there appears a nymph, whose outline Galileo attempted to correct with numerous strokes of the pen. With her right hand covering the pubic area and her left arm slightly bent, the nymph resembles the antique Venus Felix of the Belvedere court in Rome (fig. 7); with her nude body bent slightly forward,

Figure 7. Venus Felix, Roman statue. Second century A.D., Rome, Belvedere.

however, she is also reminiscent of the Medici Venus of the Uffizi Tribuna (fig. 8).[34]

In their mixture of amateurish incompleteness and ambitious effort, the pages bear witness to a conflict. The head of the male figure is not unskillful in its effect, and the face of the nymph reveals practice. At the least, the sketches bespeak a certain familiarity with the qualitative level of sixteenth-century drawing. But to be sure, the disegni of the young Galileo are by no means masterpieces, and one could argue that it was best for Galileo that he did not follow a career as an artist.

Figure 8. Medici Venus, Greek copy of the bronze Venus after Praxiteles. Florence, Uffizi (Tribuna).

---

[34] Haskell 1981, 323f., 325ss.

On the other side, even following this conclusion, two problems remain. The one lies in the question up to what degree Galileo saved his artistic ambitions and his figural thinking to his mathematical sketches. This is a subject in itself, which up to this point has not even been addressed, let alone explained. In one of the stereometric drawings,[35] for example (fig. 9), the freely drawn stroke, correcting itself on the curves, accords well with the repeatedly redrawn outline of the upper body of the nymph (fig. 6).

The second problem lies in the fact that Galileo continued to draw, and that some later pieces which have survived cannot be judged but masterly. Viviani's *vita* notes that even after his period of education, Galileo maintained "such a natural and proper inclination to the art of drawing, and in time acquired such exquisite taste, that his opinion on paintings and drawings was preferred to that of the

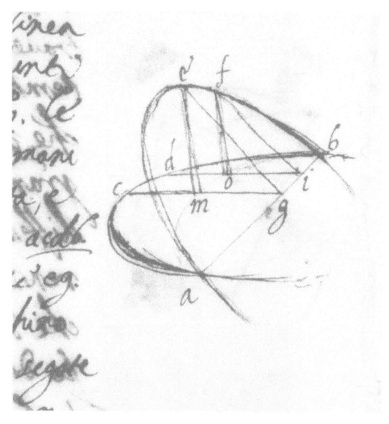

Figure 9. Galileo Galilei, stereometric drawing. Florence, Biblioteca Nazionale Centrale, MS. Gal 57, fol. 35r.

---

[35] Mss. Galileiani, 57:fol. 35.

foremost professors — even by the latter themselves."[36] These words find their confirmation in the original manuscript of the Sidereus Nuncius of 1609. Hastily sketched between calculation tables of the Medici moons are two landscapes dating from ca. 1610 (fig. 10),[37] executed in a stroke that is remarkably free and almost modern in its spontaneous assurance. In the upper stripe, a cupola projects upward to the left while buildings and trees form a staggered boundary to the right. Sails indicate that this complex is located on the water. Below, the second scene shows a sharply accentuated river course. On the opposite bank of the river lies a castle or fortified village, while in front, four sails are once again sketched in with a singularly free hand.

Figure 10. Galileo Galilei, two rivers, ink drawing. Florence, Biblioteca Nazionale Centrale, MS. Gal. 48, fol. 54v.

Whether these sketches were drawn from life or represent the free product of Galileo's imagination — a digression during his calculations for the moons of Jupiter — is difficult to say; they may be scenes from the banks of the lagoon in Venice. In any case, in their simple, confident form the sketches are convincing.

The impression of the almost transhistorical modernity that imbues the medium of drawing is likewise awakened by this view of a townscape extending across two plateaus connected by a wall running diagonally upward (fig. 11).[38] The light coming from the upper left allows the extremes of light and shadow to appear in contrasting strips.

---

[36] "Ed in vero fu di poi in lui cosi naturale e propria l'inclinazione al disegno, et acquistovvi col tempo tale esquisitezza del gusto, che 'l guidizio ch'ei dava delle pitture e disegni veniva preferito a quello de' primi professori da' professori medesmi" (Viviani 1890–1909, 602).

[37] Mss. Galileiana, 50:fol. 54ᵛ.

[38] Mss. Galileiana, 50:fol. 61ᵛ.

Figure 11. Galileo Galilei, view of a townscape extending across two plateaus connected by a wall running diagonally upward (Mss. Galileiana, 50:fol. 61ᵛ).

IV. Moon Watercolors

The drawing of a townscape explores a problem of vital importance for Galileo's study of the surface of the moon and for the wash drawings included with the Sidereus Nuncius manuscript; six of them on the recto of fol. 28 (fig. 12)[39] and a seventh one on the verso. The juxtaposition of the six drawings on a single sheet suggests that in all probability, Galileo transferred them as a group from ad hoc sketches that have now been lost.[40]

Clearly a reference to the Medici was intended, for on the verso the additional moon is included along with the horoscope for May 2, 1590, the birth date of Cosimo II de' Medici.[41] However, if Galileo sought to establish a connection here, he later abandoned the attempt, perhaps because, with the dedication of the moons of Jupiter to the Medici, he had already attained his dream of a one-man institute for advanced study, created just for him, by September 1610.

The moon images correspond to precise dates and times of day. Four drawings

---

[39] (Mss. Galileiana, 48:fol. 28ʳ). For the most recent discussion, with earlier literature, see (Whitaker 1989, 122ss.; Frieβ 1993, 121ss.).

[40] Gingerich 1975, 87f.

[41] Mss. Galileiani, 48:fol. 28ᵛ.

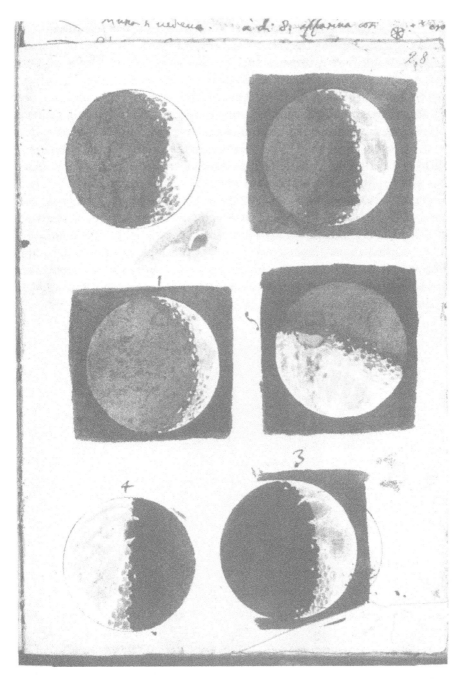

Figure 12. Galileo Galilei, six phases of the moon, ink drawing. Florence, Biblioteca Nazionale Centrale, MS. Gal. 48, fol. 28r.

show the phases of the moon between November 30 and December 2, 1609; two others represent the moon of December 17–18; while the isolated, seventh moon on the verso, with the star Theta Librae emerging to the right, is that of January 19, 1610.[42] But Galileo's drawings are remarkable not only for their precision, but also for their technique, using a brush to render the plasticity of the moon's surface. All of the circles, with diameters of 57–59 mm, are drawn with pen compasses; in each case, the center point is marked by a tiny brown dot. This area still shows light patches going down to the picture ground and varies within itself, but also presents streaks of further shadowing of the dark area of the moon. The use of brown color applied in differing densities made it possible to modulate from a deep, shadowy tone to a beige that almost fades into white.

Galileo's first moon drawing (fig. 13) uses the base color of the paper for the lighted area illuminated by the sun. In the middle of this region, a cloud-shaped area drifts to the right, darkening slightly at its right edge. In the middle of this "cloud," applied almost imperceptibly over the first layer, spots of color run from the upper left to the lower right, intensifying into a color layer of their own. A second bulge in the upper half of the illuminated area points toward the right; here,

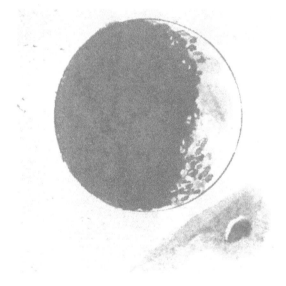

Figure 13. Lunar phase from 11.30.1609, 6–8:00 p.m. (detail from fig. 12; all times from Ewan A. Whitaker).

---

[42] The precision of Galileo's drawings has been doubted, but since Righini's conclusion that Galileo was "in fact, a remarkably faithful recorder of his visual experiences," the picture has changed completely (Righini 1975, 76; cf. Drake 1976; Righini 1978, 26–44; Whitaker 1978 [in this article Whitaker collected comparable photographs of each of the corresponding phases of the moon. He relied on the visible contours of the boundary line between light and shadow, and not necessarily on the extent of the crescent]; Whitaker 1989, 122ss.; Shea 1990).

too, similar spots and daubs were added. To the left, however, the brown grows darker in three layers of intensity. A small area attains the third color level, while to the left a "spot" lying horizontally above it is slightly darker; between them, another island deepens again by an almost imperceptible degree. This area blends into the dark side of the moon. This area still shows light patches going down to the picture ground and varies within itself, but also presents streaks of further shadowing of the dark area of the moon. In the lower right, tachistically applied daubs suggest a carpet of hills still illuminated by the sun and indentations already lying in shadow.

In order to render the growing contrast between darkness and intensifying light, additional shading is applied in the lower area of the second sphere (fig. 14), diminishing again to the left in streaks and stripes. The lighted crescent stands out from the dark ground with all the finesse of chiaroscuro, a veritable explosion of full sunlight. After the glaringly illuminated crescent and the abrupt contrast of the dividing line to the lunar night, the refined treatment of lighting continues with a diminishing of the darkness to reproduce the reflected light of the earth. Finally, the left edge of the moon becomes lighter than the transitional zone between light and shadow, so that a circular line drawn with particular strength is needed to

Figure 14. Lunar phase from 12.01.1609, 5:30 p.m. (detail from fig. 12).

distinguish it from the night sky. Both the light and the dark side of the earth's satellite shine out into space with shimmering iridescence. Daubs of an additional rust-red color appear in the upper right corner and in two places in the lower and upper dark area of the moon. In their distribution they make no geographical sense, but obviously constitute an attempt to bring further visual interest to the representation.

The dark face of the third moon (fig. 15) is a trace lighter than that of the first drawing. Strangely, the deepest dark appears outside, on the upper and lower edges of the picture, as if to provide the sphere hovering in space with a color axis

Figure 15. Lunar phase from 11.30.1609, 4:00 p.m. (detail from fig. 12).

as a kind of visual anchoring. Since the face of the fourth moon (fig. 16) is already half illuminated, the transitional zone from light to dark receives particular emphasis. A peculiarity is the giant mountain range, whose inner cliffs introduce a half-circle of shadow into the zone lighted from the left. Correspondingly there are other areas, too, that still or already lie in shadow as outposts or stragglers. Again, Galileo worked with reciprocal effects: on the dark side, the moon becomes lighter again toward the horizon, while the night sky behind darkens more intensely. Conversely, the dark of the sky diminishes on the opposite side to establish a color balance.

In the contrast zone of light and shadow, the fifth lunar disk (fig. 17) shows a stronger dark than all the previous examples. The sixth sphere (fig. 18) further intensifies the darkness of the night side against the line of contrast to the light. To the lower right, stippled flecks form a panorama of hills and craters.

Figure 16. Lunar phase from 12.17.1609, 5:00 a.m. (detail from fig. 12).

Figure 17. Lunar phase from 12.18.1609, 7:00 a.m. (detail from fig. 12).

Figure 18. Lunar phase from 12.02.1609, 5:00 p.m. (detail from fig. 12).

The finely graduated degrees of light cast over the face of the moon also serve to reveal the heights and depths of its surface. All the drawings are remarkable in the unprecedented plasticity Galileo lends to the sphere of the moon; the light plays over the irregular surface of the planet as in a film, illuminating the hills with its rays, while lower-lying regions still remain in the shadow of night. A study Galileo added to the lower right of the first moon sphere (fig. 13) demonstrates the depth effect of the light-dark contrast in an especially impressive way. In a section of the perspectively foreshortened lunar surface lies a deep crater, its dark blackness establishing a sharp contrast. The accentuation of shadow serves to create the impression of great depth. The deliberations Galileo and other astronomers had undertaken at this time, to determine the heights and depths of the moon's surface from the shadows cast on it, are here confirmed in a virtuosic modeling, constructed from the effects of light and shadow alone. The same effect, here illustrated in a single crater, characterizes all the moon drawings; the lower zone of the sixth sphere, for example, offers a three-dimensional planetary landscape of craters and mountains.

Galileo used the precision of his literary training,[43] as well, to capture in words the celestial theater of light and shadow opening up before his eyes:

> But not only does one see that the boundary between light and darkness on the moon is irregular and sinuous, but rather, what is even more amazing, there appears within the dark part of the moon a multitude of points of light completely divided and torn away from the illuminated zone and separated from it by a not inconsiderable interval. When one waits a while, they gradually increase in size and luminosity, until after two or three hours they join with the rest of the lighted part which has now become larger. Meanwhile, however, new points are continually set alight within the darkened part; sprouting upward, as it were, they grow and join themselves at last to the same illuminated surface that has extended itself still further.[44]

V. Eye and Hand

Galileo's stupendous achievement becomes especially apparent when compared with the work of the English natural scientist and cartographer Thomas Harriot, who studied the moon a number of weeks before Galileo with the help of a Dutch telescope with six-fold magnification. Harriot's drawing shows only an indistinct,

---

[43] On Galileo's literary capacity, recently: (Dietz Moss 1993, 76ss.).

[44] "Verum, non modo tenebrarum et luminis confinia in Luna inaequalia ac sinuosa cernuntur; sed, quod maiorem infert admirationem, permultae apparent lucidae cuspides intra tenebrosam Lunae partem, omnino ab illuminata plaga divisae et avulsae, ab eaque non per exiguam intercapedinem dissitae; quae paulatim, aliqua interiecta mora, magnitudine et lumine augentur, post vero secundam horam aut tertiam reliquae parti lucidae et ampliori iam factae iunguntur; interim tamen aliae atque aliae, hinc inde quasi pullulantes, intra tenebrosam partem accenduntur, augentur, ac demum eidem luminosae superficiei, magis adhuc extensae, copulantur" (Galileo 1610, 64; cf. Mann 1987, 56).

crooked figure on the illuminated sphere (fig. 19). The boundary between light and shadow appears as an unsteady line, but the amplitudes of the indentations are not pronounced enough to suggest elevations and troughs.

Two-and-a-half moon phases after Harriot, however, Galileo recognized immediately that the patterns of light and shadow on the moon had to do with its irregular surface. In view of the fact that Galileo was able to reproduce this lunar theater of light in both words and drawings, the question arises as to why Harriot did not likewise manage to capture what Galileo comprehended shortly after him. It has been suggested that Galileo was able to recognize what remained hidden from Harriot simply because Harriot's telescope was of lesser quality.[45] Galileo's telescopes were indeed better than Harriot's Dutch one,[46] but the problem can by no means be limited to the difference in technical equipment. In comparison with Harriot's poor drawing even the pictures of the moon painted with unarmed eyes by Jan van Eyck[47] and Leonardo[48] are of higher quality. William Lower, a scientific associate of Harriot, wrote in a letter of June 1610 to Harriot about his inability to see correctly when being undermined by a limiting theoretical frame-

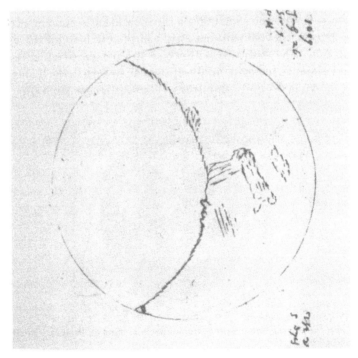

Figure 19. Thomas Harriot, lunar phase, drawing, Petworth mss. Leconsfield HMC 241/ix, fol. 26.

[45] Mann 1987, 59 n. 20.
[46] (Van Helden 1977, 26f.; van Helden 1984, 155). The early history of the invention of the telescope from the end of the sixteenth century up to Galileo has been clarified at the same time once again by Isabelle Pantin (Pantin 1992, IX–XXII and Hallyn 1992, 14–25).
[47] Montgomery 1994.
[48] Reaves 1987.

work. As if reflecting on Harriot's shortcomings, Lower confessed that when observing the "Seven Sisters" or Pleiades, he did see eight stars, but did not dare to trust what he saw: "because I was prejugd with that number, I beleved not myne eyes."[49] There was a gap between what was seen and what was perceived. Harriot, we can conclude, was limited not only by his lenses or eyes, but also by his unwillingness in seeing beyond his theoretical horizon.

For Galileo, on the other hand, the problem posed by the surface of the moon was not a new one. Since his time with Cigoli in Florence, Galileo had been confronted with the theory of "secondary light," by which the effect of bright light is reflected as an ashy shine on another surface. Thus the ashy light on the dark surface of the moon could be explained in this manner.[50] Further, the question why Galileo revived Plutarch's opinion that the moon's surface contained mountains and valleys[51] instead of confirming Averroes' theory that different densities of the smooth body of the moon created the effects of light and darkness,[52] can be explained through the lessons Galileo took out of the lectures of Ostilio Ricci. Ricci taught not only the fundamentals of geometry, Euclid and Archimedes, but also perspective, and among the texts used were the *Ludi Matematici* of Leon Battista Alberti.[53] With their sections on the mensuration and perspective representation of objects, they constituted a part of the mathematical training of the artist at that time. The virtuosity with which this perspective theory was able to calculate and visualize surface configurations can be seen in the forms, impressive even today, shown in Wenzel Jamnitzer's *Perspectiva corporum* (fig. 20).[54] These

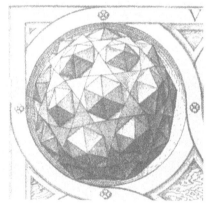

Figure 20. Wenzel Jamnitzer, stereometric body, copper engraving in: *Perspectica Corporum Regularum* (Nuremberg 1568).

---

[49] Quoted in Stevens 1900, 116; cf. Bloom 1978, 121.
[50] On the whole question of the "secondary light": (Reeves 1997, 8ss., 29ss.)
[51] On Plutrach as a source for the *Sidereus Nuncius*: (Casini 1994 and Montgomery 1996, 221f.)
[52] Ariew 1984.
[53] One of Ricci's Alberti-manuscripts is to be found among the Mss. Galileiani of the Biblioteca Nazionale Centrale in Florence (Ms.10, fol.1ʳ–16ʳ; cf. Settle 1971).
[54] Nuremberg, 1568.

figures make clear that when Galileo observed the surface of the moon through his telescope, he recognized, in the form of a planet, a problem from the basic mathematics course for artists. He was able to grasp the patterns of light and shadow on a sphere as a function of height and depth configurations. His horizon of artistic experience enabled him to render what he saw.[55]

But for all Galileo's reliance on his sense of sight,[56] he was nonetheless faced with the painful realization that his foes likewise placed their trust in their eyes. Again and again the telescope had beguiled its untrained users with mirages and deceptions. Often, for example, the earliest telescopes had produced quadratic stars that too obviously contradicted the naked eye to be taken seriously.[57] In addition, the incontrovertible objection was advanced — to which Galileo responded with enraged invective — that whoever depended on the telescope was no longer the lord of his own sense of sight. When something previously unknown was observed with the help of the telescope, it was only a matter of time until an improved or entirely new piece of equipment would confront the eye with a different reality.[58] The epistemological value of the telescopic image was thus called into question.

In view of the doubt as to the fidelity of telescopic images, the question arises as to whether Galileo's drawings were not intended to supply the power of proof denied to the telescope itself. With the help of disegno, the celestial image was transferred to a medium that corresponded to familiar visual experience and remained verifiable day or night. The drawing fulfilled ideally what the heavens could only provide in measure, and this media shift may have played a role when Galileo, in context of his drawings of sun spots, spoke of the "giudizio finale," the "Final Judgment" of the Aristotelians.[59] His statement refers directly to the sunspots; the drawings, however, are the medium and the weapon of the "final judgment." If Galileo's first phase of moon observation was empirically demonstrable, then above all in the medium of drawing.

In an almost clairvoyant way, Galileo's painter friend Cigoli affirmed the interaction of perception and drawing. When he opposed Christopher Clavius' critique of Galileo's moon observations, he found an ironic excuse in the fact that

---

[55] Edgerton 1984, 226; cf. Hallyn 1992, 55–59; Holton 1996, 185ss.

[56] Winkler 1992, 195. Winkler and Van Helden limit Galileo's trust in visual representations to the years 1610–1613. Even if so, the fundamental fact remains that it was due to Galileo's artistic experience that on the long run "astronomy became a visual science" (Winkler 1992, 195, 217; cf. the implicit critique of Winkler and Van Helden by Montgomery: Montgomery 1996, 226): "For Galilei, the image must convey its own language, apart from words."

[57] Feyerabend 1993, 148f. and van Helden 1994, 11–15.

[58] Kutschmann 1986, 149f.

[59] "Intanto gli mando alcuni disegni delle macchie solari, fatti con somma giustezza tanto circa al numero quanto alla grandezza, figura e situazione di esse di giorno i giorno nel disco solare. Se occorrerà a V.S.Ill.ma trattare di questa mia resoluzione con i litterati di cotesta città, haverò per grazia il sentire alcuna cosa de i loro pareri, et in particolare de i filosofi Paripatetici, poi che questa novità pare il giudizio finale della loro filosofia." Galileo to Maffeo Barberini, June 2, 1612 (Galilei 1890–1909, 6:304–11, [306, 311]). I follow Montgomery's interpretation of the moon drawings as a "proof," "seemingly peeled from the eye, stolen and saved from the act of perception" (Montgomery 1996, 229, 284n.13).

he could not draw, and thus was "not only half a mathematician, but a man lacking eyes as well."[60] This astounding statement presupposes that an adequate comprehension of reality involves not only its reception, but also its reproduction; not only its perception, but its construction as well. For Cigoli, Galileo could see better, because he was better prepared by his artistic training and knew how to draw. In an autodidactic process taking place between hand and eye, Galileo was better able to attain knowledge, both because he had learned to perceive the unusual and because he could demonstrate it in the medium of drawing.

Four years after Galileo's moon drawings, the list of names entering and leaving the art academy of Florence shows the following entry: "Galileo [son] of Vincenzo Galilei [has] to pay ten soldi on October 18, 1613, on account of his entrance into the academy, because the above-named has attained membership in the academy."[61] Non-artists, of course, were accepted into the academy. But Galileo was presumingly not only flattered with this appointment, but also confirmed in the methodological credo he had developed in his youth.

VI. The Empirical and Projection

There remains an irritating, irresolvable problem, which is still deserving of at least one thesis. Galileo's drawings served as the basis for four engravings, which were included in the 1610 first edition of *Sidereus Nuncius*. Because the fifth is essentially a duplicate of the third, there are only four distinct phases of the moon. They show peculiarities, which refer to the fundamental epistemological question, whether or not the belief in the interplay of perception and drawing, through consummate power of recognition, led to a strange interaction of strict empiricism, thoughtful propaganda and unwilling autosuggestion.

Two of the engravings are clear. The first shows, in accordance with the first drawing (fig. 13), the waxing moon of November 30, 1609 between 6:00 and 8:00 p.m. (fig. 21).[62] The fourth engraving depicts the waning moon (fig. 22),[63] and with its mountainous ring resembles the fourth drawing (fig. 16). Yet the middle two engravings are riddles. One of them (fig. 23),[64] shows the waxing half-moon of December 3, 1609 at 5:00 p.m. The oddity of this representation lies in the fact that while it also gives the mountain ring around the Mare Imbrium, it also shows an

---

[60] "Ora io ci ò pensato et ripensato, nè ci trovo altro ripiegho in sua difesa, se non che un matematico, sia grande quanto si vole, trovandosi senza disegnio, sia non solo un mezzo matematico, ma ancho uno huomo senza ochi." Cigoli, letter to Galileo, August 11, 1611 (Galilei 1890–1909, 11:168; cf. Edgerto 1991, 253n.41; Hallyn 1992, 58). The context of Cigoli's letter is studied by James M. Lattis in (Lattis 1994, 195ss.).

[61] "Galileo di Vinc:o Galilei de dare addi 18 di ottobre 1613 soldi 10 per sua ent[ra]ta nel Achedemia che detto disu vinto Achademico" (Florence, Archivio di Stato, Mss, Accademia del disegno No. 124, "Libro dell'Entrata e Uscita," fol. 52ᵛ; cf. Chappel 1975, 91n.4).

[62] Galilei 1610, 8ʳ.

[63] Ibid., 10ᵛ.

[64] Ibid., 9ᵛ.

crefcente parte luminofa tenebras amittunt.

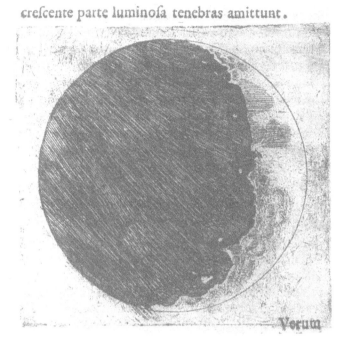

Figure 21. Galileo Galilei, lunar phase, copper engraving, in: *Sidereus Nuncius* (Venice 1610), p. 8r.

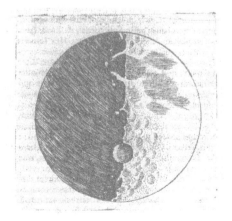

Figure 22. Galileo Galilei, lunar phase, copper engraving, in: *Sidereus Nuncius* (Venice 1610), p. 10v.

Figure 23. Anonymous, lunar surface, copper engraving, in: *Sidereus Nuncius* (Venice 1610), p. 9v.

immense crater in the middle of the lower half. However, in the drawing, recorded a day earlier, there is not even the slightest suggestion of the existence of this crater.

The crater reappears in the next engraving of the waning half-moon (fig. 24)[65] in the same position, but lit from the opposite side. Again, there is no counterpart in the drawings. Obviously, Galileo wanted to demonstrate that the great ring of the crater is only visible exactly in the half-moon stages. In this way it sits exactly in the center, thus becoming a sort of trademark. In text, Galileo described it with the emphatic tone of a revelation: "There is another thing I am unable to suppress, as I noted it with admiration: almost in the moon's center there exists a cavity, larger than all the others, whose form is entirely round. I observed it both near the first and the last quarter and have attempted to render it as accurately as possible in the second, upper drawing."[66]

The problem is, this crater does not exist; neither in reality, nor in the drawing. Galileo's illumined crescent corresponds to the appearance of the moon down to

Figure 24. Anonymous, lunar surface, copper engraving, in: *Sidereus Nuncius* (Venice 1610), p. 10r.

---

[65] Ibid., 10^r. and 10^v.

[66] "Unum quoque oblivioni minime tradam, quod non nisi aliqua cum admiratione adnotavi: medium quasi Lunae locum a cavitate quadam occupatum esse reliquis omnibus maiori, ac figura perfectae rotunditatis; hanc prope quadraturas ambas conspexi, eandemque in secundis supra positis figuris quantum licuit imitatus sum" (Galilei 1610, 67f.).

the finest details. Yet, there is no match for his large crater. One suspicion reasons that Galileo, through an act of autosuggestion or "good pedagogy" which wanted to make his point clear, enlarged one of the largest craters, the Albategnius, to such an extent, that its great, perfect roundness prominently occupied the central axis of the half-moon.[67]

An even more probable reason, as so often with Galileo, could lie in the elastic attention he afforded the perturbation which his discoveries must have fomented in many of his contemporaries. The crater would thus have served as a visual pendant to the exoteric rejection of Copernicanism. Galileo must have been cognizant that his lunar observations defied everything associated with the planet: as a round, harmonious, smooth form, a consummate sphere and celestial guarantee of platonic perfection. Yet, if Galileo was able to find a circular crater, which was positioned exactly on the central light axis of the half-moon, then the scandal must have been less acute. Though the moon was proven to be three-dimensional, it maintained its essential motif in a relatively central location, that is, the full circle; not as a real circle, but as a symbol of circular form.[68] Thus the moon was reinvested with the platonic dignity in the medium which had initially destroyed it: the irregularity of its surface.

This circle was perhaps one of the reasons for Galileo's success. In Harriot's second (17 July 1610) sketch of the moon (fig. 25), the characteristic craters

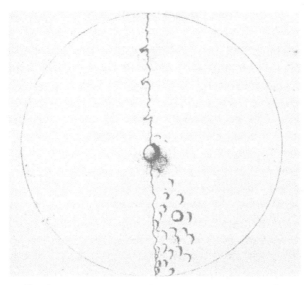

Figure 25. Thomas Harriot, lunar phase, drawing, Petworth mss. Leconsfield HMC 24/ 1ix, fol. 20.

---

[67] (Gingerich 1975, 85; cf. the comparison in Casini 1994, 57). The thesis that Galileo "exaggerated the size of what he had observed in order to bring out the salient features" was developed by Shea (Shea 1990, 56f.)

[68] Gingerich 1975, 86.

appearing lower right, demonstrate that the British scientist had come to accept
the moon's surface was not smooth, but scattered with mountains and hollows.
Apparently, it was not only his own lunar observations, but the engravings in
Galileo's publication, that prompted Harriot to see objectively.[69] Yet, his learning
process was limited once again to what he learnt, and not what he saw. In
accordance with Galileo's emphatical description, he set the fictional crater almost
exactly in the center of the moon. His drawing bridges the gap between Galileo's
text and Galileo's engraving, but is even further afield from reality than was
Galileo's crater image. Thus, in this case, it was not observation of nature, but
suggestion which won the contest of credibility.

Harriot was hardly an exception. In 1626, the Jesuit Mission in China used
Galileo's telescope pictures to extol the superiority of the Christian science to the
Chinese. One of the two engravings appears as a woodcut, turned upside-down, in
the upper-left, and situates the imaginary crater on the line of equilibrium between
lunar day and night.[70]

Galileo's images of the moon were as much a case for art history as for cognitive
psychology; in other words, he not only saw what he was able to see, but also, what
he wanted to see. Yet this realization would offer a hasty conclusion. There is an
additional facet to the problem. Beyond Galileo's projection of a circle lies the
disappearance of Kepler's ellipse.

VII. Einstein's Blind Spot

In 1925, Aby Warburg had the reading room of his Kulturwissenschaftliche
Bibliothek topped with an elliptical dome.[71] It was a gesture honoring Kepler's
discovery of the elliptical orbits of planets. The oval shape of the ceiling correlated
with figure 3 of Warburg's pictorial atlas, in which diverse models of the cosmos
lead up to Kepler's elliptical planetary orbits. In another instance, Warburg's
exhibition "Picture Collections Concerning the History of the Study of and Belief
in Stars at the Hamburg Planetarium" concluded with Kepler's ellipse.[72] With its
two poles, it seemed to symbolize a bi-polarity of creative thinking: not circling
around a single point, but oriented between the two elliptical poles of magical
thinking and Enlightenment. Warburg considered Kepler's calculations a unique
symbol of human self-liberation from the realm of the occult.[73]

In September 1928, during a visit with Albert Einstein at the seaside resort of
Scharbeutz, Warburg tried, with the aid of his pictorial atlas, "to grant [Einstein] a

[69] Bloom 1978, 121; Edgerton 1984, 227.
[70] Edgerton 1991, 269f.
[71] Jesinghausen-Lauster 1985, 216; cf. von Stockhausen 1992, 37ss.; lastly with corrections, Settis
1996, 152.
[72] Warburg 1993, fig. XVII.
[73] Jesinghausen-Lauster 1985, 215f.; cf. von Stockhausen 1992, 37ss.

glimpse of the soil in which his cosmological mathematics had originated." In a letter to Fritz Saxl, Warburg wrote later: "[Einstein] followed my pictures like a schoolboy at the movies...and (tested) the soundness of my conclusions by following up with merciless questioning. ... Only on Kepler and the ellipse, I believe, I did not get a passing grade. Otherwise he was satisfied with me."[74] In a thank-you note to Warburg, Einstein touched once more on the subject with the intent of absolving Kepler, writing that Kepler would have "been ashamed ... to earn his keep by playing such an unsophisticated game" — meaning astrological practices.[75]

Warburg was clearly unable to convince Einstein that his preoccupation with Kepler's ellipse had nothing to do with the evaluation of astrology, but instead concerned the aesthetic influence on scientific-cosmological thinking. In other words, the question whether stars move along a circle or an ellipse was not answered merely through observation and calculations, but also, and especially, through the aesthetic-visual pre-consciousness of the scientists.

Because Einstein was not prepared to accept this line of thinking, he was blinded to the problem's solution, a problem which has since become a thorn in the side of the history of science and which frustrated Einstein to the end of his life: namely, the question why Galileo rejected Kepler's 1605 publication *Nova Astronomia*. In 1953, in a mixture of puzzlement and displeasure, Einstein stated: "That this decisive advancement left no traces in Galileo's life work is a grotesque illustration that creative human beings frequently lack a receptive frame of mind."[76]

Erwin Panofsky, in Hamburg as one of those privileged scholars to have researched under the Warburg Library ellipse, was especially sensitive to such remarks.[77] Shortly after Einstein, Panofsky was awarded a place at the Institute for Advanced Studies in Princeton. In direct proximity to Einstein for almost twenty years, Einstein's remark inspired Panofsky to examine Galileo's motive for rejecting Kepler's *Nova Astronomia* and to bridge, as it were, the gap between the histories of physics and art.[78] The following year, Panofsky published his "Galileo as a Critic of the Arts" as a response to Einstein's query.[79]

Panofsky tried to demonstrate that the reasons for Galileo's disregard of *Nova*

---

[74] Einstein followed "gespannt wie ein Schuljunge im Kino meinen Bildern (...) und (prüfte) unter steten unerbittlichen Nachfragen die Stichhaltigkeit meiner Schlüsse (...). Nur bei Kepler und der Ellipse habe ich, glaube ich, nicht gut bestanden; sonst war er mit mir zufrieden" Aby Warburg to Fritz Saxl, September 5, 1928 in (Warburg 1928). This letter and the one in the following note were made available through the courtesy of Claudia Naber.

[75] Kepler would have felt ashamed "sich sein Futter durch ein so plumpes Spiel zu verdienen." (Albert Einstein to Aby Warburg, September 10, 1928, Warburg 1928). It may be surmised that Einstein was not only discomforted by Warburg's identification with Kepler, but also by his portrayal of Kepler as wrestling with astrology and magic.

[76] Einstein 1953, xvi.

[77] On the occasion of the Nobel Prize awards ceremony for his friend, the mathematician Wolfgang Pauli, in December 1945, Panofsky held one of the laudatory speeches, in which he longingly recalled those times when models of the cosmos, like Kepler's still possessed "meaning." Cited from (Ludwig 1974, 115f.).

[78] Panofsky 1954, 23f.

[79] Ibid., 24.

Astronomia were not only to be found in the close fields of physics and mathematics, but instead in a wider concept of world-view within which aesthetical values played a role as important as these. Galileo was opposed to Mannerism with its clever, trompe l'oeil, anarchic aesthetic, which had been dominating the European art world for half a century. Like his artist friends, he had had enough of this artistic direction. His affections and aversions within the arts were localized around the two poets Ludovico Ariosto and Torquato Tasso. Galileo celebrated the Rennaissance literatus Ariost as the epitome of the poetic arts, while heaping scathing criticism upon Tasso. The reasons for his rejection are art historically revealing. Galileo attacked Tasso's "distorted" verse, comparing it to anamorphosis whose meaning becomes "visible and accessible at an angle" and "which obstructs the gaze by extravagant means through a collection of fantastic, chimerical and superfluous illusions."[80]

Galileo's anti-Mannerist notion of art evoked a feeling of repulsion beyond all reason. Panofsky argued that to Galileo's sense of harmony, inextricably wedded to the movement of circles, Kepler's conclusions appeared unpalatable, even monstrous, much like anamorphosis, and that such aesthetic sensibilities created an insurmountable barrier in his mind with regard to celestial mechanics. According to Panofsky, Galileo was only capable of envisioning a harmonious cosmos based on circles, so that Kepler's compressed elliptical orbits appeared as unbearable aesthetic deformations, as though Tasso's oblique verse and anamorphosis were projected onto heaven.

This trial of understanding Galileo's silence about Kepler's elliptical orbits was appraised by Alexandre Koyrè,[81] reprinted in a revised form in *Isis* and further discussed in the history of science and art history,[82] but it must have been a great disappointment for Panofsky that Einstein ignored it. Although Einstein discussed the problem again in his last interview, in 1955, he took no account of Panofsky's conclusions.[83] Like Warburg 25 years earlier, Panofsky failed to convince Einstein of the aesthetic consequences of Kepler's elliptical orbits.

***

---

[80] "E farassi una di quelle pitture, le quali, perchè riguardate in scorcio da un luogo determinato mostrino una figura umana, sono con tal regola di prospettiva delineate, che, vedute in faccia e come naturalmente e comunemente si guardano le altre pitture, altro non rappresentano che una confusa e inordinata mescolanza di linee e di colori, dalla quale anco si potriano malamente raccapezare imagini di fiumi e sentier tortuosi, ignude spiaggiae, nugoli o stranissime chimere. (...) tanto nella poetica finzione è piu degno di biasimo che a favola corrente, scoperta e prima dirittamente veduta, sia per accomodarsi alla allegoria, obliquamente vista e sottointesa, stravagantamente ingombrata di chimere e fantastiche e superflue imaginazioni" in "Considerazioni al Tasso" (Galilei 1890–1909, IX:59–148 [129f.]).

[81] Koyrè 1955.

[82] Panofsky March 1956; Rosen March 1956; Panofsky June 1956; Fehl, 1958; Lotz 1958; Mazzi 1985; Shea 1985; Puppi 1995, 244ss.; Reeves 1997, 6f., 18ss.

[83] Cohen July 1955, 69; cf. Fölsing 1993, 243.

It was the draftsman Galileo, who was first able to recognize and record the surface of the moon. It was also the artist Galileo, who manipulated the surface of the moon into, as Hans Blumenberg described, "an interplay of revealment and veiling"[84] thereby separating it from Kepler's planetary orbits.

Yet, regardless of this condition, there persist the irresolvable epistemological questions precipitated by Galileo's disclosures. If the case of Galileo the "artist" can be generalized, then to the extent that there are art historical inquiries, whose answers would be impossible without the history of natural sciences. Just as the other way around, there are scientific complexes which remain impenetrable, as long as the visual structures of human thought, indispensable in their illumination and obscuring, are ignored.

In any event, Viviani ought to have linked not Michelangelo, but Leonardo da Vinci with Galileo. For it was Leonardo, the outstanding researcher of the natural world and occasional painter, who corresponded in a reciprocal manner to Galileo, the Medici court artist, who, by the way, worked as a researcher.

References

Ariew, Roger. 1984. "Galileo's Lunar Observations in the Context of Medieval Lunar Theory." *Studies in History and Philosophy of Science* 15:221ss.

Biagioli, Mario. 1993. *Galileo, Courtier. The Practise of Science in the Culture of Absolutism*. Chicago and London: Univ. of Chicago Press.

Bloom, Terrie F. 1978. "Borrowed Perceptions: Harriot's Maps of the Moon." *Journal for the History of Astronomy* 9:117–122.

Blumenberg, Hans. 1980. "Das Fernrohr und die Ohnmacht der Wahrheit." In *Galileo Galilei, Sidereus Nuncius [Nachricht von neuen Sternen]. Dialog über die Weltsysteme (Auswahl). Vermessungen der Hölle Dantes. Marginalien zu Tasso*, 7–75, edited by Hans Blumenberg. Frankfurt am Main, Suhrkamp.

Bredekamp, Horst. 1993. *Florentiner Fußball. Die Renaissance der Spiele. Calcio als Fest der Medici*. Frankfurt am Main and New York: Campus-Verlag.

——. 1995. "Words, Images, Ellipses." In *Meaning in the Visual Arts: Views from the Outside. A Centennial Commemoration of Erwin Panofsky (1892–1968)*, edited by Irving Lavin, 363–371. Princeton.

——. 1996a. "Galileo Galilei als Künstler." In *Übergangsbogen and Überhöhungensrampe — naturwissenschaftliche und künstlerische Verfahren. Symposium I und II*, edited by Bogomir Ecker and Bettina Sefkow, 54–63. Hamburg.

——. 1996b. "Zwei frühe Skizzenblätter Galileo Galileis." In *Ars naturam adiuvans. Festschrift für Matthias Winner zum 11. März 1996*, edited by Victoria von Fleming and Sebastian Schütze, 477–484. Mainz am Rhein.

---

[84] Blumenberg 1980, 48.

Callot, Jacques. 1971 [1592–1635]. *Das gesamte Werk*, edited by Thomas Schröder. Munich: Rogner und Bernhard.

Casini, Paolo. 1994. "Il 'Dialogo' di Galileo e la Luna di Plutarcho." In *Novità Celesti e Crisi del Sapere. Atti del Convegno Intternazionale di Studi Galileiani*, edited by Paolo Galluzzi, 57–62. Florenz.

Cavina, Anna Ottani. March 1976. "On the Theme of Landscape-II: Elsheimer and Galileo." *The Burlington Magazine* 118/876:139–144.

Chappel, Miles. 1975. "Cigoli, Galileo, and Invidia." *Art Bulletin* 57/1:91–98.

Ciampoli, Giovanni. 1978. "Discorso sopra la Corte di Roma." In Marziano Guglielminetti and Mariarosa Masoero, *Lettere e prose inedite [o parzialmente edite] di Giovanni Ciampoli. Studi secenteschi* 19:228–237.

Cohen, Bernard. July 1955. "An Interview with Einstein." *Scientific American* 93:69–73.

Contini, R. 1991. *Il Cigoli*. Soncino.

de Angelis, Paolo. (1621) imprimatur 1616. *Basilicae S. Marioris de Urbe...descriptio et delineatio*. Rome.

de Maio, Romeo. 1978. *Michelangelo e la Controriforma*. Rome and Bari.

de'Nelli, Giovanni Battista Clemente. 1793. *Vita e commercio letterario di Galileo Galilei*. Lausanne.

Dietz Moss, Jean. 1993. *Novelties in the Heavens. Rhetoric and Science in the Copernican Controversy*. Chicago and London.

Drake, Stillman. 1976. "Galileo's First Telescopic Observations." *Journal for the History of Astronomy* 7:153–168.

Dussler, Luitpold 1959. *Die Zeichnungen des Michelangelo. Kritischer Katalog*. Berlin.

Edgerton, Samuel Y. 1984. "Galileo, Florentine 'Disegno' and the 'Strange Spottedness' of the Moon." *Art Journal* 44/1:225–32.

——. 1991. *The Heritage of Giotto's Geometry*. Ithaca and London.

Einstein, Albert. 1953. "Introduction" to Galileo Galilei. In *Dialogue Concerning the Two Chief World Systems*. Translated by Stillman Drake. Berkeley.

Faranda, Franco. 1986. *Ludovico Cardi detto il Cigoli*. Rome.

Favaro, Antonio. 1880. "Inedita Galilaeiana. Frammenti tratti dalla Biblioteca Nazionale di Firenze." *Atti e Memorie dell'Instituto Veneto di Scienze, Lettere ed Arti* 21:35–43.

——. 1887. "Sul giorno della nascità di Galileo." *Memorie del R. Istituto Veneto di scienze lettere ed arti* 22:703–711.

Fehl, Philipp. 1958. "Review." *Journal of Aethetics and Art Criticism* 17:124–125.

Feldhay, Rivka. 1995. *Galileo and the Church: Political Inquisition or Critical Dialogue?* Cambridge.

Feyerabend, Paul. 1993. *Wider den Methodenzwang*, 4th ed. Frankfurt am Main: Suhrkamp.

Fölsing, Albrecht. 1993. *Albert Einstein. Eine Biographie*. Frankfurt am Main: Suhrkamp.

Frieß, Peter. 1993. *Kunst und Maschine. 500 Jahre Maschinenlinien in Bild und Skulptur*. Munich: Dt. Kunstverl.

Galilei, Galileo. 1890–1909. *Le Opere.di Galileo Galilei*. Edizione Nazionale, edited by Antonio Favaro, 20 vols. Florence: Barbèra Editrice. Reprint: 1929–1939, 1964–1966, 1968.

——. 1610. *Sidereus Nuncius*. In *Opere di Galileo* by Galilei, Galileo (1890–1909), 3/1:53–96. Florence: Barbèra Editrice.

Galluzzi, Paolo. 1994. "I Sepolcri di Galileo: o delle spoglie 'vive' di un eroe della scienza." Lecture delivered at the symposium *Galileo Galilei*. Berlin: Istituto di Cultura di Berlino.

Garrard, Mary D. 1989. *Artemisia Gentileschi. The Image of the Female Hero in Italian Baroque Art*. Princeton, New Jersey: Princeton University Press.

Gingerich, Owen. 1975. "Dissertation cum Professore Righini et Siderio Nuncio." In *Reason, Experiment, and Mysticism in the Scientific Revolution*, edited by M. L. Righini Bonelli and William R. Shea, 77–88. New York.

Hallyn, Fernand. 1992. "Introduction." In Galileo Galilei, *Le Messager des étoiles*, 14–101. Paris.

Haskell, Francis and Nicholas Penny. 1981. *Taste and the Antique: The Lure of Classical Sculpture 1500–1900*. New Haven and London.

Heikamp, Detlev. 1978. "Ammannati's Fountain for the Sala Grande of the Palazzo Vecchio in Florence." In *Dumbarton Oaks Colloquium on the History of Landscape Architecture* V, 116–176, edited by Elisabeth B. MacDougall. Washington/D.C.

Holton, Gerald. 1996. "On the Art of Scientific Imagination." *Daedalus* 125/2:183–208.

Jesinghausen-Lauster, Martin. 1985. *Die Suche nach der symbolischen Form. Der Kreis um die kulturwissenschaftliche Bibliothek Warburg*. Baden-Baden: Verlag V. Koerner.

Kant, Max Immanuel. 1923. "Entwürfe zu dem Colleg über Anthropologie aus den 70er und 80er Jahren." In *Kant's gesammelte Schriften*, edited by Königlich Preußische Akademie der Wissenschaften, 15:655–899. Berlin and Leipzig.

Kemp, Wolfgang. 1979. ". . . *einen wahrhaft bildenden Zeichenunterricht überall einzuführen" Zeichnen und Zeichenunterricht der Laien. 1500–1870. Ein Handbuch*. Frankfurt am Main.

Koyrè, Alexandre. 1955. "Attitude esthetique et pensée scientique." *Critique*, nos. 100–101, 835–847. Printed in German in Alexander Koyrè. 1988. "Kunst und Wissenschaft im Denken Galileis." In *Galilei. Die Anfänge der neuzeitlichen Wissenschaft*, 70–83. Berlin; reprinted with Panofsky's text in Erwin Panofsky. 1993. *Galilée Critique d'Art*, edited and translated by Nathalie Heinich, 81–97. Leuven.

Kutschmann, Werner. 1986. *Der Naturwissenschaftler und sein Körper*. Frankfurt am Main.

Lattis, James M. 1994. *Between Copernicus and Galileo. Christoph Clavius and the Collapse of the Ptolemaic System*. Chicago and London.

Lotz, Wolfgang. 1958. "Review." *Art Bulletin* 40/2:162–164.

Ludwig, Richard M. 1974. *Dr. Panofsky and Mr. Tarkington. An Exchange of Letters.* Princeton.

Magani, Fabrizio. 1995. "Il Collezionismo a Venezia al Tempo del Soggiorno di Galileo." In *Galileo Galilei e la Cultura Veneziana, Atti del Convegno di Studio Promosso nell' Ambito delle Celebrazioni Galileiane indette dall' Università degli Studi di Padova* (1592–1992), 1992. 137–159. Venice: Lettere ed Arti.

Mann, Heinz Herbert. 1987. "Die Plastizität des Mondes — Zu Galileo Galilei und Ludovico Cigoli." In *Natur und Kunst*, edited by Götz Pochat and Brigitte Wagner. (*Kunsthistorisches Jahrbuch Graz*, no. 23,55–59).

Masotti, Arnaldo. 1970–1980. "Ricci, Ostilio." In *Dictionary of Scientific Biography*, XI:405f., edited by Charles C. Gillispie, 16 vols. New York.

Matteoli, Anna. 1980. *Lodovico Cardi Cigoli. Pittore e Architetto.* Pisa.

———. 1992. *Disegni di Lodovico Cigoli* (1559–1613), exh. cat. Florence: Uffizi.

Mazzi, Maria Cecilia. 1985. "Introduzione." In Erwin Panofsky, *Galileo Critico delle Arti*, 7–18. Venedig.

Montgomery, Scott L. 1994. "The First Naturalistic Drawings of the Moon: Jan van Eyck and the Art of Observation." *Journal for the History of Astronomy* 25:317–320.

———. 1996. *The Scientific Voice.* New York.

Mss. Accademia del disegno No. 124. Archivio di Stato, Florence.

Mss. Galileiani. Biblioteca Nazionale Centrale. Florence.

Olschki, Leonardo. 1965 [1922]. *Bildung und Wissenschaft im Zeitalter der Renaissance in Italien*, 3:141–155, *Galilei und seine Zeit.* Vaduz.

Ostrow, Steven F. 1996a. *Art and Spirituality in Counter-Reformation Rome. The Sistine and Pauline Chapels in S. Maria Maggiore.* Cambridge.

———. 1996b. "Cigoli's Immacolata and Galileo's Moon: Astronomy and the Virgin in Early Seicento Rome." *Art Bulletin* 78/2:218–235.

Panofsky, Erwin. 1954. *Galileo as a Critic of the Arts.* The Hague.

———. March 1956. "Galileo as a Critic of the Arts." *Isis* 47/147:3–15.

———. June 1956. "More on Galileo and the Arts." *Isis* 47/148:182–185.

Pantin, Isabelle. 1992. "Introduction." In *Sidereus Nuncius* by Galileo Galilei (1610). Le Messager Celeste, pp. IX–CIV. Paris.

Perrig, Alexander. 1991. *Michelangelo's Drawings: The Science of Attribution.* New Haven and London.

Puppi, Lionello. 1995. "Galileo Galilei e la Cultura Artistica a Venezia tra la Fine del 500 e l'Inizio del '600." In *Galileo Galilei e la Cultura Veneziana, Atti del Convegno di Studio Promosso nell' Ambito delle Celebrazioni Galileiane indette dall' Università degli Studi di Padova* (1592–1992), 1992. 243–255. Venice: Lettere ed Arti.

Reaves, Gibson and Carlo Pedretti. 1987. "Leonardo da Vinci's Drawings of the Surface of the Surface Features of the Moon." *Journal for the History of Astronomy* 18/52:55–58.

Reeves, Eileen. 1997. *Painting the Heavens. Art and Science in the Age of Galileo.* Princeton: Princeton University Press.

Reti, Ladislao. 1959. "'Non si volta chi à stella è fisso'. Le 'imprese' di Leonardo da Vinci." In *Bibliothèque d'Humanisme et Renaissance* 21:7–54.

Reynolds, Ted. 1974. "The 'Accademia del Disegno' in Florence, Its Formation and Early Years, Ph.D. diss., Columbia University. New York.

Righini, Guglielmo. 1975. "New Light on Galileo's Observations." In *Reason, Experiment, and Mysticism in the Scientific Revolution*, edited by M. L. Righini Bonelli and William R. Shea, 59–76. New York.

——. 1978. *Contributo alla interpretazione scientifica dell' opera astronomica di Galileo.* Florence.

Rosen, Edward. March 1956. *Isis* 47/147:78–80.

Segre, Michael. 1989. "Viviani's Life of Galileo." *Isis* 80:207–31.

——. 1991. *In the Wake of Galileo.* New Brunswick/New Jersey.

Settis, Salvatore. 1996. "Warburg continuatus. Description d'une bibliothèque." In *Les Pouvoir des Bibliothèque. La mémoire des livres en Occident*, edited by Marc Baratin and Christian Jacob, 122–170. Paris.

Settle, Thomas B. 1971. *"Ostilio Ricci, a Bridge between Alberti and Galileo."* In *Actes du XIIe Congrès International d'Historie des Sciences*, Paris 1968, 3B:122–26. Paris.

Shea, William R. 1985. "Panofsky revisited: Galileo as a Critic of the Arts." In: *Renaissance Studies in Honor of Craigh Hugh Smyth*, edited by Andrew Morrogh et al., I:481–492. Florenz.

Shea, William S. 1990. "Galileo Galilei: An Astronomer at Work." In: *Nature, Experiment, and the Sciences. Essay on Galileo and the History of Science in Honour of Stillman Drake*, edited by Trevor H. Levere and William R. Shea, 51–76. Dordrecht.

Stevens, Henry. 1900. *Thomas Harriot, the mathematician, the philosopher and the scholar.* London.

Tabarroni, Giorgio. 1984. "I Disegni Autografi della Luna e Altre Espressioni Figurative dei Manoscritti Galileiani." In: *Novità Celesti e Crisi del Sapere. Atti del Convegno Internazionale di Studi Galileiani*, edited by Paolo Galluzzi, 51–55. Florence.

van Helden, Albert. 1977. "The Invention of the Telescope." *Transactions Of The American Philosophical Society Held At Philadelphia For Promoting Useful Knowledge* 67:part 4.

——. 1984. "Galileo and the Telescope." In *Novità Celesti e Crisi del Sapere. Atti del Convegno Internazionale di Studi Galileiani, edited by Paolo Galluzzi*, 149–158. Florence.

——. 1994. "Telescopes and Authority from Galileo to Cassini." In *Instruments*, edited by Albert van Helden and Thomas L. Hankins. =Osiris 9:9–29.

Vasari, Giorgio. 1962. *Vita di Michelangelo*, edited by Paola Barocchi. Milan and Naples.

Viviani, Vincenzio. 1890–1909. "Racconto istorico della vita del Sig. Galileo Galilei". In *Opere di Galileo* by Galilei Galileo, 19:597–646. Florence: Barbèra Editrice.

von Stockhausen, Tilmann. 1992. *Die kulturwissenschaftliche Bibliothek Warburg. Architektur, Einrichtung und Organisation.* Hamburg.

Warburg, Aby. 1928. *Warburgs Korrespondenzarchiv*, General Correspondence 1928, E. Warburg Institute, London.

Warburg, Aby M. 1993. *Bildersammlung zur Geschichte von Sternglaube und Sternkunde im Hamburger Planetarium*, edited by Uwe Fleckner et. al. Berlin.

Warnke, Martin. 1985. *Hofkünstler. Zur Vorgeschichte des modernen Künstlers.* Cologne.

Wazbinski, Zygmunt. 1987. *L'Accademia Medicea del Disegno a Firenze nel Cinquecento. Idea e Istituzione*, 2 vols. Florence.

Westfall, Richard. 1984. "Galileo and the Accademia dei Lincei." In *Novità Celesti e Crisi del Sapere. Atti del Convegno Internazionale di Studi Galileiani*, edited by Paolo Galluzzi, 189–200. Florence.

Westfehling, Uwe. 1993. *Zeichnen in der Renaissance. Entwicklung / Techniken / Formen / Themen.* Cologne.

Whitaker, Ewan A. 1978. "Galileo's Lunar Observations and the Dating of the Composition of Sidereus Nuncius." *Journal of the History of Astronomy* 9:155–69.

——. 1989. "Selenography in the Seventeenth Century." In *The General History of Astronomy*, edited by Michael Hoskin, 2:119–42. *Planetary Astronomy from the Renaissance to the Rise of Astrophysics.* Cambridge

Winkler, Mary and Albert Van Helden. 1992. "Representing the Heavens: Galileo and Visual Astronomy." *Isis* 83:195–217.

Wolf, Gerhard. 1991/92. "Regina Coeli, Facies Lunae, 'Et in Terra Pax'. Aspekte der Ausstattung der Cappella Paolina in S. Maria Maggiore." In *Römisches Jahrbuch der Bibliotheca Hertziana*, 27/28:284–336.

Humboldt-Universität zu Berlin

SARA ELIZABETH BOOTH and ALBERT VAN HELDEN

The Virgin and the Telescope:
The Moons of Cigoli and Galileo

The Argument

In 1612, Lodovico Cigoli completed a fresco in the Pauline chapel of the Basilica of Santa Maria Maggiore in Rome depicting Apocalypse 12: "A woman clothed with the sun, and the moon under her feet." He showed the crescent Moon with spots, as his friend Galileo had observed with the newly invented telescope. Considerations of the orthodox view of the perfect Moon as held by philosophers have led historians to ask why this clearly imperfect Moon in a religious painting raised no eyebrows. We argue that when considered in the context of biblical interpretation and the rhetoric of the Counter-Reformation, the imperfect Moon under the woman's feet was entirely consistent with traditional interpretations of Apocalypse 12.

In 1610, after competing with several artists, Lodovico Cardi da Cigoli (1559–1613) received the commission to paint the dome of the new Pauline Chapel of the Basilica of Santa Maria Maggiore in Rome. This basilica was no ordinary structure. It was the most important shrine in Marian worship in Western Christianity, founded in the fourth century when, according to "The Miracle of the Snow," the Virgin appeared one summer night in a vision to Pope Liberius (352–366) and a wealthy patrician and told them to build a church on the spot on the Esquiline Hill where they would find snow. The following morning they did indeed find a spot on the hill outlined in snow, and as the Virgin of their vision had instructed they erected a basilica on this spot.

Richly endowed and adorned by popes over the centuries, the basilica housed many relics, the most important of which were the *Presepio*, or the Manger of Christ, the body of St. Jerome, and the Icon of the Virgin and Child painted by St. Luke (Ostrow 1996a, 1–3, 25, 120–32). During Galileo's lifetime, two large chapels were added to the basilica: the Sistine Chapel (1585–1590) built by Pope Sixtus V to house the *Presepio* and the body of St. Jerome, as well as his own tomb and that of his predecessor, Pius V; and the Pauline Chapel (1605–1610) built by Pope Paul V to house the icon of the Virgin and Child, and his own tomb and that of his predecessor, Clement VIII.

Ostrow has analyzed Cigoli's lunar representation in the Pauline Chapel of

Santa Maria Maggiore in Rome in the context of Counter-Reformation rhetoric, and he, Marina Warner, and Eileen Reeves have documented and extensively discussed lunar symbolism in the various representations of the Virgin Mary: as the Woman of the Apocalypse, in depictions of the Assumption, the Immaculate Conception, and the Queen of Heaven (Warner 1976; Ostrow 1996a, 210–240; 1996b; Reeves 1997, 138–72). Although Cigoli's depiction of the Virgin in the dome of the Pauline Chapel has frequently been referred to as the *Immaculata* (Ostrow, 1996a, 236–40; 1996b; Reeves 1997, 138–48) and, less frequently, the *Assumption* (Edgerton 1990, 253), it is important to note that his charge was as follows:

> In the cupola is to be painted the Vision of the Apocalypse, chapter 12, that is, a Woman clothed with the Sun, and the Moon under her feet, and around her head a crown of twelve stars, opposite St. Michael the Archangel in the figure of a combatant, and [also] facing the three hierarchies, each divided into three orders. Below her is to emerge a *serpent with its head crushed as in chapter 3 of Genesis*, around [her] the twelve Apostles. (Matteoli 1980, 246; emphasis added)

Cigoli painted the Woman of the Apocalypse (fig. 1) as she was described in the instructions drawn up by the Oratorian Fathers, Tommaso and Francesco Bozio (Ostrow 1996a, 186–90), but with an interesting twist that has piqued the interest of historians of art and science alike: "the Moon under her feet" bears a remarkable likeness to one of the copper-plate engravings in Galileo's *Sidereus Nuncius* (fig. 2), and, as such, as Ostrow so aptly notes, it has been viewed as something of an "iconographic curiosity" (Ostrow 1996b, 223). It is an "astronomical moon" (hereafter referred to as a Galilean moon), clearly maculate, as opposed to a "religio-artistic moon," a perfectly smooth body without a spot. In the context of the rhetoric of the Counter-Reformation, Cigoli's painting of the Galilean moon beneath the feet of an obviously Marian figure would seem to be a serious infraction against the highly prescriptive artistic programs of the Church in which the Virgin was inevitably depicted as standing on an immaculate moon. The question we hope to resolve here is how the Moon under the Woman's feet in Cigoli's fresco functions symbolically. A related question we will also explore is why this apparent alliance between religious art and science occurred. We hope to demonstrate that the Galilean moon is not as unorthodox as it may seem, but rather a fitting part of the iconographic assemblage comprising the fresco. In short, we will argue Cigoli painted the Woman of the Apocalypse, not the *Immaculate Conception* or the *Assumption*.

Central to any discussion of Cigoli's painting is, of course, the close relationship between Galileo and Cigoli. But an exploration of other topics will also be necessary in order to arrive at a satisfactory answer to the questions posed by Cigoli's Moon: the role of *disegno* in observations and depictions of the Moon; the Moon in natural philosophy and as a subject of telescopic inquiry; its role as sign in

Fig. 1. Lodovico Cigoli, *The Woman of the Apocalypse*, 1612. Rome, Santa Maria Maggiore. Courtesy of Foto Vasari.

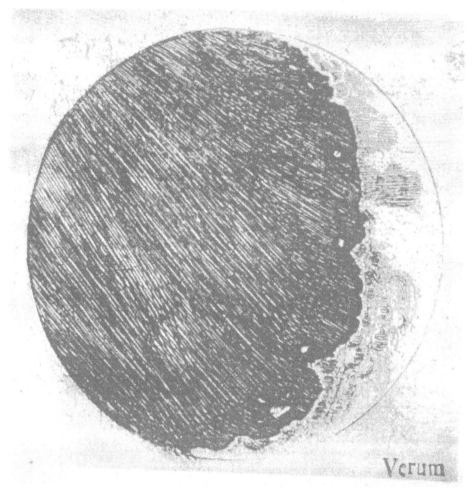

Fig. 2. Galileo Galilei, four-day old Moon. In *Sidereus Nuncius*, 1610. Courtesy of Wellesley College.

the Christian hermeneutic, or allegorical, tradition that allied Marian and lunar representations in art; and the role played by the Marian figure and the Moon in the political and artistic context in which Cigoli carried out his commission.

The Art of *Disegno* and Galileo's Moon

In his *Sidereus Nuncius* (1610), Galileo made a strong claim that, instead of being a perfectly smooth sphere, the Moon had a rough surface much like the Earth's. Although the argument was carried in the text, the four copper-plate engravings, based on Galileo's telescopic observations, provided important support for his argument. Over the past several decades, historians have proposed various theories to account for the nature of Galileo's lunar representations.

Terrie Bloom and Samuel Edgerton theorize that Galileo's training in *disegno*, which included drawing, composition, perspective, and chiaroscuro, enabled him to correctly interpret what he observed on the Moon through the telescope as relief on the lunar surface and accurately represent his observations. According to Bloom and Edgerton, the English scientist Thomas Harriot, who actually preceded Galileo in observing the Moon through a telescope, initially *could not* see this relief because Renaissance artistic conventions of perspective and chiaroscuro had not yet become common in England. However, after Harriot read Galileo's *Sidereus Nuncius* and observed the copper engravings, he could see relief and thus represent it in his own depictions of the Moon (Bloom 1978; Edgerton 1984; 1990, 235–37, 250–51). Scott Montgomery has argued that the representational conventions used by sixteenth-century map-makers, such as scalloped coastlines or shaded mountains, with which Galileo must have been familiar, helped him to *see* relief on the Moon (Montgomery 1996, 224–29),[1] whereas Mary Winkler and Albert Van Helden have argued that beginning with Galileo and Harriot, astronomers slowly developed a visual language of their own, culminating in Johannes Hevelius' *Selenographia* of 1647 (Winkler and Van Helden 1992, 1994).

However, Eileen Reeves's important book *Painting the Heavens: Art and Science in the Age of Galileo* (1997) puts Galileo's relationship with, and influence on, artists from Cigoli to Velasquez in a much wider context. She argues that, several years before the telescope Galileo, already a Copernican, looked at the Moon with the eye of an artist and realized that the "ashen" light of the Moon, i.e., the muted light seen before and after conjunction on the dark part of the Moon, was analogous to the muted light observed on objects illuminated by indirect light, the representation of which was a technique that all young artists had to learn. This realization enormously strengthened Galileo's belief in the Copernican theory in which the Earth was a planet, just like the reflecting Moon (Reeves 1997, 23–56, 91–137).

The Moon in Natural Philosophy

Although in Aristotelian natural philosophy the heavens are perfect and immutable and therefore the Moon is a perfectly spherical body, no one could argue that to the naked eye the Moon *looks like* a perfect body. It has spots. The nature of the Moon, then, had been the subject of speculation since Greek Antiquity. Among the pre-Socratic philosophers, Anaxagoras thought all heavenly bodies were fiery

---

[1] Amir Alexander, on the other hand, argues that Harriot's many lunar drawings reveal that his representations were conditioned by his practice of coastal mapping, which he had begun in the New World in the retinue of Sir Walter Raleigh (Alexander 1998).

stones. Diogenes of Apollonia thought that the planets, fixed stars, Sun and Moon consisted of red-hot pumice stone through the pores of which came rays of aether. Democritus thought that originally the Sun and Moon were not fiery, being made of the same kind of atoms as the Earth, but that later, in the process that led to the enlarging of the Sun's circle, fire came to predominate in them. He believed that both the Sun and Moon were made up of round atoms. Both Anaxagoras and Democritus believed that the Moon had mountains, valleys, and plains, much like the Earth (Dicks 1970, 59, 78, 82). It is also reported the Pythagorean philosopher Philolaus of Croton (ca. 410 BC) thought the Moon was like the Earth but had animals fifteen times larger than those of Earth because its day was fifteen times longer (Dicks 1970, 74).

However, Aristotle brushed aside the speculations of the pre-Socratics and insisted on a rigid distinction in kind between the heavens and the earthly realm. Up above, phenomena repeated themselves with perfect regularity; here below things changed: The heavens were perfect and immutable, whereas the region below the Moon was the seat of change and corruption. He argued this perfection in two ways, however: in Book I of *De Caelo* he related a heavenly body's degree of perfection to its distance from the Earth, whereas in Book II he related the degree of a body's perfection to simplicity of motion by which the sphere of the fixed stars, whose diurnal rotation was the simplest of all heavenly bodies, was most perfect (Aristotle 1930, 268.a–281.a; Grant 1985). In neither context did he discuss the Moon and its spots of which he was of course aware, because he stated that the Moon always keeps the same face turned toward the Earth (Aristotle 1930, 290.a.26). Finally, the Stoics thought that, although heavenly bodies were made up of very pure fiber or aether that pervades the entire upper region of space, the Moon was a mixture of air and gentle fire; and therefore not earthlike (Plutarch 1918, 922.d–923.e, 264–266).

Most of what we know about this subject comes from Plutarch's essay *On the Face that Appears on the Orb of the Moon*. Philosophers who believed the Moon was perfectly smooth and spherical, as befits a heavenly body, had put forward an ingenious argument to explain its spots: The Moon acts as a mirror, reflecting the ocean surrounding the earthly land mass. But Plutarch pointed out a reflection of the ocean would appear uniform without differences in shading, whereas the face of the Moon consists of lighter and darker areas (ibid., 920.f–921.d, 260–61). In addition, if the Moon were a perfect mirror of sunlight, then light would be reflected from her surface to us from one point only (ibid., 929.d–930.e, 278–80). We find this argument and its refutation again in Galileo's *Dialogue* of 1632, where Plutarch is not cited. Galileo's demonstration by means of mirrors calls to mind the eye of the artist (Galileo 1962, 69–80).

Plutarch also argued the Moon could not be composed of glass or crystal, for then the Sun's light would pass through it and solar eclipses would not be possible. On the contrary, the manner in which the Sun's light is, in fact, reflected from the lunar surface shows that its surface is like the Earth's:

Let us not then think that we offend in holding that [the Moon] is an earth, and that this her visible face, just like our earth with its great gulfs, is folded back into great depths and clefts containing water or murky air, which the light of the sun fails to penetrate or touch, but is obscured, and sends back its reflection here in shattered fragments. (Plutarch 1918, 935.c, 289)

Thus, both the Stoic notion that the Moon is a mixture of air and gentle fire, and the Aristotelian notion that it is made up of aether or the fifth element, are wrong. Plutarch puts it as follows:

It comes to this . . . Look on her as earth, and she appears a very beautiful object, venerable and highly adorned; but as star, or light, or any divine or heavenly body, I fear she may be found wanting in shapeliness and grace, and do no credit to her beautiful name, if out of all the multitude in heaven she alone goes round begging light of others. (Ibid., 929.a, 276–77)

The problem for Aristotle's followers was to reconcile several propositions. First, the Moon must be a perfectly smooth and spherical body; second, the Moon shines with light borrowed from the Sun; and third, the Moon shows dark spots. Because a perfectly smooth body, like a spherical mirror, would reflect the Sun's light from only one small area, one might conclude the lunar surface must be rough, like a wall reflecting sunlight. But because its surface *could not* be rough, the conclusion had to be that the Moon, although she shines with light borrowed from the Sun, does not reflect it. Rather the parts illuminated by the Sun absorb the light and then become self-luminous. The spots could then be explained by positing differences in what we might call optical density, whereby some parts absorb the Sun's light better than others. The parts that absorb less sunlight would thus radiate less light and appear darker to the viewer (Ariew 1984). This ingenious explanation is first recorded in the works of Alhazen (965-c. 1040), from whom it passed to Averroës (1126–1198). The influence of Averroës' ideas about the spots in the Moon can be traced from the thirteenth to the seventeenth century (Ariew 1984, 220–23), and it would therefore be an error to assume that the issue of the Moon's light and its spots was first raised in reaction to Galileo's discoveries.

The various problems of the Moon's appearance must not be underestimated. A comprehensive theory of its nature had to explain the following phenomena: the "ancient spots" visible to the naked eye; the "ashen light" before and after conjunction; the total blackness of the Moon during solar eclipses; and the reddish and sometimes grayish light during lunar eclipses. Kepler treated these problems in his *Astronomiae Pars Optica* of 1604. The arguments against the earthlike nature of the Moon, and their refutations, were all contained in Plutarch's tract, which was well known around the turn of the seventeenth century, and, indeed, Kepler appended a Latin edition of *On the Face that Appears on the Orb of the Moon* to his *Somnium*, which appeared posthumously in 1630 (Kepler 1630). What was new in Kepler's discussion was his explanation of the "ashen" light of the Moon. If this light was caused by the Moon's translucence, as Witelo had argued, or to the Moon's own light, as Erasmus Reinhold had argued, then how did one explain why the Moon appears totally dark during a solar eclipse? But in order to rule out translucence and proper light, Kepler had to explain both the "ashen light" and the reddish light during lunar eclipses. He attributed the former to reflection from the Earth and the latter to light refracted by the Earth's atmosphere into the Earth's shadow cone (Kepler 1604, 216–27, 234–47). Having disposed of these problems, Kepler could support Plutarch's argument that the Moon's surface was like the Earth's and had seas and continents, mountains and valleys.

Galileo's telescopic discoveries about the Moon (and to a lesser extent the phases of Venus) reopened the old scientific question of the nature of the Moon's surface. Galileo refuted the argument of Alhazen and Averroës by simply stating that his observations demonstrated the Moon's surface is not smooth but rough (Galileo 1989, 40–53) and that the light of the Sun is reflected from this rough surface as sunlight is from a brick wall; that is, if the Sun's light is reflected by the Moon, the lunar surface must be rough (Galileo 1962, 72–83). Of course, as Kepler's 1604 discussion shows, one hardly needed the telescope to make that argument. But for adherents to the Aristotelian cosmology who wished to preserve the perfection of the heavens, the arguments of Galileo (and Kepler) were not conclusive. All Galileo had shown was that the dark outlines of the new small spots revealed in the brighter part of the Moon changed over time as the angle of illumination from the Sun changed. If one accepted these observations, did it necessarily mean that the lunar surface is rough and uneven and that there are mountains on the Moon? Could not these new phenomena be explained by citing the old argument of "rarity and density?" (Ariew 1984, 223–25).

Galileo's own utterings in *Sidereus Nuncius* posed a question to the skeptical and suggested openings for attack. If there were mountains on the Moon, why then did its periphery appear perfectly circular? Galileo gave two reasons. First, the limb is seen tangentially across a number of mountain ranges so we see only the tops of the ridges, not the valleys; these tops merge with each other to form a periphery very close to exactly circular. Second, there was perhaps an atmosphere around the Moon, and when we observe the Moon's limb we are looking obliquely through this atmosphere, which tends to absorb light as our atmosphere does on Earth (Galileo 1989, 48–51). This last argument was subsequently dropped by Galileo when he could find no other evidence of a lunar atmosphere (Galileo 1962, 100).

Of all Galileo's initial discoveries, the earthlike nature of the Moon raised the most difficult questions. The letter of 24 March 1611 from the mathematicians of the Collegio Romano to Cardinal Bellarmine concerning Galileo's telescopic discoveries gives us some insight into how the problem of the Moon was regarded by Church mathematicians. Fathers Clavius, Grienberger, van Maelcote, and Lembo were asked for their opinions about these discoveries and replied in some detail. Regarding the Moon, they wrote:

The great inequality of the Moon cannot be denied. But it appears to Father Clavius more probable that the surface is not uneven, but rather that the lunar body has denser and rarer parts, as are the ordinary spots seen with the natural sight. Others think that the surface is indeed uneven, but thus far we are not certain enough about this to confirm it indubitably. (Galileo 1890–1909, XI: 93, idem 1989, 111)

The three younger mathematicians apparently agreed with Galileo; however, the aged Father Clavius (1537–1612), who had initially been skeptical of Galileo's discoveries, wished to preserve the traditional interpretation. Clavius had been the touchstone of astronomical orthodoxy in the Jesuit order for several generations, arguing against the homocentric spheres of Fracastoro, the Copernican theory, and the notion of a fluid heaven in which the planets moved "like birds in the air or fish in the sea" (in Lattis 1994, 94–102). Now Galileo's discoveries posed new problems. In the last edition of his enormously influential *Commentary on the Sphere of Sacrobosco*, Clavius added a brief passage about these new celestial discoveries in which he called Galileo's *Sidereus Nuncius* a "very reliable little book." About the Moon, Clavius wrote: "And when the moon is a crescent or half full, it appears so remarkably fractured and rough that I cannot marvel enough that there is such unevenness in the lunar body." And he ended the passage as follows: "Since things are thus, astronomers ought to consider how the celestial orbs may be arranged in order to save these appearances" (Clavius 1612, 3:75; in Lattis 1994, 198). Until the very end, Clavius was careful to speak as a "mathematician" only.

In the meantime, all the philosophers could do was rehearse variations of the old explanation of the Moon's spots. Father Clavius' explanation that the Moon had "denser and rarer parts" had been entirely qualitative, but Galileo's calculation of the heights of lunar mountains added a quantitative dimension to the problem. The traditional explanation now had to account for mountains four miles high, and moreover mountains covered by a transparent, perfectly spherical layer.

The alternative explanation of the naked-eye spots of the Moon — that the Moon's surface is rough like the Earth's — was revived by artists and scientists in the Renaissance. But it was Leonardo da Vinci who bridged the worlds of art and science. In the case of the Moon, he was familiar with, and rejected, the standard arguments of the philosophers to explain its spotted appearance. He argued instead that the Moon is not self-luminous; rather, it must act as a spherical mirror that reflects the Sun's light to the Earth. However, this mirror was not smooth; otherwise, sunlight would be reflected to us from only one point on its surface. Instead, the Moon's surface was made up of areas of land and seas, and the water of the seas was disturbed by waves:

The skin or surface of the water which comprises the sea of the moon . . . is always ruffled, little or much, more or less, and this roughness is the cause of the proliferation of the innumerable images of the sun which are reflected in the ridges and concavities and sides and fronts of the innumerable waves. (In Kemp 1981, 324)[2]

Thus, because the Moon shines with borrowed light, its surface must be rough. Leonardo compared this rough lunar surface with that of the Earth, arguing that the Earth's surface, too, reflected the Sun's light. The "ashen light" of the Moon was therefore reflected light from the Earth, and the Earth was in this respect no different from the planets (in Kemp 1981, 324–5).

Leonardo's arguments can be found again in Kepler's *Optics* of 1604 (Kepler 1604, 202, 223–24) and Galileo's *Dialogue* of 1632 (Galileo 1962, 67–73). Reeves has pointed out that copies of Leonardo's notebook on painting circulated in northern Italy in the late sixteenth century and it is not unreasonable to suppose that Galileo may have been familiar with Leonardo's argument from another notebook (Reeves 1997, 29–31, 114–118). Reeves also argues that the explanation of the Moon's "ashen light" was rather obvious to anyone who was thoroughly familiar, as Galileo was, with artistic techniques used to depict reflected light. Some astronomers, such as Kepler and Maestlin,[3] arrived at the explanation of the "ashen light" by *reasoning* from the Copernican assumption that the Earth was a planet and could therefore reflect light as the Moon did. Some artists, such as Leonardo (and perhaps Cigoli), started from their own *experience* in representing reflected light and on the basis of this experience concluded that the "ashen light" of the Moon was reflected "earth-shine," an indication that the Earth was like the planets. However, because of his training in both art and science, Galileo could make the argument starting from either side, and, as Reeves states, the artistic approach to the problem strengthened his Copernican convictions (Reeves 1997, 138–183).

Cigoli (1559–1613) and Galileo (1564–1642) had been friends since their youth, when they both took lessons in mathematics and perspective from Ostilio Ricci (Galileo 1890–1909, XIX: 604; Matteoli 1980, 21) and they remained in close contact throughout their lives. Unfortunately, most of the letters from Galileo to Cigoli are lost, but many of Cigoli's letters to Galileo have survived. It is clear from these that Cigoli was vitally interested in Galileo's science, and especially his astronomy. Reeves has argued that in his *Annunciation* of 1607, Cigoli showed the "ashen light" of the Moon at about the time Galileo arrived at his explanation (Reeves 1997, 91–137). But Cigoli was not a learned man in the traditional sense, although he wrote an excellent treatise on perspective, *Trattato di Prospettiva Prattica*, which has been published at last (Cigoli 1992). In October 1610, he wrote to Galileo that he had not yet seen *Sidereus Nuncius*, and that if he had seen it he

---

[2] Original text: British Library, MSS Arundel 263, 94v.

[3] In his *Astronomia pars Optica*, Kepler quotes from a (now lost) *Disputation on Eclipses* published by Maestlin in 1596, in which Maestlin explains the "ashen" light of the Moon as reflected light from the Earth (Kepler 1604, 223–24).

would not have understood it because it was in Latin.[4] He therefore urged Galileo to issue an Italian version (Galileo 1890–1909, X: 442). However, Cigoli did have a view of the relationships between "mathematics" and art. In the summer of 1611, he read a copy of the letter of the mathematicians of the Collegio Romano in which they stated their official judgment on Galileo's lunar observations (see above) and he wrote to Galileo about Father Clavius' thoughts about the Moon:

> I was most astonished by the opinion of Father Clavius about the Moon: that he doubts its unevenness because it appears to him more probable that it is not of uniform density. Now, I have thought and thought about this, and I find nothing to say in his defense except that, be he as great as he wants, a mathematician without *disegno* is not only a mediocre mathematician, but also a man without eyes. (Ibid., XI: 168)

According to Cigoli, then, one had to know drawing, or *disegno*, to be a complete mathematician. In this judgment, Galileo surely would have agreed with him. *Disegno* went together with the study of geometry and perspective, as Galileo and Cigoli had experienced it under Ostilio Ricci. Only those who did not combine these skills could be foolish enough to try to explain away the evidence of their senses by postulating invisible substances.

Cigoli, then, was intimately familiar with Galileo's argument concerning the lunar surface and the engravings of the Moon in *Sidereus Nuncius*. We may assume that before he eventually obtained his own telescope in 1612 (Galileo 1890–1909, XI: 287) he looked through the telescopes of others. And he incorporated Galileo's lunar discoveries into his own work, when, in 1610, he was commissioned to decorate the cupola of the Pauline (or Borghese) Chapel in the Basilica of Santa Maria Maggiore in Rome.

The Moon and the Allegorical Tradition

As Winkler and Van Helden have pointed out, the paucity of realistic representations of the Moon[5] before the telescope is somewhat of a puzzle (Winkler and Van Helden 1992). In the past several decades, a few pre-telescopic realistic lunar representations have been brought into the mainstream of scholarship in the history of science: In 1965, Sister Suzanne Kelly reprinted Gilbert's *De Mundo nostro Sublunari* written around 1600 (but not printed until 1651) which contains a naked-eye lunar map (Kelly 1965, 2:172–3); in 1987, Gibson Reaves and Carlo Pedretti called attention to three Leonardo drawings, one of which is a particularly realistic representation of the half-Moon (Reaves and Pedretti 1987);[6] and in 1994

---

[4] In his biography of his uncle (1628), Giovanni Battista Cardi wrote that Cigoli had been introduced to the Latin language but showed no interest in it (Matteoli 1980, 19).

[5] We are ignoring the various diagrams of the Moon in astronomical texts.

[6] The drawings are in *Codex Atalanticus*, ff. 310r and 674v, and *Codex Leicester*, f. 2r.

Scott Montgomery added the striking (if small) renderings of the Moon by Jan van Eyck (Montgomery 1994; idem 1996, 202–6). Artists had, however, painted the Moon in various non-naturalistic guises for centuries before the telescope. The most important was in the context of depictions of the Virgin Mary, a relationship derived from the exegetical interpretations, or glosses, of the Old and New Testaments. The prescriptive programs of Christian art and architecture derive from this view of Christian oratory, whose primary function was to interpret and preach the holy word.

In the Christian allegorical tradition, the Old and New Testaments were regarded as a harmonious whole; hence, the figures and events of the Old Law (Old Testament: the letter or "figura") prefigured those in the New Law (the New Testament: the spirit or "fulfillment"). In other words, the Old Testament was to be read not only as history (literally) but also as a series of signs, or predictions, of what was to be fulfilled in the New (allegorically) (Mâle 1972, 133–34).[7] For example, Jerusalem is literally the city, the "figura," in the Old Testament that prefigures the New Jerusalem, the "fulfillment," in John's vision of the Apocalypse in the New Testament. Augustine points out, however, that figurative signs are polysemic, that what they stand for can change according to the scriptural passages — the context — in which they occur. The variation of the figure can take two forms: it can be used in "a good sense, *in bono*," or "in an evil sense, *in malo*." Consequently the figurative sign "lion" is to be understood *in bono* in "the lion of the tribe of Judah . . . has prevailed" (*Apocalypse* 5.5), but *in malo*, as a sign of the Devil, in "your adversary the devil, as a roaring lion, goeth about seeking whom he may devour" (1 *Peter* 5.8; *On Christian Doctrine* III. XXV, 35–37). Thus, the Moon as sign of the Virgin's purity would signify the Moon *in bono*, whereas the Moon as sign of sublunary corruption, for example, of the infidel, or of the heresy of the Reformation, is the Moon *in malo*. As will become apparent below, this distinction is crucial to an analysis of Cigoli's Moon (see also Ostrow 1996b).

One traditional depiction of the Virgin and the Moon — The Woman of the Apocalypse — is based on various glosses of chapter 12 of the book of *Revelations*, or *Apocalypse*. The Woman of the Apocalypse, according to exegetes, is the "fulfillment" of the prophesy in the Old Testament, the "figura," in *Genesis* 3. 15: "I will put *enmities* between thee and the woman [Eve], and thy seed and her seed: she shall crush thy head, and thou shalt lie in wait for her heel" (in Reeves 1997, 142; emphasis added). The first four verses of the Twelfth Chapter of the *Apocalypse* (taken here from the Douai Bible) are as follows:

1. And a great sign appeared in heaven: A woman clothed with the sun, and the moon under her feet, and on her head a crown of twelve stars:
2. And being with child, she cried travailing in birth, and was in pain to be delivered.

---

[7] Umberto Eco notes Aquinas's statement that the authors of the Old Testament were not aware they were writing prophecy as they labored under divine inspiration (Eco, 155).

3. And there was seen another sign in heaven: and behold a great red dragon, having seven heads, and ten horns: and on his heads seven diadems:

4. And his tail drew the third part of the stars of heaven, and cast them to the earth: and the dragon stood before the woman who was ready to be delivered; that, when she should be delivered, he might devour her son. (*Douai Bible*, 288)

This passage was glossed in a number of ways. Methodius of Philippi (third century CE) in *The Symposium of the Ten Virgins* interpreted the woman as the Mother Church. The Sun in which she was clothed was Christ and His light illuminated the Church and the Moon, whose reflected light symbolized the mediating power of the Church and through its continuing cycle of waxing and waning the cycle of life. In this interpretation the Moon is thus seen as a sign *in bono*. Other exegetes identified the Woman as the Virgin Mary who crushed beneath her feet the imperfect (maculate) Moon which at various times throughout the history of Christianity signified the excessive materialism of the Church itself, the crescent of the infidel Saracens, or during the Counter-Reformation the Protestant heresy (Ostrow 1996a, 243 ff.; Reeves 1997, 139–40). In this case, then, the Moon was to be seen *in malo*. (Figures 3 and 4 show medieval miniatures of

Fig. 3. *The Woman of the Apocalypse.* From a French manuscript, c. 1320. New York: Cloisters MSS 68, f. 20. With permission.

Fig. 4. *The Woman of the Apocalypse.* From an English manuscript, c. 1250. New York: Pierpont Morgan Library, MSS 524, 7E5, 8v. With permission.

what came to be called "The Virgin of the Apocalypse," faithfully representing the *Apocalypse* 12:1–4, in which the Moon signifies the Infidel.)[8]

In the seventeenth century the doctrine of the Immaculate Conception was especially strong in Spain, where a canon for representing the Virgin as such was worked out by Francisco Pacheco, Diego Velàzquez, Bartolomé Murillo, and others, whose prescription was a melding of several traditions in which the Virgin, clothed in the Sun, was to be standing on the Moon. However, the dragon was to be omitted and eventually the Virgin's crown of twelve stars was omitted as well (Warner 1976, 246–48). Finally, there is the depiction of Mary in paintings of the Assumption. The notion that Mary, who did not suffer corruption by union of the flesh and therefore could not suffer dissolution of the body developed in the West into the notion that the Virgin's body ascended to heaven, and the feast of the Assumption became a major religious festival in the late Middle Ages.[9]

All the aforementioned images of the Moon allied with the Virgin were present at the beginning of the seventeenth century and have presented problems for historians who have dealt with how the spotted Moon in Cigoli's painting in Santa Maria Maggiore is to be interpreted. We will propose a solution below.

In many seventeenth-century representations of the *Immaculata*, the Virgin Mary is depicted as a young, beautiful maiden, arms folded across her breast. Her hands are usually pressed together in prayer, her eyes modestly cast down. Her feet, if they are visible at all, are dainty and point downward (fig. 5). In depictions of the Assumption, the Virgin Mary, her expression rapt, is represented as a lovely woman being borne aloft on a luminous cloud, sometimes by winged angels (fig. 6). But note, in Cigoli's fresco the Virgin is depicted as a mature woman. One hand gathers up her gown, while in her other hand she holds a blossoming scepter, the symbol of Jesus' power that he himself has given to her (Mâle 1949, 81).[10] There is nothing dainty or light about her feet: they are heavy and solid, as is her body, and they are planted firmly on the maculate Moon, beneath which lies the coiled Serpent. Note also the expression on her face: it is one of supplication, not ecstasy, as is often seen in depictions of the Assumption. Here, she is not only the Church Militant but Mediatrix, who "pray[s] for us now and at the hour of our death."

Given the contrast between the three Marian representations discussed above, how can we arrive at a satisfactory interpretation of the iconographic weight carried by Cigoli's Moon? First, as Arthur Danto has argued: "A [painting's] title

---

[8] The Woman on the Moon could also be seen as the Immaculate Conception. This doctrine has a very long history, beginning in the Middle Ages and culminating in Pope Pius IX's Bull, *Ineffabilis Deus* (1854), which declared as Church dogma that Mary was untainted by original sin, i.e., immaculately conceived (Warner 1976, 236–38).

[9] In 1950 the Assumption was proclaimed an article of faith by Pope Pius XII (Warner 1976, 81–102). The key text in the mass for the celebration of the Assumption, however, was Chapter 12 of the *Apocalypse* (Warner 1976, 93).

[10] Mâle makes this identification in his description of "The Coronation of the Virgin," a thirteenth-century tympanum in Notre Dame, Paris. The flowering staff can also represent the rod of Jesse: "And there shall come forth a rod out of the root of Jesse, and a flower shall rise up out of his root" (*Isaiah* 11:1).

Fig. 5. Diego Velasquez, *The Immaculate Conception*, c. 1619.
Courtesy of the National Gallery, London.

Fig. 6. Peter Paul Rubens, *The Assumption*, 1626. Courtesy of Cathedral of Our Lady, Antwerp.

is more than a name or a label; it is a *direction* for interpretation" (Danto 1981, 117). By using the very instructive example of Breugel's *Landscape with the Fall of Icarus*, he points out that a viewer who does not know the title of the painting would not understand that the legs in the painting "are the focus of the whole work ... in the sense that the whole structure of the painting is a function of these being Icarus' legs" (Danto 1981, 118). Given that the title of a painting directs the viewer's interpretation, we must exercise caution in accepting the titles, *The Immaculate Conception* (Reeves 1997, *passim*; Chappell 1975, 93, and 1992; Kemp 1990, 94; Matteoli 1980, 245) or *The Assumption* (Panofsky 1954, 5 and 1956, 3–4; Edgerton 1991, 253). If we do not, we may, like the naive viewer of Breugel's *Landscape with the Fall of Icarus*, not understand that the structure of Cigoli's painting is a function of the Galilean Moon. Furthermore, in the interpretation of scriptural passages on which paintings of the Immaculate Conception and the Assumption are based, the Virgin is compared to the Moon because both are "spotless," whereas the Moon in the Woman of the Apocalypse is compared to the mutability and evil of the sublunary world.

This is an important distinction in terms of how the Virgin and the Moon would be apprehended. In depictions of the Virgin as Immaculata or in the Assumption, for example, Mary is *like* the Moon, "as beautiful as the Moon" (Song of Songs 6:10; Reeves 1997, 142), as pure and perfect as the Moon. The two function together as what Eco in his discussion of Thomistic aesthetics refers to as a "pictorial simile." It would be difficult to imagine how Cigoli's maculate Moon and the Woman could function as a pictorial simile of either the Immaculate Conception or the Assumption. In addition, as Victor Lasareff notes, it is necessary to attend to the ideas expressed by the form, i.e., the entire iconographic program.

Clearly, the ideas expressed by the form, the iconographic program of Cigoli's fresco, are at odds with those expressed by representations of the Virgin as Immaculata and in the Assumption. The latter expresses the ideal nature of the Virgin, the former the fate of whatever is anathema to the Church and, therefore, to God. Heresy, like the serpent under the Woman's foot, will be crushed. Hence, Cigoli's Moon, as part of what Lasareff terms "a complete iconographic scheme" (Lasareff 1938, 26–28), here that of the Woman of the Apocalypse, is quite an orthodox piece of the scheme. That is to say, its Galilean character is clearly fitting within the program in which it appears.

Note also that in their charge to Cigoli (see above) the brothers Bozio refer specifically to *Genesis* chapter three, in which, according to traditional exegetical readings, Eve prefigures the Woman of the Apocalypse. Immediately after specifying Cigoli's task, they designated how the fresco was to be read, clearly referring to the Woman's prefigurement in *Genesis* chapter three, citing her own "genealogy," if you will, citing both figura and fulfillment:

> As Andrea Cesariense and St. Methodio have it, such a Woman signifies the Church; and according to St. Bernard and many other Latin writers, the Madonna literally signifies the Church no less than the Madonna who, from the beginning of the World, manifested with the Angels through the Incarnation, fights until the end of the World, triumphing in Heaven. And thus the first prophesy uttered in the creation of the World, against the Serpent who signifies the devil, "and she shall crush thy head" [*Gen.* 3:15], pertains to her (in Ostrow 1996a, 280).

Cigoli's fresco cannot be regarded as the "*tota pulchra*" type described in Reeves (Reeves 1997, 142). The Woman is clearly not depicted as "beautiful as the Moon" (Song of Songs 6:10, in Reeves 1997, 142). Nor is this a depiction of the Virgin in the fullness of her life being bodily, and peacefully, assumed into heaven. This is the Virgin as Church militant, crushing evil under her feet. We must, therefore, call this fresco, as did the Oratorian fathers, the *Woman of the Apocalypse* and read the Moon as evil and corrupt.

There is little doubt that in the Counter-Reformation climate, for Paul V and his scholars, the Moon and Serpent in this fresco depicting the Apocalypse symbolized the Protestant heretics, those who denied and ridiculed the cult of the Virgin. Cigoli's Moon may thus be viewed as a coincidental interface between his intentions to paint a Galilean Moon, and the Church's intention to depict a frightening scene. The rhetoric of the Counter-Reformation was not only designed to demonstrate the errors of the ways of the (Protestant) heretics but also to portray the fate of those who did not adhere to doctrine. The Apocalyptic Woman is, after all, part of the Last Judgment. Such pictorial programs have always been used by the Church, not only to teach, but to manipulate the faithful to act according to received doctrine. As Ostrow has demonstrated, the frescoes of the Sistine and Pauline chapels of Santa Maria Maggiore must be read in the context of the Counter-Reformation (Ostrow 1996a). In importing Galileo's spotted Moon into this religious theme, Cigoli successfully put the new astronomical Moon in the service of an apocalyptic Counter-Reformation program.

This, however, raises the question as to why no artist after Cigoli painted the Moon as he did. In the history of representations of the Virgin Mary, his depiction of a maculate Moon is unique. Indeed, lunar iconography in this context tended to disappear in Italian art, and only in Spain was it pursued through the seventeenth century. In the succession of Marian representations there, we can see how the different traditions discussed above merged into a vision of the Immaculate Conception, which jettisoned some, but also retained other symbols associated with the Woman of the Apocalypse (Reeves 1997, 184–212). Thus Giambattista Tiepolo's paintings of the Immaculate Conception (fig. 7) for Italian patrons differ radically in this respect from his one Spanish commission (Levey 1986, 274–83) in that he closely follows Francisco Pacheco's prescription for such depictions of the Virgin, who must be shown "in the flower of her age, between twelve and thirteen years old, very beautiful, with lovely and solemn eyes, a perfect nose and mouth, rosy cheeks, and with hair as close to gold as the paintbrush will allow," and with her feet on the Moon (Reeves 1997, 194). Reeves has argued that this artistic vision of the Moon in the Spanish tradition is related to the efforts by scholars such as Giulio Cesare Lagalla and the Jesuits Christoph Scheiner and François d'Aguillon, to save the perfection of the Moon (Reeves 1997, 196–212). If eventually Catholic mathematicians and philosophers had to abandon lunar perfection, their thoughts lived on, frozen as it were, in the Moons of the Immaculate Conceptions painted by the artists.

Cigoli observed the Moon through a telescope and joined a new astronomical vision of a Moon with a rough surface full of mountains and valleys to a long artistic tradition of the Virgin Mary on the Moon, just as the cult of the Immaculate Conception was becoming ever more powerful in the Counter-Reformation. At about the same time that Cigoli began painting the dome of the Pauline Chapel, Adam Elsheimer painted the *Flight into Egypt*, in which he showed an unmarked full Moon in the heavens while its reflection in the water was covered with spots (Byard 1988). But in this respect Cigoli and Elsheimer were the exceptions among artists. The maculate Moon did not become part of any artistic tradition, and even among astronomers it was some time before lunar representation was taken up seriously. Galileo's proposal to the Medici Court, in 1610, to depict and publish

Fig. 7. Giovanni Battista Tiepolo, *The Immaculate Conception*, c. 1735.
Courtesy of the Prado Museum, Madrid.

every phase of the telescopic Moon, for which he needed financial support from the Grand Duke, was never executed (Galileo 1890–1909, X: 300).

Conclusion

Galileo's Moon renewed an age-old problem about the nature of the Moon. Philosophers and astronomers committed to the Aristotelian cosmology searched for explanations of Galileo's celestial discoveries within that paradigm. Clavius and others postulated a completely transparent, that is invisible, layer of celestial material that covered Galileo's lunar mountains. But this was an inferior argument, and almost all scholars realized it. In the Christian-Aristotelian cosmology, the centrality of the Earth was supported by biblical passages. But one could not derive Aristotle's cosmology from Scripture, which said nothing about the perfection of the heavens or the Aristotelian spheres. Cardinal Bellarmine himself had, earlier in his life, constructed a cosmology entirely based on Scripture, and it looked nothing like Aristotle's universe (e.g., Baldini 1984). The question of the Moon's perfection must, therefore, not be too closely tied in our minds to the question of Copernicanism. In a biblical context the Moon could be interpreted either as a sign of purity and immaculateness or of corruption and maculateness. In the context of the interpretation of *Apocalypse* 12, the latter was the case. Regardless of whether Cigoli himself believed the maculate Moon was evidence for a heliocentric universe, the fact is that it fit in admirably with the charge written by the Brothers Bozio. Thus, his moon cannot be viewed as an "iconographic curiosity." Nor was he attempting to surreptitiously import the Galilean moon into the basilica. On the contrary, his rendition of the Woman of the Apocalypse so pleased the Pope that in 1613, just before the artist's untimely death, he had Cigoli made a member of the Order of the Knights of Malta (Matteoli 1980, 34–35).

Acknowledgments

Samuel Y. Edgerton, Jr., Steven F. Ostrow, and Eileen Reeves read a previous version of this paper. We thank them for their generous help. They are not responsible for any errors we may have made. We also thank Jet M. Prendeville, Head Librarian of the Alice Pratt Brown Library at Rice University for her help.

References

Alexander, Amir. 1998. "Lunar Maps and Coastal Outlines: Thomas Harriot's Mapping of the Moon." *Studies in History and Philosophy of Science* 29:345–68.

Alhazen. 1924–25. "Über das Licht des Mondes," translated by Karl Kohl. *Sitzungsbericht der Physikalisch-medizinischen Societät in Erlangen* 56–57:305–98.

——. 1925. *Über die Natur der Spuren [Flecken], die man auf der Oberfläche des Mondes sieht,* translated by Carl Schoy. Hannover: Heinz Lafaire.

Ariew, Roger. 1984. "Galileo's Lunar Observations in the Context of Medieval Lunar Theory." *Studies in History and Philosophy of Science* 15:213–26.

Aristotle. 1930. *De Caelo,* translated by J. L. Stocks. In *The Works of Aristotle,* vol. 2, edited by W. D. Ross. Oxford: Clarendon Press.

Augustine. 1958. *On Christian Doctrine,* translated by D. W. Robertson, Jr. New York: Macmillan.

Baldini, Ugo. 1984 [1983]. "L'astronomia del Cardinale Bellarmino." In *Novità Celesti e Crisi del Sapere: Atti del Convegno Internazionale di Studi Galileiana,* edited by Paolo Galluzzi. Monograph 7, *Annali dell'Istituto e Museo di Storia della Scienza.* Florence.

Bible. 1899. *The Holy Bible translated from the Latin Vulgate diligently compared with the Hebrew, Greek, and other editions in diverse languages. The Old Testament first published by the English College at Douay, A. D. 1609 and The New Testament first published by the English College at Rheims, A. D. 1582.* Baltimore and New York: John Murphy.

Bloom, Terrie F. 1978. "Borrowed Perceptions: Harriot's Maps of the Moon." *Journal for the History of Astronomy* 9:117–22.

Byard, Margaret M. 1988. "Galileo and the Artists," *History Today* 38 (Feb. 1988):30–38.

Chappell, Miles. 1975. "Cigoli, Galileo, and *Invidia.*" *Art Bulletin* 57:91–98.

——. 1992. "Prefazione," in Marco Chiarini, Serena Padovani, and Angelo Tartuferi, *Lodovico Cigoli, 1559–1613: tra Mannerismo e Barocco,* 11–19. Catalog of an exhibit in the Palazzo Piti, 19 July to 18 October. Florence: Amalthea.

Cigoli, Lodovico Cardi da. 1992. *Trattato pratico di prospettiva di Ludovico Cardi detto il Cigoli : manoscritto Ms 2660A del Gabinetto dei disegni e delle stampe degli Uffizi a cura di Rodolfo Profumo.* Rome: Bonsignori.

Clavius, Christopher. 1612. *Commentarius in Sphaeram Ioannis de Sacrobosco,* vol. 3. In *Opera Mathematica,* 5 vols. Mainz.

Danto, Arthur C. 1981. *The Transfiguration of the Commonplace.* Cambridge: Harvard University Press.

Dicks, D. R. 1970. *Early Greek Astronomy to Aristotle.* London: Thames and Hudson.

Duhem, Pierre. 1913–59. *Le Système du monde: Histoire des doctrines cosmologiques de Platon à Copernic,* 10 vols. Paris: Hermann.

—— 1985. *Medieval Cosmology: Theories of Infinity, Place, Time, Void, and the Plurality of Worlds*, translated and edited by Roger Ariew. Chicago: University of Chicago Press.

Eco, Umberto. 1988. *The Aesthetics of Thomas Aquinas*, translated by Hugh Bredin. Cambridge: Harvard University Press.

Edgerton, Samuel Y., Jr. 1984. "Galileo, Florentine Disegno, and the 'Strange Spottedness' of the Moon." *Art Journal* 44:225–33.

——. 1990. *The Heritage of Giotto's Geometry: Science and Art on the Eve of the Scientific Revolution*. Ithaca: Cornell University Press.

Galilei, Galileo. 1962. *Dialogue Concerning the Two Chief Systems of the World, Ptolemaic and Copernican*, translated by Stillman Drake. Berkeley: University of California Press.

——. 1890–1909. *Le Opere di Galileo Galilei*, edited by Antonio Favaro. Florence: Barbèra.

——. 1989. *Sidereus Nuncius or The Sidereal Messenger*, translated by Albert Van Helden. Chicago: University of Chicago Press.

Grant, Edward. 1985. "Celestial Perfection from the Middle Ages to the Late Seventeenth Century." In *Religion, Science, and Worldview: Essays in Honor of Richard S. Westfall*, edited by M. J. Osler and P. L. Farber, 137–62. Cambridge: Cambridge University Press.

Hevelius, Johannes. 1647. *Selenographia sive Lunae Descriptio*. Gdansk.

Kelly, Sister Suzanne, O.S.B. 1965. *The De Mundo of William Gilbert*, 2 vols. Amsterdam: Menno Hertzberger.

Kemp, Martin. 1981. *Leonardo da Vinci: The Marvellous Works of Nature and Man*. Cambridge, Harvard University Press.

——. 1990. *The Science of Art: Optical Themes in Western Art from Brunelleschi to Seurat*. New Haven: Yale University Press.

Kepler, Johannes. 1604. *Ad Vitellionem Paralipomena, quibus Astronomiae Pars Optica Traditur*. In *Johannes Kepler Gesammelte Werke*, vol. 2. Munich: C. H. Beck (1937–).

——. 1630. *Somnium seu Astronomia Lunari*. In *Gesammelte Werke*, vol. 11, part 2. Munich: C. H. Beck.

Lasareff, Victor. 1938. "Studies in the Iconography of the Virgin." *Art Bulletin* 20:26–65.

Lattis, James M. 1994. *Between Copernicus and Galileo: Christoph Clavius and the Collapse of the Ptolemaic System*. Chicago: University of Chicago Press.

Levey, Michael. 1986. *Giambattista Tiepolo: His Life and Art*. New Haven: Yale University Press.

Mâle, Émile. 1949. *Religious Art from the Twelfth to the Eighteenth Century*. New York: Pantheon.

——. 1972. *The Gothic Image: Religious Art in France of the Thirteenth Century*, translated by Dora Nussey. New York: Harper and Row.

Matteoli, Anna. 1980. *Lodovico Cardi-Cigoli. Pittore e Architetto*. Pisa: Giardini.

Montgomery, Scott L. 1994. "The First Naturalistic Drawings of the Moon: Jan Van Eyck and the Art of Observation." *Journal for the History of Astronomy* 25:317–20.

——. 1996. *The Scientific Voice*. New York: Guilford Press.

Ostrow, Steven F. 1996a. *Art and Spirituality in Counter-Reformation Rome*. Cambridge: Cambridge University Press.

——. 1996b. "Cigoli's *Immacolata* and Galileo's Moon: Astronomy and the Virgin in the Early Seicento." *Art Bulletin* 78:219–35.

Panofsky, Erwin. 1954. *Galileo as a Critic of the Arts: Aesthetic Attitude and Scientific Thought*. The Hague: Martinus Nijhoff.

——. 1956. "Galileo as a Critic of the Arts: Aesthetic Attitude and Scientific Thought." *Isis* 47:3–15.

Plutarch. 1918. *On the Face that Appears on the Orb of the Moon*. In *Selected Essays of Plutarch*, translated by A. O. Prickard, 2:246–312. Oxford: Clarendon Press.

Reaves, Gibson and Carlo Pedretti. 1987. "Leonardo da Vinci's Drawings of the Surface Features of the Moon." *Journal for the History of Astronomy* 18:55–58.

Reeves, Eileen. 1997. *Painting the Heavens: Art and Science in the Age of Galileo*. Princeton: Princeton University Press.

Trabucco, Agostino. 1957. "La 'Donna ravvolta di sole' (Apoc. 12) L'interpretazione ecclesiologica degli esegeti cattolici dal 1563 alla prima metà del secolo XIX." *Marianum* 19:1–58.

Van Helden, Albert, and Mary G. Winkler, see Winkler, Mary G. and Albert Van Helden.

Warner, Marina. 1976. *Alone of All Her Sex: The Myth and the Cult of the Virgin*. New York: Knopf.

Winkler, Mary G. and Albert Van Helden. 1992. "Representing the Heavens: Galileo and Visual Astronomy." *Isis* 83:195–217.

——. 1994. "Johannes Hevelius and the Visual Language of Astronomy." In *Renaissance and Revolution: Humanists, Scholars, Craftsmen, and Natural Philosophers in Early Modern Europe*, edited by J. V. Field and Frank A. J. L. James. Cambridge: Cambridge University Press.

University of Houston and Rice University

3. The Contexts of the Church, Patrons, and Colleagues: New Science and Traditional Power Structures

RIVKA FELDHAY

Recent Narratives on Galileo and the Church: or The Three Dogmas of the Counter-Reformation

The Argument

This article confronts an old-new orientation in the historiographical literature on the "Galileo affair." It argues that a varied group of historians moved by different cultural forces in the last decade of the twentieth century tends to crystallize a consensus about the inevitability of the conflict between Galileo and the Church and its outcome in the trial of 1633. The "neo-conflictualists" — as I call them — have built their case by adhering to and developing the "three dogmas of the Counter-Reformation": Church authoritarianism is portrayed by them as verging towards "totalitarianism." A preference for a literal reading of the Scriptures is understood as a mode of "fundamentalism." And mild skeptical positions in astronomy are read as expressions of "instrumentalism," or "fictionalism." The main thrust of the article lies in an attempt to historicize these three aspects of the Catholic reform movement. Finally, the lacunae in insufficiently explored historiographical landscape are delineated in order to tame the temptation to embrace the three dogmas, and to modify the radical conflictualist version of the story of Galileo and the Church.

Introduction

In a fairly recent book, *Galileo, Bellarmine, and the Bible*, Richard Blackwell has taken upon himself the challenging task set forward by Olaf Pedersen in the 1980s: to study the Galileo affair not only from the perspective of historians of science, but also from that of the history of theology. Blackwell has focused his gaze on one theological issue, namely the role played by the Bible in the affair, which he studied with great acumen, in great depth, and presented with acute clarity. His research, however, has not allowed him to compromise either concerning the causes of the tragic encounter between Galileo and the Church in the seventeenth century, or about future perspectives. Basically, Blackwell believes that the tendency of the Church towards increasingly centralized authority was at the heart of past events and has continued to characterize Catholicism up to the present day. He writes:

> In effect, centrally institutionalized authority tends to evolve into power.
> Human frailty being what it is, the potential for abuse increases. We begin to
> see an emphasis on obedience rather than rational evaluation, on tests of
> faith, on loyalty oaths, on intimidation, on secret proceedings, on unnamed
> accusers and unspecified allegations, on the use of the courts to suppress
> recalcitrants — and ultimately on the whole repertoire of the Inquisition.
> This is not a fantasy scenario. Rather it is precisely what happened in the
> Galileo affair. (Blackwell 1991, 176–7)

On the last page, Blackwell turns to the present and, in accordance with his basic
beliefs, assesses the situation thus:

> If we turn to the present day, the respective situations of science and
> Catholicism have changed considerably. The Catholic Church has estab-
> lished a further centralization of its religious authority in the proclamation
> of the infallibility of the pope in 1870. Simultaneously its social, political and
> cultural power has lessened considerably. Meanwhile modern society has
> evolved more and more in the direction of the democratization of political
> authority and power. Also science has replaced religion as the dominant
> cultural force, and its power has increased tremendously through its marriage
> with technology. ...Yet despite these massive changes since the age of Galileo,
> the Catholic conception of the nature of religious faith and the logic of
> centralized authority related to it seem to remain untouched. Could there be
> a second Galileo affair? What has been learned from the first one? (Ibid.,
> 179)

This last question has not been confined to the realm of rhetoric. For in yet another
paper entitled "Could there be another Galileo case?" the author's answer is stated
in clear positive terms: "it is difficult to avoid the conclusion that intellectual
honesty and freedom of thought may still not be strong enough in the Church to
prevent the recurrence of another clash between science and religion, one similar
to the Galileo affair" (Blackwell 1998, 366).

Blackwell's book, together with his article published two years ago, mark a twist
in the historiography of the Galileo affair. A new kind of energy seems to permeate
the general framework in which some of the literature of the 1990s is being written.
Blackwell's work is not unique in focusing the gaze on the authoritarianism that
characterized the Counter-Reformation Church. A seminal paper in this direction
has been William R. Shea's "Galileo and the Church," published a few years earlier
in the widely read volume entitled God and Nature. Shea has not chosen to isolate
the issue of scripture interpretation from the complex political and personal
circumstances as well as from the major theological and scientific issues of the
period. Nevertheless, like Blackwell, he tended to stress the new cultural authoritarianism that had engulfed the Counter-Reformation church and was at the root

of the affair. The authoritarian ideals, he thought, mostly manifested themselves in the insistence on the exclusive rights of theologians to interpret the Scriptures and on narrowing down interpretive options: "The Catholic church, attacked by Protestants for neglecting the Bible," he wrote, "found itself compelled, in self-defense, to harden its ground. Whatever appeared to contradict Holy Writ had to be treated with the utmost caution" (Shea 1986, 119). At the end, Shea chooses to account for the clash between Galileo and the Church in terms of an "underlying conflict between the authoritarian ideal of the Counter-Reformation and the nascent desire and need for freedom in the pursuit of scientific knowledge" (ibid., 132).

A more extreme view of Church authoritarianism has recently appeared in Marcello Pera's "The God of theologians and the god of astronomers: An apology of Bellarmine," published in the *Cambridge Companion to Galileo*. Pera understands the clash between scientific claims and truths of faith as a matter of principle that transcends the limits of specific historical circumstances. His structural analysis of the relation of science and religion leads him to the following conclusive remark about the Galileo story: "The conflict was much deeper and transcended the *dramatis personae* of the time. It was a conflict between two principles, that is, the principle that science can investigate any factual question ... and any principle that certain factual questions cannot be investigated by science because they are articles of faith" (Pera 1998, 382).

Pera's a-historical perception of the science/religion dynamics — which he attempts to back by a historical interpretation of the Galileo affair — becomes only too manifest as he casts doubt upon the sincerity of present-day Catholic strategies that claim separation between the two domains. Such separation, he contends, is not possible, because it contradicts the essential interests of any religion. Therefore:

> the fire of new Galileo affairs is still smoldering under the ashes that were thought to be cold. Such cases do not depend on historical circumstances, the imprudence of men, the transition from one tradition to another, or the power and prerogative of institutions; they are constitutive. The clash between science and religion is linked to two overlapping, although irreducible, forms of experience and the "logics" of their conceptual organization. (Ibid., 368)

The works of Shea, Blackwell, and Pera exemplify a contemporary trend in the literature on the affair, even though they differ from each other in depth, sophistication, historical orientation and important nuances. In these works the *conflict* at the heart of the relationship between religion and science is intensified either as a characteristic feature of the Catholic Counter-Reformation or as a structural and necessary feature of all religions everywhere. From the perspective of these writers, Galileo's encounter with the church became a struggle over the monopoly of the interpretation of Scripture. The results of such struggle were obviously inevitable, given the balance of power between the two sides.

While Galileo scholars, historians of the Counter-Reformation, and philo-sophers are deepening our knowledge of past events, revising our interpretations and sharpening our understanding of pre-modern Catholic culture, the Catholic church itself has initiated a project intended to foster the spirit of dialogue between science and faith. In the framework of that project Pope John Paul II established a Study Group to explore the history of the Copernican-Ptolemaic controversy of the sixteenth and seventeenth centuries and, in particular the role of Galileo in that controversy. The work of this Study Group has resulted in a series of publications by the Vatican Observatory, among them the monumental volume of Annibale Fantoli, translated into English as *Galileo for Copernicanism and for the Church*. (Fantoli 1996). Other volumes concerned with the history of the same controversy, such as Brandmüller's *Galileo e la Chiesa*, have seen light under the auspices of the Pontificia Commissione di Studi Galileiani. In addition to those studies, works related to the controversy are being published by Catholic university presses. Pierre-Noël Mayaud's volume *La Condamnation des Libres Coperniciens et sa Revocation* (Mayaud 1997), issued by the Pontificia Universitas Gregoriana, is an excellent example of the latter. The professed intention of all those writers is historical not apologetic. In his preface to Fantoli's book, George V. Coyne has succinctly clarified this point: "This is not an apologetic work in which the author takes sides. It is rather a sincere effort at an objective analysis whose purpose is to ccontribute to the good of both science and the Church." Hence, it seems natural, perhaps, that they exhibit a parallel tendency to construct their stories in the form of an *inevitable clash* between Copernicanism, Galileo, and the Church. Two short references will suffice to exemplify this point.

A. Fantoli, for example, never tires of stressing the inevitability of the condem-nation of Galileo. In spite of his initial warm relationship with the Jesuits, the enthusiastic reception — which did not necessarily mean acceptance — of his works by many clergymen, and the good will accorded to him by Urban VIII, Fantoli is convinced that Galileo could not escape the severe judgment of the Church. The root of evil, according to Fantoli, lay in the coming into being of a censuring institution like the Roman Inquisition. A grave error of its qualificators — theologians buttressed in a crumbling, but intransigent, philosophical position, and incompetent in the field of science — brought about an abuse of power which "will have its inevitable sequel in the trial and condemnation of Galileo in 1633." (Fantoli 1996, 236). Faithful to this line of argumentation, Fantoli concludes his story of the trial by claiming:

> it would have been difficult for the trial to have come to a conclusion
> different than the actual one. Galileo had, without doubt, violated a precept
> of the Holy Office (even considering only the one given to him by Bellarmine
> in a "benign" form) and had upheld, at least as probable, a doctrine declared
> to be contrary to Holy Scripture (decree of Index of 1616). As such, he had
> from the viewpoint of his judges, incurred a "serious suspicion of heresy"

from which he could not be absolved except by a public abjuration. It was likewise inevitable that, as expiation for his crime, he be condemned to the prison of the Holy Office. (Ibid., 439)

But Fantoli is not alone among Catholic writers who are convinced of the inevitability of the trial and its consequences. In the introduction to his history of the congregation of the Index, Pierre-Noël Mayaud likewise speaks of the error committed by the Church in its condemnation of Copernicanism. In this book, and especially in its first part, he aims to expose, however, the inevitability of such error:

> Nous chercherons a montrer, en particulier dans la conclusion de la l'ère Partie, les raisons de cette erreur et comment elle e'tait inevitable. Ce serait pur anachronisme en effet de negliger la profondeur, a l'epoque, de l'attachement a une lecture litterale de l'Ecriture aussi longtemps que rien d'absurde n'en decoulait et qu'aucune raison valable que s'y opposait; independamment de toutes les autres raisons avancées par des historiographes modernes, c'est essentiellement dans une fidelité a l'Ecriture, infiniment respectable, que l'Eglise a osé prendre une telle decision, et ceci a une epoque ou il n'y avait aucunes separation des savoirs." (Mayaud 1997, 2)

The return, in the 1990s, of a somewhat diluted and more sophisticated version of the "conflictualist" mode of narrating the story of Galileo and the Church should be understood against the background of the two traditional narratives that have dominated the historiography of the "Galileo affair" since the nineteenth century. As is well known, the "Conflict of Science and Religion" was first constructed by J. W. Draper and A. D. White, where the Galileo case was constitutive in laying down a whole research project. The project was designed to demonstrate the necessary and inevitable conflict between two modes of thought, two kinds of intellectual practices, and two ways of existence in the world. The trial of Galileo involved the silencing of a correct scientific theory, the humiliation of the most prestigious mathematician and philosopher of the period, and the creation of a general atmosphere of fear, suspicion, and coercion. Hence, the lesson to be drawn was of a systematic repression of human free thought by the obscurantist and authoritative church. Thus, the Galileo affair represented the culmination of long standing historical tendencies inherent in the Catholic world, a negative model for subsequent events such as the Darwinian scandal, and a constant threat for similar clashes in the future.

In spite of the enormous influence of the paradigm of necessary conflict upon the research area of "science and religion," by the end of the 1950s its overwhelming predominance eventually gave way to an alternative conceptual framework. In this framework, science and religion were separate cultural domains, each invested with authority within its own boundaries, but complementing each other's perspective on nature and man. Proper respect of these boundaries could have prevented the unnecessary clash were Galileo not so eager to convert people to the Copernican view without enough evidence, and had not the Church reacted so defensively. In the words of J. J. Langford, one of the prominent scholars in this tradition, the lesson to be drawn from the Galileo affair is completely different from that formulated by the tradition of Draper and White:

> Galileo was both a scientist and a believer; it was Galileo the scientist who wrote, Galileo the believer who recanted. But the lesson of his conflict with the Church is not that science and faith are essentially opposed. The lesson lies rather in its dramatic verification of what disaster can come to science or faith when either of these is extended beyond its proper boundaries and enters the domain of the other. A theologian qua theologian has no more authority in speaking about a matter of pure science than does a scientist in discussing Revelation and the Transcendent. (Langford 1966, 180)

Within the paradigm of separation, the story of Galileo and the Church has been interpreted mostly in contingent terms, hinging upon the personality and psychology of Galileo, or the regretful mistakes of some uninformed theologians. Galileo and the Church officials at the time tended to meddle with each other's authority. In this they both committed serious errors that brought about tragic results.

On this historiographical background, present-day conflictualist tendencies require analysis and explanation. My aim, in the following pages, is two-fold: I shall first attempt to show how the "inevitability of a conflict" is being built into the story through particular elaborations of the "three dogmas of the Counter-Reformation." In these stories, typical authoritarian attitudes of the early modern period are anachronistically interpreted as verging towards "totalitarianism." A preference for a literal reading of the Scripture is understood as a mode of "fundamentalism." And mild skeptical positions in astronomy are read as expressions of "instrumentalism." Healthy skepticism vis-à-vis these three dogmas, I shall argue, is long overdue. I shall therefore delineate the lacunae in "unexplored" historiographical landscapes that may tame the temptation to embrace the three dogmas mentioned above. In the epilogue to the paper I shall put forth the question of the "ideological undertones" which "Neo-Conflictualism" — both of clericals and of anti-clericals — carries with it and suggest ways of avoiding this position altogether. Needless to say, my comments are not offered as a detached exercise, a position I cannot claim in view of my long-term involvement in the field of Galileo studies. Rather, my suggestions should be read as an exercise of self-positioning in a dynamic field, which requires periodical withdrawals and recurrent reassessments.

Counter-Reformation Authoritarianism: A Form of "Totalitarianism"?

In recent historical literature, the attempt to prove the authoritarian nature of the Counter-Reformation church concentrates on a reading of two major decrees formulated by the Council of Trent in an early session of 1546. In these decrees the status of Catholic traditions and the monopoly of the Church over the interpretation of the bible were stated as articles of Catholic dogma. Blackwell believes that the construction of the Catholic concept of "tradition" as expressing divinely revealed truth was at the heart of the authoritarian spirit of the Counter-Reformation. Such a construction denied the "polyphonic" nature of tradition that historically allowed for a plurality of voices in biblical interpretation. Blackwell supports his reading with a quotation from the original text of the first decree, stating that both books of the Old and New Testaments and the traditions were "dictated either orally by Christ or by the Holy Spirit" (Decrees of The Council of Trent Session IV, "Decree on Tradition," 8 April 1546. In Blackwell 1991, Appendix I, 181, 9). Furthermore, according to Blackwell, the Council of Trent did not seriously concern itself with the contents of possible interpretations. Rather, its primary interest in authority restricted its concerns to protecting the identity of the class of interpreters who were perceived as guardians of Church monopolies over the holy message. This perception relies on the formulation of the second decree concerning the Holy Scriptures according to which "no one, relying on his own judgment and distorting the Sacred Scriptures according to his own conception, shall dare to interpret them contrary to the sense which Holy Mother Church, to whom it belongs to judge of their true sense and meaning, has held and does hold, or even contrary to the unanimous agreement of the Fathers" (Decrees of The Council of Trent Session IV, "Decree on the Edition and on the Interpretation of the Sacred Scriptures," 8 April 1546. In Blackwell 1991, Appendix I, 183). Blackwell contends that "this passage is not about dogma but about authority" (Blackwell 1991, 12), and concludes that the unanimous agreement of the Fathers became, during the Counter-Reformation, a "touchstone to determine the content of the Apostolic Tradition of revelation from God."

Blackwell's discussion of the Council's decrees contributes to historiography an important distinction between the *contents* of interpretation — especially whether it is considered true or false — on the one hand, and the act of *authorization* of a reading by the tradition on the other hand. He shows that an interpretation may be authorized in virtue of the status of its carriers even without a serious consideration of its contents. But Blackwell does not pursue the consequences of his own distinction. He forgets that the sheer confirmation of the place of tradition and the equalization of its status to that of the canonic text meant a recognition that the Holy Scripture is in need of interpretation in principle although it is conceived as the "voice of God." Therefore, the construction of the concept of tradition as "dictated either orally by Christ or by the Holy Spirit" does not erase interpretive pluralism in principle. Such pluralism is a necessary corollary of the recognition of the status of the tradition as equal to the canonic text that was re-established in Trent. True, the decree also testifies to the deep need of the Catholic Church for legitimization of its own position as mediating the holy message to the believers. In addition, it shows the Church representatives' striving to present a common, united front and uniformity of opinion vis-à-vis the challenge to Church authority presented by the Protestants, who denied that true believers needed Holy Scriptures to be interpreted for them. But ignoring the actual polyphony of voices in matters of interpretation, while over-emphasizing coercive means of control tends to occlude the dialectic tension between these two poles that has always characterized Catholic policy and practice of biblical interpretation. The decrees of the Council of Trent may have modified Catholic sensibilities in its quest for re-affirming institutional authority vis-à-vis the reformers, but it has not ultimately changed the basic cultural patterns that have characterized Catholicism for ages. The tension between interpretive pluralism on the one hand, and the need for control on the other is the most characteristic feature of the Catholic notion of authority. Any attempt to reduce it to "the logic of centralized authority," claiming with Blackwell that "the Catholic conception of the nature of religious faith and the logic of centralized authority related to it seem to remain untouched" (ibid., 179) distorts its true meaning and misconstrues its inner delicate fabric.

To my critique of Blackwell's understanding of the Tridentine notion of tradition, which is a matter of principle, I would like to add a historical argument. It is well known that on all major issues raised by the Council, no uniformity of opinion has prevailed in practice, although the quest for uniformity was stated again and again in many Church documents. In his monumental *History of the Council of Trent* (Jedin 1961), H. Jedin exposed opposing approaches taken by participants to the problem of biblical interpretation as well as to that of original sin, justification, and the sacraments. The careful wording of the decrees, he insists, was usually a compromise between conflicting views that left many vague areas and unclear lacunae, themselves in need of interpretation. No amount of silencing, then, was enough to suppress the plurality of voices that prevailed in practice. The most outstanding example for this state of affairs concerns the theological controversy over the interpretation of the Council's decrees on grace and free will — the controversy *de Auxiliis* which scandalized the Catholic world for almost twenty years between 1588–1607 (Feldhay 1995). During these years, in spite of the authoritarian formulations of the decrees, the Catholic elite was divided between two theological orientations. The Jesuits tended to emphasize the role of free will together with grace in the act of salvation. The Dominicans wholly rejected such an interpretation, and condemned it as heretical and opposing Catholic tradition. Nonetheless, the controversy continued to rage until Pope Paul V decided to suspend it, allowing, in fact, each order to hold to its opinions, even though public attacks on each other were prohibited. Tridentine theology thus re-assumed a monolithic front that did not, however, bring about uniformity of opinions. The controversy broke out again in the seventeenth century as the acrimonious debate between the Jansenists and the Jesuits.

This course of events exemplifies my contention that it was not possible to uproot the traditional pluralism that characterized Catholic culture *de facto* by an attempt to implement the decrees of the Council of Trent. True, during the Counter-Reformation era the Church developed new and more severe means of control than the ones known until then: the Congregation of the Inquisition and that of the Index are the most obvious examples coming to mind. But to reduce the whole cultural dynamics of the Counter-Reformation to the coercive power of these institutions is to occlude the much more complicated task of the Council of Trent in the context of which these institutions were established. In fact, the enormous work of the Council that had lasted for almost twenty years was directed towards no less than a general re-conceptualization of the relation between the realms of the transcendental and the mundane, in the context of which the problem of Church authority should be understood. This authority was not simply or even primarily invested in the exclusive right to interpret the Holy Scriptures, but actually touched upon the attempt of the Church to re-shape the relation between sacred and profane knowledge. Therefore, in trying to understand the condemnation of the Copernican books in 1616 and Galileo's trial in 1633, it is not enough to point at the Inquisition as the source of evil that embodied the whole question of Church power, as Fantoli does. Likewise, it is not enough to mark the qualificators' error as determining the whole course of events that followed as Fantoli, Blackwell, and McMullin (McMullin 1998) do in their various accounts of the Galileo affair. Rather, a deeper and more historical account of the complex concept of Church authority during the Counter-Reformation has yet to be developed. A history of the concept of Church authority would have to take into consideration long term patterns that have shaped the Catholic notion of tradition beyond a reading of two Tridentine decrees. It will have to consider the dialectic between the plurality of interpretive strategies that have constituted the tradition on the one hand, and the means of control over them on the other. And it will have to relate consensual attitudes towards the Holy Scripture to new forms of knowledge that emerged in early modernity. Only then a realistic account of Church authoritarianism will be feasible. Only then we will be able to move beyond hasty generalizations such as: "the contemporary sense of religious authority at least in the Catholic tradition, is monolithic, centralized, esoteric, resistant to change, and self protective" (Blackwell 1998, 359). Only then we will be in a position to build our stories on less essentialist notions than the all encompassing "logic of centralized authority."

Counter-Reformation Literalism: A Form of Fundamentalism?

Within the framework of "neo conflictualism," the condemnation of Copernicanism by the theologians of the Inquisition in 1616 — which led to the trial of Galileo in 1633 — is seen as a direct and necessary outcome of the Church authoritarianism first shaped in the policies of the Council of Trent. This authoritarianism gave birth to a kind of fundamentalism which did not leave room for suspending judgment over scientific theories that did not conform to biblical stances. No serious debate over the correct meaning of biblical verses preceded this decision, which was rather dictated by a "logic of centralized authority." Such logic did not leave open the possibility of modifying the traditional reading of biblical verses. Needless to say, the theologians of the Inquisition, concerned with the monopoly over the authority to interpret did not even consider the Copernican system as a candidate for a true description of the structure of the world. No theory that contradicted the literal meaning of many passages in the Bible had a chance of ever being accepted, in their mind. Therefore, when the Copernican theory was brought to the Inquisition, the theologians condemned it without much hesitation.

Both Blackwell and McMullin (McMullin 1998) are convinced that in its dealings with Copernicanism and with Galileo's discoveries the church authorities had no real interest in scientific theories, which were much lower in status than theology in the context of seventeenth century culture. Therefore, when a clash occurred, scientific claims could not but lose. Thus, McMullin, in the opening pages of his essay, frames his story by stating:

> What these consultors showed themselves committed to defend was not primarily a cosmology. In their own eyes, they were vindicating the authority of Scripture in regard to the truth of its literal content. The Copernican theses about the Earth's motion and the Sun's stability were, in their view, clearly at odds with specific passages in the Bible. To affirm such theses, therefore, was equivalent to calling the authority of Scripture into question. It was that, and not a presumed link between Aristotelian cosmology and the content of Christian doctrine, that led them to condemn the Copernican claim about the Sun as "formally heretical. (Ibid., 273)

And he continues:

> The Galileo affair ought not then be construed, as it so often has been, as primarily a clash between rival cosmologies. ... What called them [the theologians] into action was a perceived threat to the authority of Scripture as well as to their own authority as its licensed interpreters. ... Once *they* entered the lists, the ground of battle shifted, as Galileo very quickly saw. He realized that if he were ever to get a hearing for the new cosmology on its philosophic (scientific) merits, he would have to defend himself on an entirely different front first. And it was on *this* front that the battle was lost

before it was ever really joined on the side of cosmology. (Ibid., 275; emphasis in orginal)

In a similar vein, Blackwell concludes his investigation with the following sentences:

If we can assume that the Church officials clearly perceived the alternatives sketched above, then one main factor may have been that they were convinced that heliocentrism is a Category III claim, that is, impossible to prove! The reason simply is that what is false cannot be proven to be true. And they were convinced that Copernicanism is false because the Bible, and therefore God, asserts the opposite. (Blackwell 1991, 172)

Such claims are supported by two kinds of evidence. The first relates to the well-known tendency of Counter-Reformation theologians to prefer a literal interpretation of the holy text over an allegorical one. Blackwell thinks that this tendency conformed to the concept of tradition constructed by the Council of Trent and designed to present it as monological. Both the adherence to a literal interpretation and the construction of tradition as monolithic were meant to erase a variety of interpretive voices in order to support the exclusive authority of the Church institutional elite. Blackwell concludes that no matter how ingenious an interpretation, an individual who suggested it "was always in jeopardy ... if he undertook an actual reinterpretation of the Church's traditional reading of a particular problematic text" (ibid., 37).

The second kind of evidence relates to the positions of Cardinal Bellarmine. In Blackwell's and McMullin's studies the figure of Cardinal Bellarmine looms much larger than that of one human agent, one protagonist, one actor in an extremely complex historical drama. For them — as for many other writers on the Galileo affair — the Cardinal's personality and opinions somehow encapsulated the official position of the Counter-Reformation church on matters of biblical interpretation and its relation to profane knowledge. Bellarmine's views in these matters had been formed through long years of polemics with Protestants and crystallized in the three monumental volumes *Disputationes de controversiis christianae fidei adversus hujus temporis haereticos* (1586–93). Blackwell points out that this work contained Bellarmine's reflections on the work of the Council of Trent. The structure of the work is compared to the sequence of sessions of the Council, a comparison that supports the claim that the work represents the reception of the Decrees and their application in practice. Blackwell cites passages from the *Controversies* that point out that for Bellarmine "everything in the Scriptures is true ... this truth guarantee *applies not only to matters of faith and morals...but to both general and even specific claims made in the Scriptures*" (ibid., 31; emphasis added). This truth condition, Blackwell continues, "was clearly destined to clash with Galileo's scientific standard of truth" (ibid., 32). Adopting an even narrower approach to the biblical text than the Council itself, Bellarmine prepared the structural conditions for the following course of events —

the condemnation of the Copernican books, and the trial of 1633. Moreover, in his eyes, the authority of the interpreter was more important than the meaning of the interpretation. Thus Blackwell insists that "issues relating to the meaning of Scripture are subordinated to the question of *who* is to judge what the true meaning is" (ibid. 36; emphasis in original).

The understanding of Bellarmine's views as fundamentalist is strengthened by a reading of his famous letter to the Carmelite Antonio Foscarini where he had warned both Foscarini and Galileo not to hold Copernican views as absolutely true in nature. In that letter Bellarmine invoked his exegetical principle according to which anything concerning the empirical world which was stated in the bible should be considered a "matter of faith," if not in relation to the subject discussed, then in relation the speaker. Blackwell and McMullin believe that the statement of this principle in itself meant that there was no way to re-interpret the Scriptures in order to accommodate their meaning. No theory could be taught and defended, let alone developed and finally demonstrated if it contradicted the meaning of the Holy Scriptures as commonly and unanimously understood by the tradition. "Foscarini and Galileo had no possibility of a reply to this pronouncement from the most powerful cardinal of the day. Checkmate!" writes Blackwell (ibid., 106).

This interpretation, however, seems to me to be a retrospective reading of the consequences of the trial of 1633. Much of its conviction lies in the coherence of the story that is told by constructing a necessary causal chain from the authoritative Tridentine decrees of 1546, through Bellarmine's letter to Foscarini of 1616 and up to the sentence of the trial in 1633. However, in reality Bellarmine's letter cannot be deduced from the Tridentine decrees, nor is the sentence deducible from Bellarmine's letter. This presumably inevitable chain of events is at the heart of the "neo-conflictualist" interpretive strategy. I shall hence sketch three arguments against it, although their full development requires more research than has yet been done on the Counter-Reformation background of the affair. First I shall show that the Tridentine decrees of scriptural interpretation are far less "fundamentalist" than presumed by the "neo conflictualists." Second, I shall use some of the materials brought by Blackwell himself concerning principles of interpretations developed by Catholic theologians after Trent. These testify to the persistence of traditional broad approaches to scripture interpretation that can hardly confirm the impression of a growing fundamentalism among Catholic interpreters. Last, I shall suggest an alternative reading to Bellarmine's letter to Foscarini, a reading that would point out the limits of Bellarmine's fundamentalism, in spite of his preference for literal interpretations and his insistence on the exclusive authority of the consensus of the Fathers in exposing the tradition.

First, it is worth emphasizing that although two of the decrees of the Council of Trent dealt with the re-confirmation of the status of the tradition and the authority to interpret the Holy Scriptures, the decree restricted such authority to matters of faith and morality. It then follows that concerning the interpretation of other facts mentioned in the bible but not specifically related to faith — cosmological facts, for example, which touch upon the verse in Joshua — the Church does not have such exclusive authority. This formulation testifies to recognition of the autonomy of reason and its judging faculties in the realm of nature, in contradistinction from the duty to fully accept Church interpretation in what concerns the supernatural realm. No doubt the Copernican theses did not touch upon the supernatural realm. Hence, the need to intervene in cosmological matters did not directly follow from the Council's official policy.

Beyond the question of authority, the decrees did not formulate any principles of interpretation. Therefore, the preference for literal interpretation was not part of the Church official position. The most important Catholic theologians of the Tridentine era continued to develop the distinction first formulated in Trent between "matters of faith" on the one hand, and other facts mentioned in the Bible on the other hand. These other facts, they maintained, did not necessarily require adherence to the literal meaning of the text, or to the consensus of the tradition. Blackwell himself quotes two great theologians who explicitly warned against burdening theology with too much authority in things that do not touch upon "matters of faith." Thus, in his *De locis theologicis* (1563) Melchior Cano, considered among the most illustrious participators of the Council of Trent, argued that in matters concerning the realm of nature, the authority of theologians is not superior to that of the philosophers. Thus he wrote that: "When the authority of the saints, be they few or many, pertains to the faculties contained within the natural light of reason, *it does not provide certain arguments but only arguments as strong as reason itself when in agreement with nature*" (Melchior Cano, O.P., "De locis theologis," in *Opera*, vol. I, Rome 1890, VII, 3, quoted by Blackwell, 18; emphasis added). Likewise the Jesuit Benedictus Pereyra, one of the giants of biblical interpreters of his time, who had always showed preference for literal reading of the Scriptures also rejected a fundamentalist approach to the text: "in dealing with the teachings of Moses," he wrote, "do not think or say anything affirmatively and assertively which is contrary to the manifest evidence and arguments of philosophy or the other disciplines," and more generally he asserted: "Scripture is clearly very broad by its very nature and is open to various readings and interpretations" (Benedictus Pererius Valentini (Pereyra), *Commentariorum et disputationum in Genesim tomi quatuor*, Romae 1591–95, quoted in Blackwell 20).

In the light of these quotations, the identification between literalism and fundamentalism seems doubtful, and probably wrong. Moreover, if one takes into consideration that the very term "literalism" in the sixteenth century excluded allegorical interpretations but not metaphorical ones, there is all the more reason to doubt all approaches which attempt to attach fundamentalism to the great Church interpreters of the period.

Last but not least are Bellarmine's positions. Needless to say, Bellarmine's insistence on the need to interpret the Scriptures literally, and his life-long involvement in buttressing the Church's exclusive authority to interpret vis-à-vis Pro-

testant attacks echo through the formulations of the letter to Foscarini (Finocchiaro 1989, 67–69). In contrast to the Church demand for authority in "matters of
faith" alone, Bellarmine invoked the sacredness of the Holy Scriptures, from
which he inferred their truth-value not only in theological and moral matters but in
all other factual and empirical things. "Nor can one answer that this is not a matter
of faith," he wrote, "since if it is not a matter of faith as regards the topic ... it is a
matter of faith as regards the speaker" (Finocchiaro 1989). In this Bellarmine
deviated from the decree and broadened its scope of application. He surely wished
to suspend any attempt of re-interpreting scriptural verses in accordance with the
Copernican theory, as long as this theory has not been proven. However, throughout the letter Bellarmine's voice is not dogmatic but pragmatic. Thus he appeals to
Foscarini's "practical reason" — not to any exegetical principle — in order to
convince him to recognize the difficulty which lies in the attempt to accommodate
the literal sense contrary to the opinion of all Church Fathers and traditional Latin
and Greek commentators. The crucial evidence, however, lies in Bellarmine's own
words by which he delineates the limits of literalism in Scripture interpretation: "if
there were a true demonstration that the sun is at the center of the world and the
earth in the third heaven, and that the sun does not circle the earth but the earth
circles the sun," Bellarmine writes, "then one would have to proceed with great
care in explaining the Scriptures that appear contrary, and say rather that we do
not understand them than that what is demonstrated is false" (ibid.). McMullin
thinks that "in context, one can see that he was not conceding this allusion to the
traditional Augustinian principle to be a real possibility." It is his innate courtesy
to his correspondent, a respected theologian, that leads him to add the qualified
"until it is shown to me" to the assertion: "I will not believe that there is such a
demonstration." He has already indicated that he thinks such a demonstration to
be permanently out of reach" (McMullin 1998, 283). Nowhere, however, does
Berllarmine make the contention that demonstration should not be sought for
since it is unattainable in principle. Denying his actual words with the hypothesis
that he did not really mean them seems unconvincing. In fact, Bellarmine here
joins the opinion of Melchior Cano, that of Benedictus Pereyra, and the general
attitude of Catholics that recognize the inherent opacity of Scripture and the need
for interpretation that would not violate the truth of natural reason. He certainly
wished to suspend attempts at re-interpretation in a period of great sensitivity to
the authority of the tradition, and in a state of uncertainty concerning the validity
of Copernicanism as a scientific theory. These pragmatic considerations, rather
than any imaginary fundamentalism, however, pushed him to broaden the application of the Tridentine decrees, indicating at the same time their limits as well.

The Scientific Status of Astronomical Theories: A Form of Instrumentalism?

If Copernicanism had been considered a truly demonstrated theory according to scientific canons of proof at the beginning of the seventeenth century, it would have been impossible to judge it as a heresy, in spite of its being contrary to the literal meaning of the Scripture. This conclusion directly stems from the basic principles of interpretation accepted even by a literalist of the stature of Bellarmine, who shunned, as we have seen, any suggestion that "what is demonstrated is false." Therefore, the Inquisition decree of 1616 that condemned Copernicanism included the implicit assumption that the Copernican statements were meant as mathematical hypotheses, not as absolute truths about the universe. This assumption was supported by Osiander's preface to the *De Revolutionibus*, even though it did not conform to Copernicus' own intention. Consequently, the congregation of the Index, in what seemed to be a move in conformity with the decision of the Inquisition, decreed that Copernicus' book should be *suspended until corrected*. It is important to emphasize that in 1616 only those books that attempted reinterpretation of Scripture in order to facilitate an accommodation of Copernicanism were condemned. Later on — only in 1620 — the congregation also made suggestions for specific corrections. Practically the corrections located places in the book where the Copernican theory was explicitly presented as natural truth, and suggested ways of presenting it as a mathematical hypothesis. This state of affairs presents researchers with an interpretive dilemma concerning the true meaning of the decision and the position of the Church. Was it in fact legitimate to read Copernicus' book only for its practical uses such as the calendar reform? Was it actually forbidden as a scientific-theoretical text? Or maybe the very decision of the congregation of the Index to suspend — not prohibit — the book actually left the limits of its possible uses opaque, so that in spite of many constraints an attempt to prove the theory was not completely forbidden?

All neo-conflictualists are of the common opinion that in the Catholic world Copernicanism had been buried already in 1616, with the condemnation that the Copernican theses were contrary to Holy Scripture. To the explanation hitherto mentioned for such an ominous decision — the authoritarianism of the Tridentine Church which entailed interpretive fundamentalism — is added another type of argument. This argument touches upon instrumentalism in astronomy, which is inferred mainly from Bellarmine's letter to Foscarini and from Mosaic astronomy developed in the Louvain lectures which he had given in the 1580s and which he continued to hold throughout his life. And indeed, Bellarmine opened his letter to Foscarini with a distinction, common among contemporary scholars, between two types of scientific proofs — demonstration *ex suppositione* and absolute demonstration. The first, he argued, is well known to mathematicians and satisfies the norms of the profession ("and that is sufficient for the mathematician"). The second kind of demonstration is commonly practiced by natural philosophers and theologians. Then Bellarmine contended that when applied to Copernicanism, the use of the first kind of demonstration presents no danger whatsoever, while the second is indeed dangerous:

> to say that the assumption that the earth moves and the sun stands still saves all the appearances better than do eccentrics and epicycles is to speak well, and contains nothing dangerous. But to wish to assert that the sun is really located in the center of the world and revolves only on itself without moving from east to west, and that the earth is located in the third heaven and revolves with great speed around the sun, is a very dangerous thing ... not only because it irritates all the philosophers and scholastic theologians, but also because it is damaging to the Holy Faith by making the Holy Scriptures false. (Finocchiaro 1989, 67–69)

Focusing the gaze on Bellarmine's view of astronomy, and interpreting his distinction between mathematical-hypothetical discourse and a philosophical discourse of truth as expressing an instrumentalist position adds a third layer to the story of an inevitable conflict between Galileo and the Church. "The firm conviction that mathematical astronomy could not *in principle* provide a demonstration of the Earth's motion and without such a demonstration the literal sense of Scripture ... could not be challenged, seems to have been Bellarmine's guiding line throughout," writes McMullin (McMullin 1998, 282; emphasis in original), and even more emphatically he delineates Bellarmine's view by stating: "Bellarmine is not merely pointing to the fact that the Copernicans have not yet come up with a proper demonstration of the Earth's motion. He is, in his own mind, at least, giving reasons to believe that they never *could* (ibid., 283). Such a view of astronomy that denied it any independent claim for truth, adds Blackwell, well suited the logic of "centralized" authority which Blackwell assigns first to the Tridentine church but then to Catholicism at large, and which he sharply contrasts with the structure of authority in modern scientific discourse. On the other hand Galileo made broad use of the principle of accommodation which could not be adopted by the authoritarian Tridentine Church and was explicitly rejected by Bellarmine. In spite of the warning of the Cardinal, he did not discard his attempts to find a physical proof for Copernicanism. Occasionally, he was even tempted to present his arguments in favor of Copernicanism as proofs. In all this he challenged the explicit position of the Church represented by Bellarmine. No wonder, then, that he was eventually tried under the pretext of vehement suspicion of heresy.

A healthy dose of skepticism towards this story is, however, needed. Just as it is necessary to cast doubt upon the totalitarian tendencies of the Tridentine church, and just as attaching fundamentalist stances to Bellarmine is exaggerated, so it is possible to interpret differently the distinction between hypothetical and absolute discussion of cosmological issues. In making the distinction between demonstration *ex suppositione* and true demonstration, and in demanding that Foscarini and Galileo limit their claims to the field of mathematics, Bellarmine shows his awareness of the major divisions that split the academic world of his period. In order to understand his position in its proper historical context, it is worth quoting at length from N. Jardine's study on the status of astronomical science in the sixteenth century. A great number of astronomers at the time, Jardine argues,

> without openly committing themselves to radical skepticism, doubt or deny the capacity of astonomers' planetary models to represent the disposition and motions of the heavenly bodies and insist on a strict distinction between the proper concerns of the mathematical astronomer and those of the natural philosopher. ... Such a combination of doubt or denial of the reality of planetary models with insistence on the strict demarcation of a celestial physics concerned with the nature of the cosmos from a mathematical astronomy concerned only with saving the phenomena, without regard to the truth of the hypotheses employed, becomes increasingly prevalent in the course of the sixteenth century. (Jardine 1984, 237–38).

Jardine deems this position pragmatic and cites various reasons for its popularity. Mainly, however, he stresses the need felt by many scholars to avoid conflict between theologians and philosophers as well as between astronomers and naturalists and to allow the continuation of a working tradition at a transitional stage in its development. Moreover, Jardine is especially concerned to avoid the identification of such a position with any kind of modern fictionalism. "No protagonist of the pragmatic compromise expounded the strict instrumentalist view that truth and falsity are not predicable of astronomical hypotheses. And even the more relaxed instrumentalism which claims only that predictive success rather than truth is the goal of astronomy can rarely be attributed without qualification" (ibid., 239).

   A reading of the letter to Foscarini in the context of the positions held by the majority of astronomers of his time reveals that Bellarmine's advice to Galileo and Foscarini was not meant to bury the discussion of Copernicanism, as Blackwell and McMullin claim, but rather to enable its continuation.

Epilogue

In this paper I have tried to show that the reconstitution of Galileo's involvement with the Catholic Church as a narrative of *inevitable conflict* is common to many historians writing in the last decade. But whereas many share the concept — and thus structure the story in similar ways — they do not necessarily share the meaning that the story attempts to convey. Thus, it appears to be the case that the concept of the conflict serves many purposes for different people from different cultural and ideological milieus. For Blackwell, casting the story in terms of an inevitable conflict directly leads to a more general statement of the inherent antagonism between Catholic authoritarianism and modern science not only in the past but also in the present (Blackwell 1998, 359). For Pera (Pera 1998), religion and science are in principle irreconcilable in a transhistorical, transcultural sense. In the context of the Catholic project that encourages a dialogue between religion and science the inevitablity of the conflict is very differently understood, however. For Fantoli and Mayaud, for example, shifting the emphasis of the story to the contradiction between Scriptures and the Copernican theses helps to focus the gravity of the historical error committed by the church, but also to limit its boundaries and plea for a historical understanding of its roots. In their view the church was defending Scriptures and the right of their interpretation because it was the core of its collective identity severely challenged by its Protestant enemies. Thus, the philosophical-scientific controversy is re-constructed in terms of identity politics, where criteria of truth and falsehood are less relevant than the quest for coherence. Both groups, however, are stuck with a simplistic story that still fails to provide real historical explanations for the well-known facts.

Indeed, the three dogmas of the Counter-Reformation as defined in this paper are built around three key-concepts that may serve as the nuclei of a revised story, once criticized and thoroughly historicized. Historical research has yet to show how the boundaries of church authority crystallized not simply in the encounter between Bruno, Campanella, Galileo, and the Inquisition. Rather, it crystallized in a series of cultural struggles between Catholics and Protestants on the one hand, but also between traditionalist and more openly modernizing intellectual elites within the Catholic world itself on the other hand. Simultaneously, philosopher-scientists of the type of Bruno, Campanella, and Galileo also strove to break through the traditional position of commentators of great authors such as Aristotle, Plato, or Thomas Aquinas, and build up their voice as speakers in an independent and authoritative discourse on nature. Thus, the force that was indeed brutally exercised on Galileo in the conclusion of the trial of 1633 did not directly and necessarily result from his disobedience to Bellarmine's warning. Neither should Bellarmine's warning be read as a necessary deduction from the Tridentine decrees. Rather, it was force exercised at a moment of loss of control in a long and complex historical process where different notions of authority — religious, scientific, philosophical — competed for cultural hegemony in a field that has not yet been differentiated into clear bounded spheres.

Likewise, the preference for literal interpretation manifested by Bellarmine, should be examined not only in the context of the decrees that directly concerned Holy Scriptures but also in association with the rest of the Tridentine decrees and the doctrinaire developments after Trent. Moreover, practices of interpretation should also be investigated in the wider contexts of contemporary rules of inter-preting classical texts, and not least of all the interpretation of nature. Last, it should be remembered that at the time of Galileo, the Church had its own science, developed especially by the Jesuits who tried to implement the Tridentine reform through educating the whole Catholic population and through assimilating new types of knowledge. Investigations of this kind will show that at the time of the

Counter-Reformation the Church itself experienced struggles between different options for the re-organization of culture. The traditionalists tended to adhere to the old boundaries between disciplines and the hierarchy between them and to maintain their authority through traditional means. The modernizers attempted to assimilate new areas of knowledge but were exposed to the danger of losing traditional authority to the new disciplines. Galileo was implicated in this struggle and was not always able to exploit it for his own benefit. This cultural struggle had intellectual and ethical aspects, as well as force-oriented aspects. The only hope to research it historically is by renouncing the anachronistic use of dichotomous, simplistic categories in its representation.

References

Blackwell, Richard, J. 1991. *Galileo, Bellarmine, and the Bible.* University of Notre Dame Press.

——. 1998. "Could there be Another Galileo Case?" In *The Cambridge Companion to Galileo*, edited by Peter Machamer, 348–366. Cambridge: Cambridge University Press.

Fantoli, Annibale. [1994] 1996. *Galileo for Copernicanism and for the Church.* Translated by George V. Coyne, 2nd ed. Vatican Observatory Foundation.

Feldhay, Rivka. 1995. *Galileo and the Church: Political Inquisition or Critical Dialogue?* Cambridge: Cambridge University Press.

Finocchiaro, Maurice. 1989. *The Galileo Affair: A Documentary History.* University of California Press.

Jedin, Hubert [1957] 1991. *A History of the Council of Trent.* Translated by D. Ernest Graf, 2 vols. London: Thomas and Sons.

Langford, Jerome, J. 1966. *Galileo Science and the Church.* Ann Arbor: University of Michigan Press.

Mayaud, Pierre and S. J. Noël. 1997. *La Condamnation des Livres Coperniciens et sa Révocation.* Rome: Pontificia Universitas Gregoriana.

McMullin, Ernan. 1998. "Galileo on Science and Scripture." In *The Cambridge Companion to Galileo*, edited by Peter Machamer, 271–347. Cambridge: Cambridge University Press.

Pera, Marcello. 1998. "The God of Theologians and the God of Astronomers: An Apology of Bellamive." In *The Cambridge Companion to Galileo*, edited by Peter Machamer. Cambridge: Cambridge University Press.

Shea, William R. 1986. "Galileo and the Church." In *God and Nature*, edited by David C. Lindberg and Ronald I Numbers. University of California Press.

The Cohn Institute for the History and Philosophy of Science and Ideas
Tel Aviv University

PAOLO GALLUZZI

Gassendi and *l'Affaire Galilée* of the Laws of Motion*

The Argument

In the lively discussions on Galileo's laws of motion after the Pisan's death, we observe what might be called a new "Galilean affair." That is, a trial brought against his new science of motion mainly by French and Italian Jesuits with the substantial adherence of M. Mersenne. This new trail was originated by Gassendi's presentation of Galileo's *de motu* not simply as a perfectly coherent doctrine, but also as a convincing argument in favor of the truth of Copernicanism.

L'affaire Galilée that I will be dealing with does not concern the condemnation of the Pisan scientist in 1633 and the reactions which this event produced in Europe. Even if for many reasons it still remains problematical, that *affaire* has been the object of thorough studies, even recent ones, to which I have nothing new to add. There is, however, another Galilean *affaire* that began immediately after the conclusion of the first, upon the death of Galileo in 1642. In this second episode the scandal was represented by the Galilean laws of motion, submitted to a series of severe censures between 1642 and 1648, especially in France.

At the origin of the *affaire* of the Galilean science of motion is the articulated presentation that Gassendi gave of it in the *De motu impresso a motore traslato* published in 1642 (Gassendi 1642),[1] but materially drafted in 1640, two years after the publication of Galileo's *Two new sciences*, that Gassendi had read with considerable attention.

---

* This paper reproduces, with a few changes, the essay published in Italian in 1993 (Galluzzi 1993). Since then a few studies have been published which discuss the themes dealt with in this work, shedding new light on the protagonists of this capital discussion. Among these new contributions appear of special interest the long awaited critical edition of Baliani 's *De motu*, edited by Giovanna Baroncelli, and the essay by Carla Rita Palmerino (Palmerino 2000). This last work presents an insightful examination of the reasons behind the evolution of the attitude of Father Mersenne in the lively European discussions on the Galilean science *de motu*.

[1] I am using the version included in the edition of 1658. In this edition the title of the work was changed and a third *epistula* "ad venerandum senem Iosephum Galterium" was added. The third *epistula* contains a reply of the Canon of Digne to accusations moved to him by Morin (n. 11; Gassendi 1642, 3:478–563).

The conclusion of this *affaire* which, unlike the first, did not end with any formal convictions, can be made to coincide with the death of Marin Mersenne, in 1648. During these seven years, Mersenne was in fact the active and able director of a controversy to which he also offered a significant personal contribution. Since 1633, Mersenne had manifested interest in the Galilean science of motion, having inserted in his *Harmonie Universelle* a description of it, based on the *Dialogues* (Mersenne 1637, 1:85–92, 125–128).[2] Nor must it be forgotten that in 1634 he published a version of the Galilean treatise on mechanics (Mersenne 1634a),[3] while in 1638 he gave to the press *Le nouvelles pensées de Galilée*, a free translation of part of the *Two new sciences* (Mersenne 1639).[4] In France, thanks to Mersenne and to his epistolary network, the Jesuits Pierre Le Cazre and Honoré Fabri were involved in the *affaire*, as well as Boulliau, Roberval, Le Tenneur and, more marginally, Descartes and Pierre Fermat. In Holland, the solicitations of Mersenne caused the intervention of the "enfant prodige" Christiaan Huygens, who would give eloquent proof of his own mathematical talent. Lastly, during his journey to Italy between the end of 1644 and the first months of 1645, Mersenne succeeded in involving in these discussions Evangelista Torricelli and Michelangelo Ricci. Moreover, after his return to France, he kept up an intense epistolary exchange on these same topics with Galileo's disciples in Florence and Rome, as well as with the Genoese Giovan Battista Baliani.

In the *De motu impresso* of 1642, Gassendi, for the first time, clearly laid out the problematic issues that would become the core of the new *affaire*.

The brilliant pages by Alexandre Koyré on the *De motu impresso* by Gassendi in the *Études Galiléennes* (Koyré [1939] 1961)[5] and in the *Newtonian Studies* (Koyré 1965),[6] a point of reference that is still important today. While stressing Gassendi's mathematical limits, Koyré exalted his transformation of Galilean kinematics into dynamics and his bold insertion of the Galilean theory of motion into an organic philosophical framework.

Interested above all in reconstructing the exquisitely intellectual aspects of the process of definition and assimilation of the key concepts of the new conception of the physical world, Koyré has rightly recognized the *De motu impresso* by Gassendi as a significant stage in the process of progressive geometrization of the cosmos. Koyré nevertheless showed scarce interest in the reactions that the *De motu* provoked, especially in France, in the years immediately following its publication.

However, contemporary readers of Gassendi's *Epistulae* were provoked, not so much by the considerable conceptual novelties on which Koyré has insisted, but

---

[2] For the intricate editorial genesis of the edition of this monumental work, see Lenoble [1943] 1971, XXI–XXV.

[3] A critical edition of Mersenne's translation was published by B. Rochot in 1966, with an insightful introduction.

[4] For the selective and interpretative character of Mersenne's translation of the Galilean *Two New Sciences*, see Costabel and Lerner 1973, 1:7–51.

[5] I am using the Italian translation by M. Torrini (Koyré 1976, 311–324).

[6] I am using the Italian translation by P. Galluzzi (Koyré 1972, 194–197, 204–206).

rather by the Galilean definition of natural motion, accepted without reserve by Gassendi, with all the consequences which derived from it in terms of relations between spaces traversed from rest in equal successive times, of the paths of projectiles, and of the relationship between the prescriptions of Galileo's laws and the motions of "real" bodies.

Although it may seem paradoxical, in the many reactions following Gassendi's work — the speculations on the indifference of bodies to motion or to rest, on the violent character of free fall, on gravity and its external cause, on the relativistic conception of motion, and on many other similar key concepts — assumed much less importance than the careful examination of Galileo's definition on the acceleration in natural motion and of the laws that the Pisan had deduced in the *Two new sciences*.

To understand this phenomenon, we have to focus carefully on the decisive role that the Galilean laws of motion played in the *Epistulae duae de motu impresso* by Gassendi. He had in fact proposed a bold integration of the Dialogue and the *Two new sciences*, accomplishing that organic objective which had been in Galileo's mind since 1609, but which the anti-Copernican sentence of 1633 had prevented him from achieving.

To the reader of the *De motu impresso*, the Copernican structure of the Gassendian universe appeared evident, even if the heliocentric theory was declared in purely hypothetical terms. Gassendi's universe appeared dominated by simple laws of universal validity, which expressed the rigorous convergence between the rational principles of the new physics and the mechanical action of forces: a universe that could be understood through measurement, thanks to mathematical procedures and instruments. It was a universe in which absolute reference points were missing and in which, as a consequence, there were no privileged directions of movement; a universe wisely regulated by its Creator — the most refined mechanic — who, due to the perfection of the original plan, could remain discreetly detached from the everyday operation of the great machine of the world, where occult "qualities," "impulses," and internal virtues disappeared. In this universe, there was no room even for gravity — it being replaced solely by infinite and empty spaces traversed by atoms in continuous movement. The movement that dominated this universe and that guaranteed its orderly conservation throughout time was governed by laws which Galileo had defined and described.

In the *De motu impresso*, Gassendi strived to show with acute speculations and with convincing experiments the absolute unsustainability of the integration of Ptolemaic cosmology and Aristotelian dynamics, which had been very effective for a long time. The criticism of traditional dynamics implied the necessity of abandoning geocentric cosmology and paved the way for an entirely new science of motion which Gassendi derived from Galileo, while transforming it, particularly through deducing the effects described by the Pisan scientist from precise physical causes. Gassendi, in short, gave an eloquent demonstration of the formidable allegiance between the Copernican cosmology and the Galilean doctrine of motion, vigorously emphasizing their congruence and full compatibility. It was above all through stressing the indisputable superiority of the Galilean science of motion over traditional dynamics, and its capacity to sustain experimental tests, that Gassendi intended to effectively undermine traditional cosmology, which found its fundamental support in the Aristotelian concept of nature and motion.

In effect, the Galilean laws of motion seemed to account perfectly for those "real" motions that the principles of Aristotelian physics were unable to explain. Gassendi also pointed out the dramatic conceptual fragility of ancient dynamics, cleverly emphasizing how such fragility undermined the foundations of that same traditional cosmology that, by the admission of its own supporters, was closely tied to the Aristotelian science of motion. In fact, those who had tried to elaborate a series of presumed confutations of the motions of the Earth had relied on the principles and on the laws of the traditional theory of motion.

It begins to be clear why the *De motu impresso* by Gassendi gave rise to a new Galilean *affaire*. It must be stressed that this new episode is closely connected to the Galilean *affaire* of 1633 of which it constitutes a natural appendix. The passionate defense of the Galilean laws of motion implied for Gassendi the sound promotion of the philosophical and conceptual foundations of Copernican cosmology. Such a defense, moreover, did not entail any obligation to explicitly declare his own adhesion to the theory of the Earth's motions, thus avoiding the risk of a confrontation with the ecclesiastic authorities who had condemned it.[7]

The First Reactions to Gassendi's *De motu*

The intrinsic alliance between Copernican theory, new philosophy, and Galilean science of motion appeared evident to the contemporary readers of the *Epistulae de motu impresso a motore translato* as a result of the immediate and instinctive reaction by the notorious astrologer Morin (Costabel 1974)[8] and by the Jesuit Pierre Le Cazre,[9] Rector of the Dijon College.

---

[7] Although the *De motu impresso a motore translato* contains many points that are worthy of investigation, an exhaustive analysis of this text is still lacking. Besides the above quoted pages of A. Koyré, useful remarks on Gassendi's *Epistulae* are to be found in Bloch 1971 (190–201).

[8] As is known, Morin heavily attacked Galileo before and after the sentence of 1633 (Morin 1631 and 1634). At the conclusion of his work of 1634, Morin explicitly informed the reader that the *Dialogue* had been condemned by the Holy See and that Galileo had been forced to abjure his doctrines. In the letter to Jean de Beaugrand, of 11 November 1635, Galileo vigorously resented the gratuitous malignity of Morin stating that Morin would have offended him much less by publishing the sentence and the abjuration than by keeping silent about it: "soggiunge che havrebbe aggiunta la sentenza e abiurazione fatta in Roma, ma ha stimato meglio il tacerla per sostentar la mia fama .... Assai ... meno m'havrebbe offeso il Morino publicando che tacendo mie sentenze e abiurazioni" (Galilei 1890–1909, 16:341ff.).

[9] Biographical data on Father Le Cazre (Rennes 1589; Dijon 1664) are extremely scarce. Rector of the Colleges of Dijon, Metz, and Nancy, he was Provincial of Champagne and Assistant of France. For Sommervogel he "scripsit docte et accurate multa de disciplinis philosophicis, theologicis, mathematicis, physicis" [he wrote many erudite and accurate things on philosophical, theological,

In his very fierce *Alae Telluris fractae* (Morin 1643),[10] Morin, going directly to the heart of the matter, maintains that Gassendi had just bared his true face as a follower of Copernicus, notwithstanding the formal condemnation of heliocentrism by the ecclesiastic authorities. According to Morin, who had already distinguished himself in the denouncement of Galileo's Copernicanism, Gassendi's concept of free fall guided by purely mechanical attractive forces and performed according to the space-time proportions rigorously described by Galileo, constituted, in fact, a fundamental component and a distinct characteristic of the Copernican system: attraction and the Galilean laws of naturally accelerated motion conferred therefore an indelible stamp of infamy.

Gassendi avoided replying publicly to this attack of unprecedented violence. However, he immediately wrote a bitter answer that would see the light, unbeknownst to him and with his apparent disapproval ("me invito"), on the initiative of some of his friends in 1649 (Gassendi 1649).[11] In his rebuttal, Gassendi reasserted the purely hypothetical nature of his references to the Earth's motions, but he energetically defended the definition of acceleration in free fall and the laws of natural motion that Galileo had described in the *Discourses*. He took care, moreover, to emphasize that that concept of motion, as well as the causal model of attraction, did not necessarily presuppose the Copernican system, but would have also been admissible in the hypothesis of an immobile Earth (Gassendi 1658, 3:562–563).

Even the Jesuit Le Cazre, in his first private letter, sent to Gassendi from Dijon on November 3 1642 (ibid., 6:448–452) through Senator Filbert de la Mare,[12] stressed the evident integration in the *De motu impresso a motore translato* between the Galilean analysis of the laws of motion and the Copernican system although without emulating Morin in his verbal affronts and explicit threats.

It is worth remembering that Gassendi had established in his *Epistulae* of 1642 the most evident link between the laws of motion and the Copernican system through his endorsement of the Galilean explanation of the tides. Le Cazre formulated radical criticism of Galileo's concept of motion, which depended, as he

---

mathematical and physical disciplines]. Moreover, he affirms that in the National Library of Paris there is a manuscript treatise by Father Cazre, *De descensu gravium* (Sommervogel supposes it is a copy of one of the letters to Gassendi); the signature that he gives (Ms. Lat. 61, 40A) is unfortunately wrong and does not consent retrieval of the manuscript (Sommervogel 1890–1932, 2:col. 934–935). When the *De motu impresso* was published, Father Le Cazre was Rector of the important Jesuit College of Dijon (Tannery et al. 1945–1988, 9:321–323, 323 n. 5).

[10] For the later attacks by Morin against the corpuscular philosophy of Gassendi, see Sortais 1921–1923, 2:167–173.

[11] L'*Apologia* also contains the so-called "Copernican letters" of Galileo. Gassendi's reply — a letter to his friend Gaultier — was published unbeknown to him on the initiative of some of his friends, who were worried that his not answering could harm the reputation of the Canon of Digne. One year later a new edition was issued which contained the letter in which Gassendi protested his own complete extraneity, or rather contrariety, to the publication of the *Apologia* (Gassendi 1650). For Gassendi's letter of justification, see Gassendi 1650, 5–6. Gaultier's answer to the *Alae Telluris fractae* was republished in Gassendi 1658, 3:520–563.

[12] De la Mare, who had received a presentation copy of the *De motu impresso* from Gassendi (see his letter of thanks in Gassendi 1658, 6:447), lent it to Le Cazre (Gassendi 1658, 4:448–452).

stated, upon false principles "ex principiis falsis" (ibid.). He refused to admit that the parabola — which, according to Galileo and Gassendi, resulted from the composition of a uniform horizontal motion with the acceleration of free fall — would not have brought about an increase in the time of fall and an impact of greater intensity (ibid., 448–449). He also contested the increase of spaces traversed in equal times from rest, according to the series of odd numbers, that had been proposed by Galileo and reaffirmed by Gassendi (ibid.). Lastly, he denied that a heavy body projected perpendicularly would have taken the same amount of time in its ascent, as it would in its descent (ibid.).

After having hurriedly detached himself from the fundamental Galilean-Gassendian conceptions of motion, the Jesuit launched an attack upon philosophical-cosmological matters. He emphasized that atomism and the reduction of qualities to single local motions proposed by Gassendi in the *De motu* would not be appreciated by learned and pious men "viris eruditis ac piis minus placitura videantur" (ibid., 449). In fact, as a necessary consequence, it resulted that accidental forms do not exist; then they cannot be separated from substance, with serious consequences for the Eucharist mystery "formae accidentariae nullae sunt, multoque minus inveniri et esse possunt ab omni substantia separatae. Quid sanctioribus igitur nostrae Religionis mysteriis fiet?" (ibid., 450). This distressing interrogative on the destiny of the dogma of the Eucharist sustained the "friendly" recommendation to renounce the publication of the work on the philosophy of Epicurus: "Audio certe tibi etiamnum in manibus esse iustum volumen aliud, quo haec Epicuri ac Democriti somnia illustrare labores... Vix per Deum immortalem ne tui nominis authoritate, infirmioribus quidem errandi, caeteris vero prae conceptam ingenj tui ac iudicij sagacitate opinionem imminuendi praebeas occasionem" (ibid., 450–451).

Le Cazre insinuated that, in the light of his purely hypothetically declared assumption of the Earth's motion, the care and insistence with which Gassendi continually proposed arguments, observations, and experiments in favor of the Copernican system, appeared somewhat strange:

> Cur adeo studiose caetera quoque omnia Copernicanorum argumenta congeris ... Cur ... de Telluris eiusdem circa Solem motu tam prolixam nec ad praedictum finem necessariam adjungis disputationem in qua Copernicanorum argumenta omnia et rationes quae potes et vales dicendi facultate stabilire et confirmare, sed aliorum quoque obiectiones infirmare, pari studio et contentione moliris? Quid amabo te amplius eras facturus si eam sententiam animo destinato asserendam propugnandamque assumeres? (Ibid., 451)

And before manifesting his own surprise at Gassendi's subscription to the Galilean theory of the tides,[13] Le Cazre reminded Gassendi that the anti-Copernican verdict

---

[13] Gassendi had reproposed the Galilean theory of tides and was shown to consider it as convincing

was issued not simply by some Cardinals, but by the Pope himself "non Cardinales tantum aliquot (ut ais), sed supremum Ecclesiae caput; Pontificio decreto in Galilaeo damnaverit, et ut ne in posterum verbo aut scripto doceretur sanctissime prohibuerit."[14] It had been, according to Cazre, a very healthy edict that impeded the followers of that wrongful theory from sustaining absurd consequences "portenta propemodo infinita ... excolantes culicem et elephantos deglutientes"; and he recalled at the same time that philosophers were obliged to submit to the dogmas of faith.[15]

Considering the Jesuit's threatening tone, the specification of his friendly intentions by Senator de la Mare in his letter of transmission to Gassendi of Le Cazre's writing, were certainly read with some relief on the part of the Canon of Digne: "de quo apud te praestare possum eum esse qui haec ad te scripserit non severiori Catholicae fidei tuendi studio adductus ... sed unius veritatis inquirendae ratione" (ibid., 452).[16]

Gassendi's reply to the Jesuit was published in 1646, as the last of three letters (Gassendi 1646)[17] in response to the *Physica demonstratio* (Le Cazre 1646)[18] given to the press by Le Cazre in 1645, in order to refute the Galilean conception of motion. Gassendi responded to the Jesuit's accusations point by point, claiming firmly the *libertas philosophandi* in physical investigation and supporting the compatibility of atomism with the Christian faith.[19] Furthermore, he insisted on the extraordinarily rational and experimental congruence exhibited by the organic body of doctrines formed by the Copernican system, by the conception of atoms in

---

in the *Epistola II De motu impresso a motore translato* (Gassendi 1658, 3:517–519). He was even more explicit when he proposed again the Copernican proof deduced from tides (*de aestu maris deque defensa a Morino titubatione Telluris*) in the *Pars secunda* of his reply to the *Alae Telluris fractae* of Morin (Gassendi 1658, 3:531–41).

[14] Gassendi — like Peiresc and other French Galileans — remained convinced for a long time that the sentence of 1633 had been issued against Galileo "ad personam." Thus the sentence did not imply, in his view, adherence to geocentricism as an article of faith. In the *De motu impresso*, the Canon of Digne affirmed that only Cardinals had approved the sentence: "Cardinales aliquot approbasse Terrae quietem dicuntur" clearly showing that he did not consider these positions in favour of the immobility and centrality of the Earth as "articulum fidei" (an article of faith), nor as a dogma "apud universam Ecclesiam promulgatum ac receptum" (Gassendi 1658, 3:519).

[15] "Nempe meminisse semper oportet nos non philosophos tantum esse, sed etiam Christianos, philosophiamque nostram nec debere, nec vero etiam posse a Christiana Fide discrepare" (Gassendi 1658, 6:451).

[16] In his letter of 7 November 1642, De la Mare defined Le Cazre as "vir alioquin maxime inter suos extimationis et philosophicis mathematicisque disciplinis large imbutus, ut tu melius aliquando ex illius scriptis noveris, quae non pauca habet de motu."

[17] The *Epistola tertia*: "quod, tametsi tempore primam, visum est tamen postponere, quod praeter argumentum cum superioribus commune, contineat etiam explicationem plurium aliarum difficultatum" (Gassendi 1658, 3:625–650), dated December 6 1642, contains the reply to criticisms by Le Cazre in his letter of 3 November 1642. The first two *Epistulae* confuted Le Cazre's arguments (Le Cazre 1645b and 1646) against the Galilean-Gassendian science *de motu*.

[18] The work was dedicated to Gassendi.

[19] Replying to the accusation of defending a "rash" philosophy, Gassendi defended *libertas philosophandi*: "ad quae mere sunt naturalia, quod attinet, non nego quidem ea me philosophari libertate, ut non uni alicui Sectae eruditorum, ut vocas, haerescam" (Gassendi 1658, 3:627). Gassendi devoted a long paragraph to the compatibility of the philosophy of atoms with Christian faith in general and the sacrament of the Eucharist in particular (Gassendi 1658, 3:636–638).

movement in empty spaces and by the Galilean conception of motion with its relativistic structure and its rigorous laws, which appeared to Gassendi the direct consequence of the mechanical attractions between the atoms. He lastly emphasized that the superiority, at least on a hypothetical level, of this organic body of doctrines was further highlighted by the possibility it offered of conceiving and constructing working models of the world machine. As is well known, the conception of knowledge as a tool which allows man to reproduce natural phenomena is continuously insisted upon by Gassendi.

In an extraordinary passage of his letter to Le Cazre of November 1642, Gassendi in fact described a physical model, a sort of hydraulic planetary, capable of simulating the formation and the orderly functioning of the cosmos. If a new Daedalus or an Archita could construct this type of system, he stressed, basing themselves on the new principles of the Galilean-Gassendian dynamics, why should we exclude the possibility that Divine omnipotence would be able to create the universe in which we live and make it work perfectly using these same principles?:

> Heinc proinde dico, et unumquemque Siderum globum in ea parte mundani spatij, quam Deus ab initio ipsi praescripsit, circumgyrari, et globum Telluris in ea parte mundani spatij quiescere, in qua Deus ipsum initio constituit. Rem ita esse intelligo, ut si quis plureis apparet globos ex ea materia, quae sub pari mole, sive ambitu tantum ponderet, quantum aqua, et ipsos in aqua quiescente constituat. Quilibet enim eorum globorum, ubicumque fuerit constitutus, ibi conquiescet; et neque ex summo imum petet, neque ex imo summum; neque ex utrovis extremo medium; neque ex medio utrumvis extremum; neque ex medio, extremove locum interceptum, neque ex loco intercepto medium aut extremum. Etsi fingas Daedalum, Architam, aut alium artificem adeo ingeniosum, ut uno eorum alicubi intra aquam constituto efficere possit ... varios illis circa ipsum obeundos motus indere; ii globi peragent suos motus quamcumque ad partem instituti fuerint; nempe seu prope superficiem, seu prope fundum; seu sub medium, seu prope medium, seu procul a medio; scilicet tam ille, quam isti, ob ipsam cum aqua ...aequilibritatem neque graves erunt, ut subsidant, neque leves ut avolent.
>
> Subinde ergo comparo cum immoto globo Tellurem, cum circum-ductis Sidera; et dico, sicut globus ille emoveri potest e loco in quo est, et promoveri versus alium; sic posse quoque Tellurem emoveri e loco in quo est et promoveri versus alium, sic posse ipsam quoque Tellurem emoveri a loco in quo est et promoveri versus Lunam; addoque ut globus ille in quocumque alio aquae loco reponatur, in eo pari modo quiescet, neque priorem repetet; sic et Tellurem, in quocumque loco constituta fuerit, in eo mansuram, nec pristinum repetituram. Et dicis tu quidem *attolli*, quod ego heic simpliciter dico *emoveri*. (Gassendi 1658, 3:631–632)

As in the "Platonic myth" of the Galilean *Dialogue*, acceleration, the properties of motion, and attraction with its purely mechanical model of action, constituted the fundamental elements of cosmogony. At the same time, the presumed privilege of centrality invoked for the Earth by Le Cazre and by the supporters of traditional ideas appeared simply as the effect of Man's natural tendency to make himself the measure of everything:

> Translata Terra versus Lunam ad Antipodas existentem nos non propterea avolaturos in derelictum a Terra locum; ut neque etiam antipodas, translata Terra versus Lunam factam nobis ad verticem; sed et illos et nos perinde in eadem antiqua sede versaturos, tanquam simul translato Terrae centro, respectu cuius et comparate ad situm capitis pedumque nostrorum, censebimur semper ascendere et descendere, sive locum sursum deorsumque habere; non autem simpliciter respectu loci in quo Terra aut erit, aut fuerit, et qui seu centrum Mundi sit, seu non sit, nihil ad ascensum aut descensum faciat. (Ibid., 632).

Gassendi would later on insist upon the inevitable and necessary congruence between the principles of motion introduced by Galileo and the structure of the real world. Inertia and the conservation of motion, the composition of motions, and acceleration as a continuous process constitute, in fact, the major components of a new natural philosophy and of a new conception of the universe (ibid., 632–636).

After the violent reactions of Le Cazre and Morin, the debate on the *De motu* by Gassendi assumes a different tone.

Insinuations regarding the Copernican implications attenuated, while the adversaries' attention concentrated on the critical examination of Gassendi's presentation of the Galilean laws of motion.

The reasons behind the attenuation of the cosmological polemics can only be hypothesized. Gassendi was an influential person, a respected man of the Church, esteemed with irreproachable behavior. He had besides, on many occasions, stressed that the theories that he was outlining belonged to the sphere of purely hypothetical doctrines, which he proposed only in trying to account for phenomena; in any event, he was absolutely ready to submit to the decisions of the Church.[20]

Under these circumstances, an explicit accusation of heresy directed at Gassendi had little chance of success and could even backfire against whoever proclaimed it. Therefore, it was more prudent and opportune to try to confute the new conception of motion, demonstrating its intrinsic weakness and thereby undermining the

---

[20] He answered Le Cazre's insinuations, for example, by confirming that, even though, as far as he knew, the Sovereign Pontiff had not confirmed the sentence of Galileo, making it "universal," he was very ready to submit to the decisions of the Church: "quod me attinet, me vel sola fama, habitaque fides tuis literis ita movet ut non expectem promulgationem [of the sentence], sed statim prorsus exosculer et plane caeca, ut dicitur obedientia ipsum excipiam" (Gassendi 1658, 3:641). However, he had presented arguments that seemed not only to make plausible, but also necessary and diffusely practised a non-literal interpretation of the Holy Scriptures.

structure that Gassendi had proposed as the fundamental evidence in favor of the Copernican hypothesis. This option is evident in the *Physica demonstratio* by Le Cazre (Le Cazre 1646). Here the Jesuit avoided proposing strong insinuations on the connection between the Galilean dynamics exposed in the *De motu impresso a motore translato* and the Copernican system. The Jesuit's attack was concentrated exclusively on the Galilean definition of motion in free fall, which he considered wrong. Le Cazre (ibid, 36ff.) resolutely refuted the odd numbers law (1, 3, 5, 7, 9 etc.) suggesting that it be substituted with the continuously double geometric proportion (1, 2, 4, 8, 16, etc.). He also contested that a heavy body moving naturally from rest, would pass through — as Galileo had stated — all the infinite minor degrees of velocity before reaching any given velocity. This different opinion derived from Le Cazre's conviction that velocity increases not according to time, but according to distance (ibid., 26ff.). Lastly, he denied the validity of the postulate proposed by Galileo at the opening of the *De motu naturaliter accelerato*,[21] and he also rejected the experiment of the interrupted pendulum introduced by the Pisan scientist to confirm this postulate.[22]

Le Cazre's reasoning constitutes an emblematic example of the objective difficulties of assimilating the new concept of motion, offering a whole array of paralogisms, misunderstandings, deductions from weak principles, and pseudo-experiments, such as that of the scale with which he believed to have demonstrated that impact (and the velocity of heavy bodies in natural motion), increases with the height of the fall.[23]

Even without the insinuations regarding Gassendi's Copernican sympathies, the discussion of the new laws of motion is still characterized by a strong *vis polemica*. For Le Cazre, Galileo had proclaimed himself author of a "new science," while his construction rests upon false principles "non modo suspicionibus meris, vixque probabilibus coniecturis ... sed ex principiis aperte falsis evidentibusque paralogismis omnia concludi: ex quo consequens est novam illam evanescere scientiam" (ibid., 5–6). And he continued accusing Galileo and his "tyrones" of insolence, of intentional mystifications to sustain at all costs their pseudo-science, and of unheard-of stupidities, giving a sinister laugh at the paradoxical blindness of a philosopher who proclaimed himself to be "linceo": "Lynceus Philosophus ac Mathematicus, Lynceorum princeps, in tam aperta luce caecutiat" (ibid., 9).

The spiteful tone indicates that the objective had not changed, even if Le Cazre had now chosen to open fire, not directly upon Gassendi, who is indeed considered an accomplice, but upon Galileo himself.

Gassendi replied with a letter to Le Cazre in March 1645, later distributed as the

---

[21] "Altera quoque erroris causa in Galilaei placitis inde etiam manavit quod sibi dari et gratis concedi postulat: gradus velocitatis eiusdem mobilis super diversas planorum inclinationes acquisitos tunc esse aequales, cum eorundem planorum elevationes ponuntur aequales" (Le Cazre 1646, 9ff.).

[22] "Experientia qua Galilaeus suum postulatum confirmat renititur" (Le Cazre 1646, 11).

[23] "Experientia nova et admiratione digna, modum mensuram ac rationem accelerationis motus in naturalium gravis descensu evidenter exprimens" (Le Cazre 1646, 18–26; see also Tannery et al. 1945–1988, 12:122–123).

first of the *Epistolae tres de proportione qua gravia decidentia accelerantur*, published in Paris in 1646 (Gassendi 1658, 3:564–588).[24] He enhanced Le Cazre's misunderstandings, showing the weakness of his reasoning, reversing the interpretation of his scale experiment (ibid., 575–579),[25] and above all, demonstrating with a careful geometric analysis that Le Cazre's continuously double geometric proportion resulted in absurd consequences. Gassendi passionately defended the Galilean theories, particularly insisting on the conception of acceleration as a continuous process and on the fundamental and close connection between velocity and time (ibid., 582–583, 565–566).[26] Regarding the Galilean postulate refuted by Le Cazre, Gassendi informed the interlocutor that it had been demonstrated by Torricelli in the *De motu* of 1644 (ibid., 569–572).[27]

Le Cazre responded with the *Vindiciae demonstrationis physicae*, sent privately to Gassendi from Metz on 6 April 1645.[28] In this writing he meticulously reasserted his own theories, accusing the interlocutor of having falsified or, at the very least, having misunderstood his thought. The Jesuit reproposed the scale experiment (ibid., 604–607), contested the admissibility of the Torricellian demonstration of the postulate,[29] and obstinately insisted upon the falsity of the Galilean conception of motion, drawing arguments from the experiments of both Mersenne in the *Cogitata* (Mersenne 1644a), and the reflections by the Jesuit Onorato Fabri on the real nature and laws of acceleration in free fall.[30]

The second of Gassendi's *Epistulae* of 1646 (Gassendi 1658, 588–625) contains a mordant reply to the *Vindiciae*. It is worth remembering the explicit self-criticism to which Gassendi submits himself: *De quodam lapsu emendando circa causam accelerati gravium motuum* (ibid., 621–623). In the *De motu impresso*, Gassendi had attributed acceleration to a twofold cause: constant attraction from the center, and the air displaced by the descending body which flowed behind it, giving it a further push (ibid., 497–498). This twofold cause (curiously, the second one reproposed the Aristotelian theory of antiperistasis to explain the motion of heavy bodies once separated from their motor; it seemed, furthermore, to presuppose a full universe, which is surprising in an atomist like Gassendi) had been introduced by Gassendi in order to account for the increase according to the series of odd

---

[24] "Epistola prima admodum Reverendo et religiosissimo doctissimoque Viro P. Petro Cazraeo Societatis Iesu."

[25] "De experimento in bilance facto ac aliud revera probante quam velocitates esse sicut spatia."

[26] Egidio Festa, whom I thank, has brought to my notice that in some texts of the *Syntagma* (Gassendi 1658, 1:341), Gassendi, confirming the "continuous" character of acceleration in natural motion, proposes an analysis from which it follows that this continuity is not intrinsic, but only apparent ("ad sensum"). See Festa, in Gassendi 1994, 2:355–364.

[27] "De postulato Galilaei circa motum super aeque altis non aeque inclinatis planis." Gassendi had had from Pierre Carcavy a copy of the *De motu* of Evangelista Torricelli (Torricelli 1644) where "Galilaei successor eximius demonstraverit in eo istud postulatum" (Gassendi 1658, 3:570).

[28] The *Vindiciae* were sent to press later (Le Cazre 1645b).

[29] "Quae vero de libro Torricelli postea adjungis (etsi ea non viderim) partim vera, partim falsa et saltem incerta esse non dubito" (Gassendi 1658, 3:601).

[30] At least in one case, Father Le Cazre echoes the "physical" analysis of motion of the confrère Honoré Fabri (Gassendi 1658, 3:616).

numbers of spaces traversed in equal times from rest. The new reflections developed in the effort to reassert, against Le Cazre's objections, that acceleration is a continuous process, have now convinced Gassendi that the odd number law can be deduced by the simple hypothesis of an attraction from the center. The recognition of the error is proposed without embarrassment. Indeed, it becomes a further argument to use against Le Cazre's wrong theories. Gassendi admits in fact to having fallen into error because he had not taken into account the continuous process of acceleration "quia enim non attendi velocitatis gradum primo momento acquisitum ita integrum maneret in secundo, ut ad superandum duo spatia valeret, ipsumque habui quasi valeret solum ad superandum unicum" (ibid., 621). He had himself experienced the difficulties implicit in assimilating the "integration" of continually increasing degrees of velocity and the implications of this key concept of Galilean kinematic for the proportion of the spaces traversed in equal times from rest. Ingenuously confirming his *lapsus* he made Le Cazre know at first-hand the decisive importance geometric competence played in the understanding of the new concept of motion and the crucial importance of the close relationship between velocity and time that was contested by Le Cazre.[31]

The *Epistulae* of 1646 marks Gassendi's formal exit from the dispute on the Galilean conception of motion. But his retreat from the scene did not at all bring about a conclusion to the Galilean *affaire* of the laws of motion.

New Interlocutors

Other characters were on the scene long before, proposing new and subtler strategies. The key character was undoubtedly the Jesuit Onorato Fabri who still needs to be studied thoroughly in order to illuminate the many apparently contradictory aspects of his personality and his scientific production.[32]

---

[31] "Iam lapsus fuit, quatenus proinde velocitates ut spatia habere se admisi imprudens. Quia enim non satis attendi velocitatis gradum primo momento acquisitum ita integrum manere in secundo ut ad superandum duo spatia valeret, ipsumque habui quasi valeret solum ad superandum unicum" (Gassendi 1658, 3:621).

[32] Father Fabri appears like an ambiguous personality, on whom his contemporaries expressed contrasting opinions. He surely represented one of the most important participants in the dialogue which authoritative representatives of the Company of Jesus in Italy and in France held with exponents of the new science. At the same time, Fabri conceived new apologetic strategy views. He tried to incorporate the main scientific novelties, both astronomical and mechanical, into the body of Aristotelian natural philosophy, which he updated substituting syllogistic logic with mathematics. This attitude may help to explain the opposition of the more intransigent representatives of the Jesuit Order against Father Fabri. For example, Thibout wrote to Mersenne, on 17 June 1647, that Fabri "a ce que m'asseure Mons.r Mousnier, il est traversé par les Peres de sa Compaignie, et croit on qu'il font tout ce qu'il peuvent pour le faire sortir, comme ils ont faict leur possible pour empecher l'impression de ses oeuvres" (Tannery et al. 1945–1988, 15:245). The sending to Rome of Fabri, at the end of 1646 at the decision of his superiors of the College of Lyon, where he resided, has indeed been interpreted as punishment inflicted because of his innovative teaching of philosophy and science (Tannery et al. 1945–1988, 15:234–236). On the other hand, the constant contraposition of Fabri to the mechanical and astronomical conclusions of Galileo and Galileans provoked the resolute opposition of many

The debut of Father Fabri in the second Galilean *affaire* goes back to 1643, and appears immediately marked by extremely ambiguous characters. On August 9, 1643, encouraged by a Jesuit brother whom he does not name, Fabri sent a long letter to Gassendi, still in part unpublished, in which he sketched a sort of autobiography (Tannery et al. 1945–1988, 12:275–279).[33] He strongly emphasized his own choice of submitting natural phenomena to rigorous mathematical treatment, in consideration of "rerum physicarum et mathematicarum communio" (ibid., 276).[34] Furthermore, he clearly kept his distance from the natural philosophers of the Scholastic, full of litigious and often purely verbal disputes "Porro, cum Physicam appello, nolim, quaeso, intelligas litigiosam illam quam vulgo in scholis nostri philosophi docent... sed iucundam illam quae naturales effectus primo explorat sensu tum vero ad suas causas reducit" (ibid., 276).

Fabri then revealed to Gassendi his own theories about the physical causes of the motion of the heavenly bodies, specifying that he followed the common opinion as to rest of the Earth and motion of the Sun (ibid., 278). He outlined a system of evident Tychonian structure, but full of Keplerian reminiscences and founded on the explanation of planetary orbits as the consequence of the composition of uniform circular motions and accelerated straight motions. On this basis Fabri came to delineate a purely mechanical cosmogony that echoed the "Platonic myth" of Galileo's *Dialogue*, a work which he appears to have read very attentively (ibid., 279). He also specified that he had conceived a series of arguments against the Copernican system, about which he solicited Gassendi's opinion (ibid., 277–278). Lastly, he informed Gassendi of having written a treatise *de motu locali* based on hundreds of rigorous geometric propositions (ibid., 277).[35] Curiously, Fabri supplied Gassendi with very brief information about the results of his work

---

authoritative followers of the Pisan scientist. In particular, Giovanni Alfonso Borelli will vigorously denounce the ambiguous new apologetics of the Jesuit of Lyon on several occasions. He considered him an enemy much more dangerous than the declared opponents of the new scientific ideas (Borelli 1667). Borelli had already opposed Fabri's *Brevis adnotatio in Systema Saturnium* in 1659–1660 (Fabri 1659; Galluzzi 1977). As confirmation of the ambiguity of this important personage, there is the different and much more favorable disposition of Prince Leopoldo towards him. The Prince generously received and effectively protected him when, in 1660, Father Fabri was examined and imprisoned by the Holy Office. Fabri had indeed declared the necessity of a not literal interpretation of the passages of the Holy Scriptures where the immobility of the Earth is affirmed, in case an indisputable proof for the movement of the Earth should be produced. Moreover, Michelangelo Ricci presented him to Torricelli with flattering words (Galluzzi et Torrini 1975, 1:381). See Sommervogel 1890–1932, 3:col 510–521; 9:col. 309) and Fellmann 1959, 1971, and 1992; for his polemics with Gassendi, see Sortais 1920–1922, 2:38–401). For an essential biography, see De Vregile 1906, 5–15. For useful information about the development of his reflections *de motu*, see Drake 1970b, 1973, 1974, 1975; Caruso 1987. For Fabri's reflections on mathematics, see Fellmann 1959. A Ph.D. dissertation has been dedicated to Fabri by Lukens (1979).

[33] In this work are published only some paragraphs of the letter. The letter is in the National Library of Paris, Fonds Lat., Nouv. Acq. 600:19–31. This letter is not published in Vol. 6 (which contains the correspondence) of Gassendi's *Opera omnia*.

[34] "Nullus fere sit in Physica tractatus qui mathesi carere possit."

[35] He also added that he had written a treatise on secondary qualities: "motui locali succedunt qualitates sensibiles: color, lumen, sonus, et caet. quarum omnium explicationi quantum mathesis conferat tuo judicio relinquo." Mersenne will give a survey of Fabri's theory of colors (Mersenne 1644b, f. 5r not numbered).

on motion, limiting himself to declaring that acceleration along different inclinations increases according to the inverse sines, a theory which corresponded to that previously proposed by the resolute Copernican Ismael Boulliau (ibid., 3:527, 626–629; Mersenne 1644a, 49–50).[36]

Gassendi's answer, on 20 August of the same year (ibid., 12: 282–284; Gassendi 1658, 6:167–168), was full of appreciation[37] and cautious reserve (especially regarding the Jesuit's anti-Copernican arguments).[38] As to local motion, Gassendi emphasized that the increase of acceleration according to inverse sines is practically equal to the proportion affirmed by Galileo, Mersenne and himself "nonnihil differt ab ea quae Galilaeus, Mersennusque et, si fas dicere, etiam ego, observitare visi sumus." However, the difference "circa ipsa initia motus" was so modest that it could be considered insignificant (ibid., 283–284).[39]

On 21 August of the same year, just twelve days after the letter to Gassendi, Father Fabri addressed a long letter to Mersenne in French (ibid., 285–302), the first known document of a correspondence which had begun some time earlier. The Jesuit illustrated to Mersenne his own ideas regarding the cause and nature of local motion. Curiously enough, he proposed a substantially different theory to that communicated to Gassendi. In "natural" motion, according to Father Fabri, spaces traversed in equal times from rest did not increase according to the Galilean proportion of odd numbers, nor according to inverse sines, but according to natural numbers. He had derived such a proportion from a rigorous causal analysis of motion. Movement for the Jesuit was produced by *impetuosité*: degrees of equal impetus were acquired in single instances of time and the velocity of motion increased by the summing of this impetus.[40] The reasons that induced Father Fabri to propose the increase of spaces according to the series of natural numbers were based on his conception of the "instant." Fabri talked of "physically" finite instants,[41] that is of atoms of equal times, to each of which corresponds the production of a degree of impetus, and, as a consequence, a traversed space. In the

---

[36] For Boulliau's deduction of the law of the acceleration of natural motion from the Copernican hypothesis, see Koyré [1955] 1973, 55–66.

[37] In particular, Gassendi shows curiosity about the anticipation that he had received on Fabri's treatise of colors: "Miratus iam fueram tuum illum tractatum *de Coloribus* tametsi neque integrum, neque nisi cursim legendum concessum" (Tannery et al. 1945–1988, 12:282).

[38] "Quas te adversus Copernicum excogitasse rationes dicis, dignae erunt haud dubie tua illa solertia; siquidem, tametsi non satis percipio quid ex tarditate ejus partis rotae quae adversam axi viam tenet possit in ipsum colligi; auguror tamen te exinde rationem tenuisse validissimam vel ex eo quod terram quiescentem supponis, ubi de motu coelestium physica causa sermonem instituis" (Tannery et al. 1945–1988, 12:283).

[39] Gassendi verified the extreme similarity between the two propositions in terms of spaces traversed in equal successive times from rest.

[40] "Sur quoi je dis — que ma proportion double arithmetique suppose necessairement des instants et des points, puisque chasque production nouvelle de mon impetuosité se fait dans un instant consequif" (Tannery et al. 1945–1988, 12:286).

[41] "Je compose le continu de points ou d'indivisibles, ce qui est contre tout le 10e d'Euclide ... a mon advis ... le continu, un ligne par exemple, n'est pas divisible actuellement jusques a l'infiny, mais seulement en puisance" (Tannery et al. 1945–1988, 12:290–291).

second instant, the first impetus being conserved, the degrees of impetus become two and therefore the space traversed is doubled. And so on.[42]

Mersenne objected to him that experience demonstrated that spaces in natural motion increased in equal time from rest according to the Galilean proportion of odd numbers. Fabri replied that those results were derived from the fact that "la commune mesure du temps che l'on prent pour mesurer tout les temps de la cheute et la proportion des accroissements de vistesse n'est jamais un instant, et qu'elle en contient presque une infinité."

If "l'on reduit les parties des temps que l'on a prises aux instants, l'on trouvera que l'espace acquis respond a peu près aux experiences de Galileo et de V. R." (ibid., 286). Father Fabri stressed the substantial equivalence on the level of experimental verifications between the two hypotheses, insisting on the fact that the more indivisible physical instants were contained in the equal parts of time taken as measure, the less would result the differences between the values foreseen by the two different proportions.[43] It is not surprising then for Father Fabri that the experiments seemed to substantially confirm the Galilean proportion, notwithstanding it was wrong. The small differences between experiments and theoretical previsions that Mersenne had underlined in the fourth of his *Question theologique, physiques, morales et mathematiques* (ibid., 289)[44] were to be imputed to the slight differences between the mistaken Galilean proportion and Father Fabri's physically correct one.

The Jesuit was careful to explain that, if on the practical and experimental level, both proportions were basically equivalent, on the level of truth their difference was considerable. His hypothesis of "finite" and indivisible instants implied acceleration as a discontinuous process, that is the integration of instantaneous and ever increasing degrees of uniform velocities. Thus, a falling body starting from rest did not pass through all minor degrees of velocity — as Galileo had affirmed — but initiated motion with a determined degree of velocity. The proportion of the growth of spaces traversed in equal times from rest according to natural numbers

---

[42] For a discussion of Fabri's analysis, see Drake 1970b, 1975.

[43] "D'autant plus que l'on prendra de partyes de temps toutes esgales a 2 instants, comme celle que j'ay prise, durant laquelle le corps qui descent fait les trois partyes de l'espace AB; et d'autant plus aussy que l'on prendra d'espace, ma proportion arithmetique s'esloignera tousjours plus de la vostre, quoyqu'inegalement. Par exemple: sy je ne mets que deux partyes de temps dont chascune soit esgale a deux instants, la proportion de Galilee me donne 12 points d'espace, supposant que durant la 1ere partye de temps le corps descendant face l'espace susdit AB, qui contient 3 points; et ma proportion arithmetique donnera 10 pointz, parce que j'ay 4 instants; et sy je metz trois partyes de temps, la proportion de Galilee me donne 27 et la mienne donne 21, parce qu'il y a 6 instantz; et sy je prens 4 partyes de temps, Galilee me donne 48, et moy qui ay 8 instantz, je donne 36. Par ou il appert que d'autant plus de partyes de temps que l'on prendra successivement, que la proportion de Galilee a la mienne croistra en inegalité majeure, comme il est aysé de voir dans les exemples donnés; car quand j'ay pris seulement 2 partyes de temps, la proportion estoit 1 1/5; quand j'en ay pris trois, elle estoit 1 2/7; quand j'en ay pris 4, elle estoit 1 1/3; et sy j'en prens 5, elle sera 1 4/11" (Tannery et al. 1945–1988, 12:287–288).

[44] "Il ne faut pas s'estonner sy la proportion trouvèe par l'experience respond a peu prés a celle de Galilee, et en la veritable raison du peu qui s'en manque, ce que vostre R. a fort bien observé en son livre de *Questions curieuses* Q. 4" (Mersenne 1634b, Question 4,16).

was therefore the only true one, depending on a rigorous causal explanation of natural motion (impetus and the atoms of time). Father Fabri concluded by mentioning the imminent publication of his *Metaphysica*, wherein the conformity of the causes of motion with metaphysical principles would have appeared evident (ibid., 293–295).[45]

Illustrating the analysis of motion of his teacher Fabri, Mousnier, in the *Tractatus physicus de motu locali* of 1646, confirmed that although usable on practical grounds, the proportion adopted by Galileo was intrinsically false: "igitur haec esto clavis huius difficultatis; progressio simplex principium physicum habet, non experimentum; progressio numerorum imparium experimentum non principium, utramque cum principio et experimento componimus" (Mousnier 1646, 108).

Therefore, on the level of truth, the Galilean theory could not be admitted: motion had to be explained "per causas" and physics presupposed metaphysics.

Father Fabri's position undoubtedly presented many novelties and differences in comparison to those of Le Cazre (Father Fabri will, in fact, reject his continuously double geometric proportion).[46] It appeared, above all, much more subtle and articulated. Fabri did not expect Gassendi and the supporters of the new ideas to recognize Galileo's doctrine of motion as absurd and abandon it. Galileo's theory was instead acceptable as a perfectly useful instrument for physics research. Even if he is never mentioned in the letter from Fabri to Mersenne, it is absolutely clear that Gassendi is the main target of his reflections. Gassendi had proposed the Galilean analysis of motion as an essential ingredient of the Copernican cosmology. And the consequences of this operation menaced, besides the dogmas of faith, the whole foundation of traditional knowledge.

It is probably not by chance that Fabri avoided presenting these subtle analyses in the letter written a few days before to Gassendi, whereas, for quite some time, he had been keeping Mersenne informed about them. Mersenne had in fact shown some hesitation in accepting the Galilean theory and laws of motion. Besides, he was not at all willing to follow Gassendi in using the reformed Galilean dynamics as the supporting structure of a new philosophy and cosmology.[47]

Father Fabri privileged his relationship with Mersenne not only in order to bring him over to his side, but also to use the Father Minim as an effective channel of communication and transmission. He knew that Gassendi would have been

---

[45] The subordination of *physica* to metaphysical principles is emphasized as necessary: "Or comme l'extension peut estre plus parfaite ou plus imparfaite, je dis le mesme de la duration mais cecy est une pure metaphysique, a laquelle il appartient d'expliquer tous ces effects formels des raisons universelles, c'est â dire qui conviennent esgalement a l'estre materiel et au spirituel. Par exemple, estre, estre substance, estre accident, quantité, qualité, rapport, estre au lieu, au temps, se mouvoir, ou plus viste ou plus tardivement, agir, patir, et caetera, ce que la veritable physique doibt supposer."

[46] "reicitur sententia illorum qui volunt hanc progressionem fieri inxta proportionem geometricam, quam vides in his numeris 1.2.4.8.16 quae licet initio minus recedat a vera, in progressu tamen multum aberrat, nec est ulla ratio quae pro illa faciet" (Mousnier 1646, 111).

[47] For Mersenne's position in reference to the Copernican hypothesis, see: Lenoble [1943] 1971, 454ff.; Hine 1973; and Dear 1988, 32–34, 113–114.

immediately informed by Mersenne. Moreover, he counted on testing the reactions of the main European interlocutors in correspondence with Mersenne, before making public his own hypothesis.

The Controversy Reaches Italy

Mersenne did not betray Father Fabri's expectations. On the occasion of his journey to Italy between the end of 1644 and the first months of 1645, Mersenne had the opportunity to intensify his ties with Torricelli with whom he had been corresponding since 1643.[48] To Galileo's successor in Florence and to Michelangelo Ricci, his customary interlocutor in Rome, he described the animated discussions that the Galilean laws of motion were causing in France and the criticism they were subjected to, not only by the Jesuits, but also by Roberval. He did not even refrain from informing the most authoritative of the Galilean disciples, Evangelista Torricelli, who had just finished publishing a *De motu* which was completely "Galilean" in its principles (Torricelli 1644), about the doubtful results of the tests to which he had submitted the Galilean laws of motion. Mersenne had made public the tests in the *Ballistic* section of the *Cogitata* of 1644.

From Italy, and later, after his return from Paris, Mersenne continued to keep Torricelli informed of Le Cazre's oppositions (Mersenne to Torricelli, 15 March 1645, in Galluzzi et Torrini 1975, 1:222–224; Tannery et al. 1945–1988, 13:399–404) of the passionate defense of Galileo held by Gassendi[49] and of the objections that he himself, incited by Roberval's severe criticisms, had formed against Galileo's doctrine of motion, even in the Torricellian reformulation (Mersenne to Torricelli, 13 December 1645, ibid., 553–556).[50] Lastly, he communicated to Torricelli that the Jesuit Onorato Fabri had entered into the controversy about the Galilean proportion of acceleration: "putat se demonstrasse proportionem accelerationis motus gravium Galilaei geometrice falsam esse, licet in praxi bonam; suam vero per numeros naturales 1, 2, 3, 4, 5 esse veram" (Mersenne to Torricelli, 1 March 1647, ibid., 350–352; ibid., 15:116–120). And he stated that the new interlocutor was learned and talented and deserved to be taken seriously: "vocatur Honoratus Fabri, estque Gesuita ... Est ad modum acutus et totam philosophiam quatuordecim voluminibus pollicetur" (Tannery et al. 1945–1988, 15:118–119).

Galileo's disciples in Florence and Rome probably suspected from the very beginning that the attack on the laws of motion was a renewal, or an appendix, to the Galilean *affaire* of 1633. This explains their extremely cautious attitude.

---

[48] The first letter of Mersenne to Torricelli is of August 1 1643 (Galluzzi et Torrini 1975, 1:71; Tannery et al. 1945–1988, 12:268–269).

[49] "Gassendi strenue refellit scriptum Jesuistae Cazrei" (Mersenne to Torricelli, 10 October 1645, in Galluzzi et Torrini 1975, 1:287–289; Tannery et al. 1945–1988, 13:492–496).

[50] Together with the letter of 13 December 1645, Mersenne sent to Torricelli the letter in which Roberval explained his own radical reservations on the Galilean science of motion and on Torricelli's *De motu* (Roberval's letter in Torricelli 1919–1944, 3:349–356).

Torricelli, in particular, tried in every way to avoid the insistent solicitations of Mersenne, whom he did not trust, above all because of the Minim's active personal participation in the critical discussions on the Galilean laws of motion.[51]

Compelled by his official position to assume the role of Galileo's defender, Torricelli moved with extreme caution, continuously inviting Mersenne not to divulgate his rebuttals as he specified at the end of his letter of the end of June, 1645, where he stated that Le Cazre's hypothesis was absurd: "Oro Paternitatem Vestram ne quis videat hanc epistulam; neque enim respondere est animus neque talia me movent."[52] When transmitting to Mersenne in Rome an earlier letter with the illustration of two experiments that confirmed the Galilean laws of *de motu*, Torricelli took care to specify that he would not have responded to further objections. And he begged the Father not to make public his thoughts except "solis tuis amicis" (Torricelli to Mersenne, 25 December 1644, in Galluzzi et Torrini 1975, 1:174–175). Faced with Mersenne's asphyxiating insistence, surprised as he was at the disciple's hesitation to enter the field of battle in the venerated Master's defense, Torricelli ended up removing the very object of contention, bitterly declaring that he did not care whether the principles *de motu* were true or false; in any case, it was legitimate for a geometer to assume them as true principles and derive from them all consequences ("che i princìpi della dottrina *de motu* siano veri o falsi a me importa pochissimo — si sfogò col Ricci — poiché, se non son veri, fingasi che sien veri conforme abbiamo supposto, e poi prendansi tutte le altre specolazioni derivate da essi principi, non come cose miste, ma pure geometriche" (Torricelli to Michelangelo Ricci, 10 February 1646, ibid., 276).

It was a retreat dictated by reasons of opportunity, not the renunciation of his own convictions. Not by chance, Torricelli will insist with force on the truth of the Galilean principles of *de motu*. Just a few months later in a letter of August 8, 1647 to his trusted friend G. B. Renieri he carefully rejected Fabri's proportion according to natural numbers (Torricelli to Renieri, 8 August 1647, ibid., 391–394).[53] It is worth noting that Torricelli carefully avoided sending his confutation of Fabri's proportion to Mersenne.

To the Italian disciples of the Pisan scientist it was evident from the beginning that the discussion on the Galilean laws of motion was not a candid intellectual

---

[51] Torricelli and the other disciples of Galileo had little sympathy for Mersenne, probably because of his diligence in collecting and transmitting critical remarks on the Galilean science of motion. The highly-colored representation of Mersenne that Carlo Dati gave later is symptomatic of the Galileans' suspicious attitude. Mersenne is represented as one who is more fit for collecting and promoting other people's inventions than for communicating his own: "come quei mercatanti che per iscarsezza di loro avere, malamente potendo far negozi, sfogano il genio loro guadagnando pure assai nel contrattare e mettere in vendita le merci altrui" [like those merchants who, because of lack of their substance, being able to do business with difficulty, express their mind in earning still much by bargaining and putting other people's goods up for sale] (Dati 1663, 6).

[52] Torricelli also declared the untenability of the experiment of the balance, proposed by Le Cazre and considered by him as conclusive (Galluzzi et Torrini 1975, 1:247–249).

[53] For Torricelli's attitude towards critics of the Galilean doctrine *de motu* and, especially, towards Mersenne, see Galluzzi 1976, 73–84.

debate, but rather an attempt to summon a new lawsuit against Galileo.[54] And this awareness explains their hesitations and caution. Torricelli and Ricci presumably ignored the role played by Gassendi at the beginning of these discussions with the publication of the *Epistulae de motu impresso*. They probably had not read this work, notwithstanding that Mersenne had punctually and repeatedly informed them of the noble and effective defense of Galileo's ideas assumed by Gassendi.[55]

It has also to be noted that the Italian situation was substantially different, because of the more effective capacity of control on scientific debate by the ecclesiastic authorities. Besides, as we shall see, the Italian Jesuits will soon follow the example of their French colleagues, directing severe criticism against the Galilean science of motion.

From France to the Netherlands

1. Le Tenneur

Father Fabri's theory of motion, already anticipated with explicit reservations by Mersenne in the *Ballistic* section of the *Cogitata* of 1644 "sententia philosophi subtilissimi qui statuit accelerationem pro diversis temporibus in eadem ratione qua numeri serie naturali disponuntur" (Mersenne 1644a, 52), caused animated reactions after the publication, by his pupil Pierre Mousnier (Mousnier 1646)[56] in 1646, of the collection of the Jesuit's lessons. It was of course Mersenne who in timely fashion informed the supporters of the Galilean theory *de motu* of the Jesuit's work and about his criticism of Galileo's conception of motion: "Insinuavit mihi Mersennus noster — Jacques-Alexandre Le Tenneur wrote to Gassendi on January 16, 1647 — esse aliquem alium quem totius Societatis acutissimum vocant, qui aliquid etiam adversus Galilaeum molitur, gloriaturque se demonstrasse spatia aequalibus temporibus in descensu gravium emensa esse inter se ut series naturalis numerorum 1.2.3.4. etc." And Mersenne had obviously urged Le Tenneur to reply: "monuit autem bonus ille noster ad illam me pugnam accingerem." The invitation to contend was passed on to Gassendi by Le Tenneur, who was convinced that — as in the case of the earlier attack by Le Cazre — this was a new attempt by the Jesuit Order, not only against Galileo but also against the *Epistulae de motu*

---

[54] By sending from Rome to Torricelli a summary of Le Cazre's *Physica demonstratio* (he had had a copy of it from Mersenne), Ricci emphasized the weakness and the absurdity of the remarks directed against Galileo. However he denounced the animosity of the French Jesuit towards the Pisan scientist: as he said, the Jesuit shows himself as light in his behaviour as he is in his doctrine: "con mille vanti di sé medesimo e scherno di Galileo si dimostra non meno leggero nei costumi che sia nella dottrina" (Ricci to Torricelli, 26 March 1645, in Galluzzi et Torrini 1975, 1:229).

[55] It is possible that Torricelli did not know Gassendi's *De motu impresso a motore translato*. Torricelli will declare to Mersenne in July 1646 that he had not read any of Gassendi's work: "Nihil enim ex eius [Gassendi] operibus vidi praeter vitam D. De Peiresc" (Galluzzi et Torrini 1975, 1:309).

[56] The *Liber secundus* was dedicated to discussions "de motu naturali" (Mousnier 1646, 74–132).

impresso by Gassendi: "fac igitur ut cum novo illo hoste dimicare victoriamque reportare possis" (Tannery et al. 1945–1988, 15:49; Gassendi 1658, 6:505). In his reply of March 11, Gassendi, however, declined the invitation for reasons of health and because of his engagement in other studies.[57] However, he urged Le Tenneur to assume the role of Galileo's paladin: "tradita tibi lampas iam est ut hunc quasi cursum absolvas" (Gassendi 1658, 6:266).

Mersenne's decision to invite Le Tenneur to reply to Father Fabri, was probably the consequence of his awareness of Gassendi's unavailability. Probably disturbed by the exhausting dispute with Father Le Cazre and seriously preoccupied by the relentless attacks coming from the Jesuit Order, Gassendi decided to retire officially from the controversy. However, from his letters we learn that he went on soliciting other authoritative friends, as, for example, Fermat,[58] to enter the field. Moreover, he did not tire of encouraging and offering advice to Le Tenneur.[59]

The latter had already had occasion to distinguish himself in the *affaire Galilèe* of the laws of motion. At the end of 1646 he had sent to Gassendi his *Disputatio physico-mathematica* (Le Tenneur 1646),[60] in which he offered a new criticism, based on acute mathematical analyses, of Father Le Cazre's continuously double geometric proportion. In the *Disputatio*, which is still unpublished, Le Tenneur stressed the necessary convergence of new Galilean mechanics and Copernican cosmology as had Gassendi, and dedicated considerable attention to the problem of the plausibility of recourse to the texts of the Sacred Scriptures in the disputes about the systems of the world. He skillfully demonstrated in many places that the Scriptures seem to belie even the Ptolemaic hypothesis. He asked, on this part of his writing, the opinion of Gassendi who hurried to send him his own approval.[61]

With a series of observations and demonstrations circulated in Mersenne's epistolary network, Le Tenneur showed that he had earned the trust granted him

---

[57] Gassendi was completing the *Institutio astronomica iuxta hipothesim tam veterum quam Copernici et Tychonis*, published in 1647 (Gassendi 1647).

[58] The Senator of Toulouse, solicited by Gassendi, set up a rigorously geometrical deduction of the Galilean proportion of the acceleration of natural motion almost certainly in 1646. Gassendi's solicitation must be related with his polemic with Father Le Cazre, as clearly emerges from the end of Fermat's letter: "Haec succinte et familiariter, Clarissime Gassende, scripsimus ne tibi in posterum fascescant negotium aut Cazreus, aut quivis alius Galilei adversarius et in immensum excrescant volumina quae unica demonstratione vel fatentibus ipsis authoribus aut destruentur aut inutilia et superflua efficientur" (Gassendi 1658, 6:543).

[59] No specific work has been devoted to this important personage. He animated Mersenne's correspondence after 1646, especially about topics *de motu* and in relation to debates on the existence of vacuum. For Le Tenneur, see Drake 1970b, 1973, 1974, 1975. For his participation in the discussion about vacuum, see Middleton 1965, 40.

[60] The manuscript, dated 1 November 1646, is in the National Library of Paris, Ms. Fonds Lat. 6740. The codex, which belonged to Melchisedec Thevenot, contains some letters of Father Le Cazre.

[61] In his letter of 24 November 1646, Le Tenneur had informed Gassendi of his having written some reflections on the presumed incompatibility between the Copernican doctrine and the Holy Scriptures: "Visum autem haec qualiacumque sint tibi communicare et censurae iudicioque tuo submittere, ut pote quorum author praecipue extitisti et quibus ortum dedit incundissima scriptorum tuorum lectio" (Tannery et al. 1945–1988, 14:628; Gassendi 1658, 6:504). Gassendi replied, expressing his own enthusiastic approval of the *Disputatio*; see Gassendi to Le Tenneur, 1 December 1646, in Gassendi 1658, 6:260–261).

by Gassendi, whom he kept constantly informed about the developments of the controversy. We are, in fact, in the presence of an important figure unjustly neglected. Le Tenneur is one of the many talented gentlemen of science who populate Mersenne's correspondence not only in France but also in Italy, England, and in the Low Countries.

His notable worth as a mathematician emerges in the *Traité des quantitez incommensurables* (LeTenneur 1640)[62] and in the *De motu naturaliter accellerato* (Le Tenneur 1649).[63] In a letter to Mersenne full of autobiographical references written on July 9, 1647 (Le Tenneur to Mersenne, 9 July 1647, in Tannery et al. 1945–1988, 15:287–299), Le Tenneur would proudly proclaim himself an amateur and self-educated person. He has carefully studied Galileo's *Dialogues* and *Two new sciences*, while he acutely and passionately defends the *De motu* by Torricelli from Roberval's objections (Le Tenneur to Mersenne, 9 July 1647, ibid., 289).[64] Le Tenneur appears as a convinced "Galilean" full of admiration for Gassendi's atomism. Of Gassendi he particularly appreciates the insertion of scientific ideas into an organic philosophical background as well as his talent as experimenter. As many others in those years, Le Tenneur was fascinated by Descartes' physics, but, as he wrote to Mersenne, he admired Descartes' insightful arguments, but did not find them at all convincing "je demeure plustot esblouy de la lumiere des ses raisonnement que je ne m'en trouve exclairé" (Le Tenneur to Mersenne, 9 July 1647, ibid., 295).

Thanks to these qualities Le Tenneur ended up assuming a central role in the conclusive phase of the *Galilée affaire* of the laws of motion. In a long letter-treatise to Mersenne on April 13, 1647, Le Tenneur in fact produced an effective rejection of Father Fabri's hypothesis, strongly reaffirming the truth and absolute accuracy of the Galilean analysis of motion (ibid., 173–199).[65] First of all, Le Tenneur demonstrated the fragility and impracticality of resorting to "physical" instants, or atoms of time, as suggested by Fabri (ibid., 175–177). He also vindicated the full legitimization of the Galilean definition of accelerated motion in the *Two new sciences*, insisting on the concept of acceleration as a continuous process and on

---

[62] For an analysis of this work, see Drake 1973, 267.

[63] See Drake 1973, 268.

[64] Le Tenneur is particularly severe towards Roberval and criticizes Mersenne because he was constantly appealing to his presumed authority and, above all, seemed to consider Roberval as Le Tenneur's teacher: "Vous prenes plaisir a apeler le braue Roberual mon ancien maistre. Qui donc vous en a tant dit? Se vante t'il de m'auoir apris beaucoup de choses? Certes sy cela est, il a grand tort, car je vous proteste que je ne tiens quoy que ce soit de luy. Ce n'est pas qu'en effect, il ne m'en peut monstrer beaucoup et que je ne me tienne bien inferieur a luy en science, mais il faut parler des choses comm elles sont. Et a fin que vous scachies comme la chose s'est passee. Il est vray qu'il me prit fantaisie vn jour de me faire expliquer par luy quelque theorique de planetes que je n'entendois pas. Mais, ayant recogneu apres quelques peu de visites qu'il auoit sy bien le don de s'expliquer que j'en faisois tout autant moy seul et que je n'auancois pas plus pour l'entendre parler, ie luy enuoiay de l'argent et le priay de ne plus prendre le peyne de venir" (Le Tenneur to Mersenne, 9 July 1647, in Tannery et al. 1945–1988, 15:291).

[65] The confutation concerns the presentation of Fabri's theses *de motu naturali* by Mousnier in the *Liber Secundus* of the *Tractatus Physicus* (Mousnier 1646).

the direct proportionality between acceleration and time (ibid., 182–186). He furthermore pointed out with clever demonstrations the absurd consequences that came from Father Fabri's hypothesis, offering at the same time an intelligent illustration of the Galilean theory of acceleration (based on the integration of infinite ever increasing degrees of velocity with the passing of time) and showing that from that theory derived necessarily the proportion of odd numbers, opposed by Father Fabri;[66] Le Tenneur proclaimed the hypothesis of the Jesuit as being simply false, while reasserting his full confidence in Galileo's conception of motion (ibid., 194–195). He finally declared that he had entered into the controversy on the insistence of a few friends and of Mersenne, and not because he had any faith in the possibility of changing the opinions of his stubborn interlocutors: "Quod scilicet de sua fama detrahi arbitraretur, si aliquando recantarent" (ibid., 174). In fact, Father Fabri replied defending his own theory point by point.[67]

Mersenne's correspondence records in detail the development of this phase of the *affaire* of the Galilean laws of motion. Le Tenneur nurtured the conviction that both Fathers Fabri and Le Cazre, even if with different approaches, were motivated by a common goal: to discredit Galileo and Gassendi, his prophet in France.[68]

On the other hand, Le Tenneur was aware that a too passionate intervention in defense of the Galilean laws of motion could have produced dangerous consequences. Thus, it is not by chance that he implored Mersenne not to over-emphasize his role as Galileo's paladin: "Ca esté en effect ma seule intention de decouvrir les erreurs de ces deux personnages [Le Cazre and Fabri] et non pas d'estaller l'opinion de Galilée" (Le Tenneur to Mersenne, 9 July 1647, ibid., 292).

The letters from Le Tenneur to Mersenne in 1647 (unfortunately Mersenne's replies are lost) have considerable importance because they allow us to follow the evolution of Mersenne's behavior in the *Galilée affaire*. It is evident that the fascination with the theory elaborated by Father Fabri, together with his reservations on the bold reproposal of Galileo's conceptions expressed in the condemned *Dialogo* (that is, the full integration between mechanics and cosmology) in the *Epistulae di motu impresso* by Gassendi, pushed Mersenne to reconsider the

---

[66] "Monendum autem est Galilaeum per omnes tarditatis gradus aliud nihil intelligere quam infinitos" (Tannery et al. 1945–1988, 15:196). Moreover, he confuted Fabri's thesis on the increase of velocity as a discontinuous process.

[67] Le Tenneur's letter was transmitted by Mersenne to Fabri and to Mousnier without revealing the author. In a letter to Mousnier of May 1647, Fabri recognizes the worth of the anonymous interlocutor and of his arguments. He left the task of answering to his own disciple, since the *Tractatus*, against which Le Tenneur had railed, "invito me, in lucem edidisti" (Tannery et al. 1945–1988, 15:235). The long answer, prepared by Mousnier but surely elaborated or, at least, approved by Father Fabri, was sent by Mousnier to Mersenne on 1 October 1647. It was inserted later in Mousnier-Fabri's *Metaphysica demonstrativa* (Mousnier Fabri 1648, *Appendix*, 3:609–659).

[68] In his letter to Mersenne of 21 May 1647, for example, Le Tenneur unites Father Le Cazre and Father Fabri because of their common engagement in demolishing Galileo's reputation: "Car l'on reproche a Galilée que sa définition du mouvement acceleré n'est pas bonne et luy objecte que l'acceleration ne se doit pas faire dez le commencement; l'autre dit, que c'est mal a propos qu'il suppose une infinité d'jnstants et que le grave tombant doive passer par tous les degrez de tardité" (Le Tenneur to Mersenne, 21 May 1647, Tannery et al. 1945–1988, 15:227–228).

favorable attitude towards the Galilean laws of motion manifested since 1633 (Mersenne 1637, 1:85–92, 125–128)[69] and substantially reaffirmed in the *Cogitata* of 1644 (Mersenne 1644a).

His initial timid doubts on the plausibility of the Galilean definition of naturally accelerated motion, on the sustainability of the principles on which it depended, and on the possibility of experimental verification of Galileo's laws of motion were, with the passing of time, progressively increasing. Le Tenneur was fully aware of this trend and for this reason, in letters to Mersenne in 1647, he took every possible step to avoid the Minim ending up by radically denying the Galilean analysis of movement. Such fears appear particularly evident in the letter to Mersenne on September 13, 1647 (Tannery et al. 1945–1988, 15:417–424). Mersenne had sent Le Tenneur the proofs of the *Tomus III* of his own *Novarum observationum* requesting his opinion on the book. Le Tenneur insisted on the necessity of attenuating the criticism that the Minim addressed to various aspects of the Galilean theory of motion. And he explicitly beseeched him not to mix with the group of the Pisan scientist's detractors: "voila sans doute ce qui se peut dire la dessus sans nier absolument l'opinion galiléenne comme font nos beaux docteurs sourcilleuz" (ibid., 420).[70]

2. Huygens

The letters that Mersenne exchanged with the young Christiaan Huygens between 1646 and 1647 confirm that Le Tenneur's impression of a progressive retreat by Mersenne from Galilean positions was anything but unfounded. Mersenne had been informed by Constantin Huygens of the demonstration that his son Christiaan had given of the Galilean odd numbers proportion (Mersenne to Costantin Huygens, 12 October 1646, ibid., 15:527–529). The demonstration was contained in the very important *De motu naturaliter accelerato* of 1646 (Huygens 1646),[71] in which Christiaan, just seventeen years old, criticized the theories exposed by Caramuel in his work on motion of 1644 (Caramuel 1644).[72] This text shows that Huygens had read very carefully the *De motu impresso* by Gassendi from which he had most probably obtained information about the Galilean theories and demonstrations on motion which he will be able to learn directly from the *Two new sciences* only later on.[73] From the *De motu impresso* by Gassendi, Huygens also

---

[69] See also n. 2.

[70] The two "doctors" are obviously Father Le Cazre and Father Fabri.

[71] See D'Elia 1985, 33–46.

[72] Caramuel rejects the Galilean proportion of space traversed in equal times from rest. He opts indeed — like Fabri — for the series of natural numbers that he finds confirmed by repeated and accurate experimental verifications. Caramuel's tone towards Galileo was particularly disdainful.

[73] The first of the two parts into which the *De motu naturaliter accelerato* is divided was written by Huygens without having been able to consult the *Two new sciences*. He read this work at the end of 1646, drawing inspiration for the second part of his work (Huygens 1888–1950, 11:68 n.1).

drew the theory of attraction as the cause of motion, from which he had inferred with rigorous geometrical procedures the necessary truth of the odd numbers proportion.[74]

On 13 October 1646, Mersenne manifested to Christiaan his own scepticism on the possibility of giving a rigorous demonstration of the Galilean proportion. With a resolute tone he affirmed that the principles which Galileo used "dans tout ce qui'il a dit du mouvement ne sont guère fremes." He then challenged the young correspondent: "si nonobstant cete consideration vous croyez que votre demon-stration soit ancore valable vous me ferez plaisir de me la communiquer" (Tannery et al. 1945–1988, XIV, 538–541; Huygens 1888–1950, 1:558–559).[75] In his letter of 28 October 1646, Christiaan Huygens showed how weak were the foundations on which Father Mersenne's confidence relied. He dissolved, point by point, with shrewd demonstrations, Mersenne's doubts (ibid., 567–573; ibid., 24–27). He peremptorily concluded that, in the vacuum, the Galilean laws of motion would prove absolutely true. And this was for Huygens decisive evidence in favor of Galileo, given that the rigorous and precise knowledge of motion had to be founded not on uncertain experiments, as Mersenne claimed, but only upon mathematical reasoning.[76] Mersenne replied on November 16, declaring himself stupefied at the shrewdness of the analysis. I believe, he wrote to Christiaan, that "Galilée eust esté ravi de vous avoir pour garand de son opinion." And yet, he reaffirmed that he still had many doubts (ibid., 612–614; ibid., 30–31) regarding the Galilean hypothesis.

These doubts are still, in fact, expressed in the *Tomus Tertius* of the *Novarum observationum* (Mersenne 1647, chap. 15–19:131–169) published in 1647. How-ever difficult it is to distinguish Mersenne's personal views from the tangle of opinions of other authors that he records with scrupulous faithfulness, it emerges as evident the effects of his progressive distancing between 1646 and 1647 from the Galilean science of motion. In the *Novarum observationum*, Mersenne inserted a lengthy review of the diverse conceptions of motion, recalling his initial position in favor of Galileo's doctrine, especially as a consequence of a series of substantially positive experimental verifications of it (ibid., chap. 15:131). He alluded to Gas-sendi's *De motu impresso* and to Le Cazre's objections "quem Gassendi copiosis-sime refutavit" (ibid.). He recalled the discussions between Le Tenneur and Father Fabri;[77] he introduced Roberval's objections and Torricelli's counter-deductions (ibid., 132); he furthermore faced the problem of the cause of motion, informing

---

[74] See Gassendi's demonstration of acceleration (founded on attraction) adopted by Huygens in the *De motu* (Huygens 1888–1950, 11:69).

[75] See also Dear 1988, 210.

[76] "Et ie ne trouve point d'autre progression qui ayent quelque regularité et la proprieté requise que cellecy. Et pour cela je croij qu'il n'y a point d'ordre du tout, ou que c'est celuy de ces nombres impairs" (Tannery et al. 1945–1988, 14:572; Huygens 1888–1950, 1:27).

[77] "Est et alius vir ingenio praeclarissimo qui numeros ab unitate naturaliter consequentes 1.2.3.4. etc. maluit adhibere gravium casibus ... quam etiam Clariss. Tennerius amicus singularis scripto nondum vulgato refellit" (Mersenne 1647, chap. 15:131–132).

the reader of the Gassendian proposal of attraction as the cause of motion (about which he expressed doubts), and of the Cartesian explanation of motion as the consequence of the pressure of subtle matter in a "full" universe.[78] He then put forth a relevant series of reservations on the Galilean definition of motion and on the admissibility of the postulate of the *Two new sciences* (ibid., 133–141), stressing how experimental verifications showed only approximate correspondence between Galilean laws and real phenomena. On the other hand, he resignedly admitted that to effect precise observations and experiments in this field was extremely difficult, especially because of the impossibility of having an exact measurement of time, even using pendulums, to the perfection of which Mersenne himself had contributed considerably (ibid., chap. 19:152–159).[79]

The conclusions that he deduced were bitter: without knowing its cause, motion cannot be an object of science. Since the cause of motion is unknown,[80] it can only be affirmed that the proportion of odd numbers proposed by Galileo describes with some approximation the real conduct of heavy bodies, and only in the initial part of motion from rest. Whoever tried to affirm Galileo's theories as real science went beyond legitimate limits, given that of motion we can at most claim a "docta ignorantia."[81] For Mersenne, the issues of motion once again called for the same embarrassing situation experienced with cosmological issues: "quemadmodum neque rationes quae hactenus allatae sunt in gratiam utriusque motus Terrae quidquam demonstrant ... etiam si plures vellent eam moveri ob rationum praestantiam quod id innurere videtur" (ibid., chap. 15:135). It was like admitting that in the *affaire* of the laws of motion, exactly as in the *affaire* of the world systems, someone had cheated, claiming to possess the truth without however putting forth necessarily demonstrative arguments.[82]

This analysis and these conclusions ended up by putting the good Father Mersenne objectively on the side of those who contrasted the operation proposed in Gassendi's *De motu impresso* and developed by his emulators. In short, what occurred was exactly that which Le Tenneur had intuited and tried to avert. If he did not explicitly side with Father Fabri, Mersenne reinforced however objectively his position; be it because he refused to consider Galilean doctrines of motion as a theory capable of accounting for real phenomena, be it because he vindicated the

---

[78] "Quid si [gravia] neque trahantur a Terra, neque propria gravitate ferantur, sed expellantur ab aere aut alia materia subtiliore, eo fere modo quo suber et alia corpora leviora sub aquam immersa expelluntur ab aqua, non aliqua peculiari aquae vi aut qualitate, nisi ea quae locum suum repetit" (Mersenne 1647, chap. 15:132).

[79] For Mersenne's experiments on pendulums, see Koyré 1953.

[80] "Vides igitur de his casibus corporum, quae vulgo gravia dicuntur, nihil penitus demonstrari posse donec innotescat principium, seu vera et immediata causa ob quam versus centrum haec et illa corpora suum iter instituant, quantunque iuventur aut impediantur in toto itinere ab omnibus aliis corporibus occurrentibus aut circumstantibus" (Mersenne 1647, chap. 15:133).

[81] "Quod si dixeris nos igitur ea ratione nullam scientiam istorum motuum habituros, quidni doctam ignorantiam ignoranti scientiae praeponas?" (Mersenne 1647, chap. 15:134).

[82] "Hinc fit ut in isto negotio, aliisque similibus, etiamnum cum D. Paulo possimus asserere: si quis autem se existimat scire aliquid, nondum cognovit quemadmodum oporteat eum scire" (Mersenne 1647, chap. 15:141).

necessity of that explanation per causas, on which the Jesuit had particularly insisted.

Not at random, exactly as happened in the Copernican affair, also in this case the proposal was made for a purely "hypothetical" conception of the Galilean doctrine of motion, to be considered as a practical tool — again it was Fabri who suggested it first — but not as a principle to act as a pillar of the new conception of the universe, delineated by Galileo in the condemned *Dialogo* and reproposed by Gassendi in the *De motu impresso a motore translato*. One could therefore say that for the theory of motion as well a kind of "Osiander argument" was put forth. Mersenne had more than a marginal part in it, certainly to the full satisfaction of the Fathers of the Company of Jesus. The explosive potential of the *Galilée affaire* of the laws of motion could, in this way, be finally defused. Mersenne might have been induced to accentuate his criticism by his hesitations before the audacious welding of Copernicanism and Galilean dynamics proposed by Gassendi. He could, moreover, have felt the fascination of the more "philosophical" explanations proposed by Fabri, or been affected by pressures of the Jesuits, or, finally, have been influenced, in the last years of his life by Descartes' strong reservations on Galileo's science of motion.[83]

It appears evident, in any case, that many authoritative members of the Company engaged in a process of systematic demolition of the Galilean theory of motion. With the exception of Father Riccioli, who had confirmed the Galilean laws via experiments but deriving from them an anti-Copernican argument (Koyré 1955; Galluzzi 1977), the Jesuit scientists systematically and neatly marked their distances from the principles and theorems illustrated by Galileo in the *De motu naturaliter accelerato* of the *Two new sciences*.

Some of them, like Le Cazre, and, later on, Father La Loubère (La Loubère 1658),[84] tried to force the "organic" supporters of the Galilean ideas, like Gassendi, into silence, with violent and threatening attacks, underlining the absolutely false character of those doctrines and their inevitable heretical implications. Others, like Father Fabri, assumed a more skillful and subtle attitude, by limiting themselves to emphasizing that the proportions between spaces and times in free fall proposed by Galileo could be replaced by other rules that had obtained in

---

[83] The final accentuation of the critical tones of Mersenne towards the Galilean science *de motu* has been stressed by Peter Dear. Dear remarks in Mersenne the passage from an initial pragmatic adherence to a refusal of the "principles" on which the Galilean vision was founded (Dear 1988, 215). For Dear it is evident that Mersenne suffered the influence of Father Fabri. However such an influence is not enough to explain Mersenne's change of attitude (Dear 1988, 216–217). Dear's opinion is that Mersenne was pushed to search for an explanation of motion "per causas" because of the influence of Descartes' *Principia philosophiae* (Dear 1988, 218–219). Dear also attributes Mersenne's sceptical attitude towards Christiaan Huygens (Mersenne to Costantin Huygens, 12 October 1646, in Tannery et al. 1945–1988, 14:527–529) to the influence of the explanatory model of physical phenomena contained in the *Principia* of Descartes. A different view of the reasons behind the evolution of Mersenne's attitude towards Galileo's laws of motion has been recently proposed by C. R. Palmerino (see Palmerino 2000).

[84] La Loubère again proposed Le Cazre's hypothesis against the then late Gassendi.

experimental tests a comparable degree of confirmation. These rules — like the continuously double proportion of Father Fabri — had the added value of not compelling that the whole structure of traditional knowledge be unhinged. The supporters of this second line of action legitimated the practical use of the Galilean laws generally avoiding open insinuations regarding their connection to the Copernican system. Moreover, they tried to win over some of the most influential and illustrious protagonists of the Galilean *affaire* to their cause — stressing the reasons of opportunity and caution. This strategy, however, did not produce successful results with Gassendi who after 1646, kept silent on these matters. Much better results were obtained with Mersenne, who in the last years of his life came substantially to satisfy their expectations.

Back in Italy

Precise evidence shows that continuous pressure was exerted by Fabri and other Jesuit Fathers on another illustrious protagonist of research on motion, the Genoese patrician Giovanni Battista Baliani, to distance himself from the Galilean position. It has to be stressed that, just between 1646 and 1647, Baliani proceeded with a relevant revision of the first edition of his *De motu* (Baliani 1646),[85] published in the year 1638 (Baliani, 1638), the same year as Galileo's *Two new sciences*. The first edition of the work presented an analysis of motion substantially convergent with that of Galileo, as Gassendi himself had underlined, recognizing however that Baliani achieved analogous results to the Pisan scientist in an independent way.[86] Serge Moscovici, in a book on Baliani written over 30 years ago, tried to free the Genoese patrician from the accusation, broadly recorded in historical tradition, that he, animated by a strong spirit of emulation, had deliberately distinguished his position from Galileo's in the second edition of the *De motu* (Moscovici 1967).[87] It is, in any case, indisputable that the second edition of Baliani's work introduced noteworthy modifications (Baliani 1646 and 1998).[88]

---

[85] On Baliani, see Drake 1970a; Grillo 1963. See also Moscovici 1967; Costantini 1969; Baliani 1998 (*Introduction* by G. Baroncelli).

[86] On 11 October 1640, Gassendi wrote to Girolamo de' Bardi of his having received Baliani's *De motu* one year before. He had, above all, appreciated Baliani's method of demonstration. Gassendi compared it with Galileo's *Two new sciences*: "Si Balianus solo ratiocinio eam proportionem [of the acceleration in natural motion] invexerit, quam primus, quod sciam, Galilaeus est experiundo assecutos" (Gassendi 1658, 6:100). To be noticed at first, here, is the clear position of Gassendi in favor of Galileo as to the priority of the discovery. Moreover, Gassendi — still referring to a friend of Baliani — made it explicitly clear that he did not subscribe to Baliani's statements on the unreliability of the experiments: "Et postulat quidem concedi nonnulla quae quispiam forte abnueret, quod Naturae subtilitas hebetudinem sensus non sequatur" (Gassendi 1658, 6:100).

[87] For the reception of Baliani's *De motu*, see: Moscovici 1967, 79–84 and Baliani 1998 (*Introduction* by G. Baroncelli). Traditionally a severe opinion has been expressed about the way in which Baliani behaved towards Galileo. This attitude has been attributed to the envy and to the spirit of emulation of the Genoese patrician.

[88] The work that was published in 1647 was considerably increased by the insertion of two books on "solids" (*De impetu* and *De motu super pluribus planis diversimode inclinatis*) and three books on

Claudio Costantini's important book, *Baliani e i Gesuiti* (Costantini 1969), provides a series of documents and remarks that help us to understand the real reasons for the reformulation of the laws of motion made by Baliani. Costantini has shown that a group of Jesuit scientists, among whom are Fathers Grassi, Confalonieri and Cabeo, tried from 1648 onwards, to put every possible pressure on Baliani to induce him to join the Company in the battle against the supporters of the vacuum (ibid., 92–94). In the eyes of the Jesuits the goal of obtaining explicit support for their cause from Baliani came to be of a very special value, since the Genoese patrician was considered one of the most authoritative representatives of the new scientific ideas (ibid., 75–76). Even more important, it would have therefore appeared that Baliani, abandoning his original positions, would have assumed a critical attitude against Galileo's laws of motion.[89]

In effect, the revision of 1647 radically changed Baliani's original approach, transforming the book into a work in which the hypothesis of Father Fabri took an absolutely central position. Baliani sustained, in fact, that in free fall the spaces grow in equal times from rest according to the natural numbers from the unity. He followed with an explanation of Fabri's theory of the "indivisible instants," or atoms of time; he declared that on the experimental ground there was perfect equivalence between his own hypothesis and the Galilean one, by referring to the same arguments as the French Jesuit (Baliani 1646, 108–140).[90] Again, like Father Fabri, and in full accord with the final phase of the evolution of Mersenne in the *de motu* debate, Baliani claimed, in opposition to Galileo, the necessity of a causal explanation of motion, as a consequence of the unreliability of experimental verifications.[91]

It was not a case of plagiarism, but rather the acceptance of the pressure and constant suggestions received from the Jesuits with whom he was in direct and friendly contact. And in fact nobody accused Baliani of having presented, as flour from his own sack, a hypothesis taken from an author, Fabri, who was not even quoted in the Genoese's *De motu*.

"liquids." A critical edition of Baliani's *De motu* has been recently published by Giovanna Baroncelli (Baliani 1998).

[89] Moscovici — who defends the good faith of Baliani — admits that he was the object of instrumentalization of some Fathers of the Company of Jesus: "Certain Jésuites ont peut-être utilisé les remarques de Baliani comme un machine de guerre contre Galilée" (Moscovici 1967, 80).

[90] Baliani's analysis follows step by step that of Mousnier-Fabri, sometimes also in the terminology. Baliani proposes the argumentation of the Jesuit of Lyon both in reference to the reasons that induce to prefer the series of natural numbers instead of the Galilean odd numbers proportion, and in remarking the impossibility of experimentally determining which of the two hypotheses is true (Baliani 1646, 110–111). The close correspondence between the illustration of the laws of motion in the second edition of Baliani's work and Fabri's exposition has escaped Moscovici. On the contrary, S. Drake has remarked the strict analogy between the two positions, but he has considered it a purely accidental coincidence (Drake 1974, 50–51). On the basis of Drake's works, Dear has pointed out the convergence between the Jesuit of Lyon and the second edition of the *De motu* of the Genoese patrician, but without giving any explanation for it (Dear 1988, 215–216).

[91] "Hic pariter peragere libuit, videlicet naturam motus pro viribus investigare, causas nimirum et principia a quibus hae demum motus passiones proveniant" (Baliani 1646, 98). And again, in criticizing the Galilean proportion of the odd numbers: "Hanc ... propositionem inniti experimentis sensui deceptioni obnoxiis; quibus insensibilis error detegi nequit" (Baliani 1646, 110).

In Baliani's letter to Mersenne of 13 March 1647, from which it is possible to deduce that at that time the printing of the second *De motu* was still not completed (Tannery et al. 1945–1988, 15:145–151), we find very important evidence of the Jesuits' pressure on the Genoese patrician. In this letter Baliani informed Mersenne of having received, in preceding weeks, a visit from Father Fabri who was en route to Rome: "E' stato qui il P. Onorato Fabri con cui ho trattato con molto gusto e mi pare un uomo molto dotto e vedo che in molte cose habbiamo dato nell'istessi pensieri" (The honorable Father Fabri was here and I have talked with him with great pleasure, and he seems, to me, a very learned man and I see that in many things we share the same thoughts) (ibid., 147). Father Fabri, for his part, would declare triumphantly that his own hypothesis had met with the approval of Baliani.[92] Regarding Fabri's hypothesis, moreover, we have a long letter, not dated but most probably from the year 1647, from the notorious Father Orazio Grassi no less, which, on a request of Baliani's, comments upon some aspects of the *de motu* theory of Father Fabri,[93] as reported by Mousnier in the *Tractatus Physicus*.

Moreover, in the year 1646, another Italian Jesuit, who was also very familiar with Baliani, Father Nicolo' Cabeo, who had for a long time taught in the Jesuit Seminar of Genoa,[94] introduced in his monumental comment to the *Meteorologica* of Aristotle a long and resentful review of the Galilean doctrine of motion. In this bitterly critical review are echoed the objections Le Cazre made to Gassendi's *Epistulae*, and the analyses and arguments of Father Fabri (Cabeo 1646). Cabeo insisted energetically that Galileo was preceded in his presumed discoveries by Giovanni Battista Baliani. Cabeo praised the modesty of the Genoese patrician, modesty that had induced him not to follow Galileo in the pretension of putting forward as *scientia* a series of propositions dependent upon clearly false principles.[95]

---

[92] In the answer of Mousnier to Mersenne, inserted into the *Metaphysica demonstrativa*, the underlining of the adherence of Baliani to Fabri's theory was skilfully matched with the praise on his scientific and moral authority: "vir certe maxima apud omnes gloria, sive res praeclare ab illo gestas in publicis numeribus, quibus egrege defunctus est, sive luculentissime ingenii et doctrinae monumenta, quae in publicam lucem edidit, sive demum singularem humanitatem consideres, ad quam natura illum munifice finxit" (Mousnier-Fabri 1648, 587).

[93] The letter, of February 5 but without indication of the year, has been published by Moscovici with the date 1649 which is surely incorrect (Moscovici 1967, 256–263). Father Grassi's references to Fabri's propositions are, in fact, constantly referred to the *Tractatus physicus* (Mousnier 1646). Hence, the letter seems to mirror Baliani's effort to assimilate Fabri's hypothesis on natural motion with the help of Father Orazio Grassi. The second edition of the *De motu* was not yet completed at the beginning of 1646. Between the end of January and the beginning of February the Genoese patrician was still setting up the pages of the *Liber Quartus*, where he inserted the hypothesis of Fabri. Father Grassi's letter brings to light that the difficulties of Baliani in reference to the *Tractatus physicus* concerned Fabri's concept of *impetus*. It has to be underlined that Baliani, being fully aware of the notion of conservation of movement (that he calls "naturalis motus continuatio," a principle not taken into account in Fabri's theory of motion) did not propose impetus as the cause of acceleration. (Baliani 1646, 101–108).

[94] For the relations between Baliani and Father Cabeo, see Costantini 1969. For Cabeo, see Frajese 1971; Ingegno 1972.

[95] Father Cabeo inserts his own violent criticisms against the Galilean doctrines *de motu naturali* in the context of the explanation of the dynamics of "driving rains." A "driving rain" was the fall of a great quantity of water, kept in the firmament, caused by the will of Providence: "intolerabilis

Thus the Jesuits flattered Baliani and urged him to oppose Galileo.[96] As Baliani's intense exchange of letters with Mersenne between 1646 and 1647 shows, it seems that even the Father Minim had an active role in encouraging the Genoese to distance himself from Galileo's position. Moreover, in 1645 Mersenne stayed for a long time with Baliani in Genoa, and most probably the two had occasion to discuss Fabri's hypothesis. Mersenne's correspondence helps us to understand why Baliani was tempted by these interested pressures. In fact, he was flattered by the prospect of gaining an important position in the gallery of the heroes of science, removing himself from the shadow projected by the personality of Galileo, by which his first edition of the *De motu* had remained substantially darkened. This spirit of emulation and of vindication is evident in the conclusive statements of the introduction to the *Liber Quartus* of the second edition of his *De motu* in which he had introduced as his own the hypothesis of Fabri regarding the arithmetic progression of spaces in accelerated motion "Augetur, igitur, ni fallor, motus iuxta progressionem arithmeticam, non numerorum imparium ab unitate ... sed naturalem. Ego ... detexisse spero causam ... a qua huiusmodi proportio emanat aperuisse et insuper quales errores fuerint in suppositionibus et experimentis huc usque habitis ... neque enim is sum qui tantum mihi tribuam ut rerum arcana intimius caeteris rimari nihi videar ... Nec inutiliter me laborasse existimavero si credar vitam silentio non pertransisse" (ibid., 114; Baliani 1998, 187).

A few months later, Baliani would make a lively protest against Mersenne who, in the Tomus III of the *Novarum observationum* of 1647 (Baliani to Mersenne, 1 October 1647, in Tannery et al. 1945–1988, 15:462–465), had presented him as a disciple of Galileo. Invoking Father Cabeo's testimony registered in his commentary to the *Metereologica*, Baliani proudly claimed, not only his priority regarding Galileo, but also gave evidence of his substantial autonomy, emphasizing that his

---

quaedam Galilaei iactantia, qua se solum ab orbe condito hanc turpem ignorantiam ab hominibus sustulisse profitetur ... stomachum enim ne dum risum excitant ejusmodi jactantiae. Certe illo, ipso anno quem prodiere Dialogi Galilaei, dum essem Genuae, narravit mihi Ioannes Baptista Balianus, nobile genuensis, vir ingenio et eruditione illustris, se incrementum velocitatis multis ab inde annis quam quidquam de Galilaeo audiret" (Cabeo 1646, 423–424). Later, Cabeo, denying that the propositions *de motu* of the *Two new sciences* demonstrate anything, since they were founded on purely experimental evidence, praised Baliani, who did not consider such propositions as demonstrated, but simply as "suppositum ex experientia, ex qua deinde, tamquam ex principio experimentali deducit pulcherrimas consequentias ..." (Cabeo 1646, 424). In fact, Cabeo opposed animatedly even the hypothetical admissibility of the Galilean laws of motion, by reproposing the arguments of Father Le Cazre and Father Fabri: that is the untenability of the postulate of the *Two new sciences*; the increase of velocity according to the series of odd numbers; acceleration as discontinuous process (beginning with a determined velocity) and proportional to the distance from rest; and, finally, the necessity of submitting physical discussion to metaphysical principles.

[96] In the field of cosmological ideas as well, the Jesuits contrasted the *pietas* and the opportune steadiness of Baliani with the conceitedness and the lack of responsibility of Galileo. The same was done by Father Riccioli, in the *Almagestum*. He affirmed that he exposed the selenocentric hypothesis "non tam ut viri iudicium contra Galilaeum adiungerem, quam ut exemplo ipsius discant reliqui conceptus mentis propriae, si forte sacrarum literarum aut ecclesiasticarum sanctionum authoritati minime congruant, aut non parere, sed pio ac prudenti ideoque foecundo abortu eos comprimere, aut si forte peperint intra merae hypothesis fascias et incunabula coercere" (Riccioli 1651, 1, *Pars Secunda*: 381).

De motu did not depend on the disputable principles of Galileo: "Valde gauderem quod dicas viros doctos censere me optime demonstrare de gravium motu ne adderes ex hypothesi positionum Galilaei, nam aliud est me cum Galilaeo in pluribus convenire, aliud meas demonstrationes ex eius principiis pendere, quod significari videris qua ductus ratione non percipio" (Baliani to Mersenne, 1 October 1647, ibid., 463). And he concluded, adding that at most he could be considered an admirer of Galileo — Fabri defined himself in the same way[97] — but not his *sectator* (Baliani to Mersenne, 1 October 1647, ibid., 464).

Mersenne promised to give him satisfaction as soon as possible. And in fact, the Minim's authoritative certification of Baliani's independence and autonomy from Galileo will be inserted, in Italian translation, in the edition of the *Works* (Baliani 1666, 10)[98] of Baliani of 1666:

> Ho gran gusto che V.S. mi habbia imparata per l'ultima sua che Galileo non sia il primo che ha osservato la proportione del moto de i corpi gravi che cascano giù, perché io pubblicherò a tutti quanti che in ciò siete stato il primo Osservatore, come l'ha confermato il P. Cabeo nel luogo citato da voi nelle sue Meteore. [I was very pleased in reading in your last letter that Galileo was not the first one to observe the proportion of the motion of falling heavy bodies, and I will tell the fact publicly that you were the first observer of this phenomenon, as Father Cabeo has confirmed in the statement quoted from you in his Meteore.] (Tannery et al. 1945–1988, 15:504–505)

So Baliani, flattered by interested friendships and spurred by a strong sense of emulation towards Galileo, made a mistake which will seriously compromise his reputation, as the generally and constantly negative judgments of the historians on the second edition of his *De motu* confirm (Caverni 1891–1900, 5:28; Moscovivi 1967, 79–84). Baliani was, therefore, the involuntary victim — as a matter of fact it was an intellectual suicide — of the Galilean *affaire* of the laws of motion.

The affair of the laws of motion has by now assumed the aspect of a second "trial," in which — along with legitimate and sometimes well-founded reservations on the coherence and plausibility of the *De motu naturaliter accelerato* in the *Two new sciences* — played a decisive role intellectual incompatibility and obscure intrigues. The awareness of the role played by the Jesuits in the *affaire* of the trial and of the condemnation of Galileo, induced many of the protagonists — in Italy as well as in France — to intuitively recognize, beyond the attacks directed at *de motu* doctrines, a new subtle intellectual strategy of the Jesuit Order. We lack definite proof that would allow us to state that the polemical initiatives of the

---

[97] In the *Metaphysica demonstrativa* Mousnier stressed Fabri's admiration for Galileo.: "Galilaeum vestrum magnifacimus et saepius a nostro Philosopho [Father Fabri], tum in privatis colloquiis, tum in publicis praelectionibus accepi magnum Galilaeum (sic enim illum vocabat) eo ingenio praeditum fuisse cui vix simili a multis retro saeculis extitisset" (Mousnier-Fabri 1648, 658–659).

[98] The original of Mersenne's letter, dated 25 October 1647, is lost.

different Jesuits were inspired by an homogeneous and coherent plan, as Tenneur and the Galileans in Tuscany took, however, for granted. More probably, Father Le Cazre on the one side, and Father Fabri on the other, developed two different strategies, which nevertheless had in common the purpose of contrasting the integration of Galileo's *Dialogo* and the *Two new sciences* — that is of the Pisan dynamics and of Copernican cosmology — clearly proposed by Gassendi in the *Epistulae de motu impresso* of 1642.

As soon as the Copernican *affaire* came to an end, with the death of Galileo, a second one was opened, which was less dramatic in its results, but no less delicate or important. The close alliance, suggested by Gassendi, between acceleration, the analysis of the infinite, attraction, atoms and the void, on one side, and Copernican cosmology, on the other, in the framework of a rigorously mathematical treatment, but open to experimental controls, produced extraordinary consequences in natural philosophy; at the same time, it represented an extremely effective vehicle of promotion of Galileo's image in the European context. This alliance, and the affair to which it gave rise, stimulated, in fact, the maintenance of an intense interest in Galileo in the second half of the century, up to Newton and beyond.

The awareness of the decisive role played by Gassendi in the transmission of Galilean ideas found clear expression in the introduction to the Florentine reprint of the Lyonnais edition of the works of Gassendi (Gassendi 1727). Niccolò Averani, a later and not exceptional heir to the Galilean tradition, who had promoted and edited the work,[99] exalted in the introduction with its solemn expressions, the continually effective defense of Galileo's scientific reputation by Gassendi, proudly emphasizing — and not without foundation — that Gassendi also owed much to Galileo:

> Pace enim Gallorum hoc sit dictum, quidquid laudis in Gassendum confertur, in Florentinorum laudem reflecti videtur; nam qui Gassendum laudant et immortalem Galilaeum totius Italie lumen extollunt ... Quod libenter doctissima Gallorum natio concedere non detractabit, quum Gassendus ipse primas concederet viro de se optime merito; cum quo tantae amoris caussae studiorum similitudine altae intercedebant, virtutisque aeternum suspiciendae appellaverit ("Tipographus" to "philosophiae studios" 1713, ibid., I, VII).

---

[99] For Niccolò Averani and for the motivations behind the Florentine reprint of the *Opera omnia*, see Carranza 1962 and Ferrone 1982, 155–160.

References

Baliani, G. B. 1638. *De motu naturali gravium solidorum.* Genuae: ex Typographia Io. Mariae Farroni.

——. 1646. *De motu naturalium gravium solidorum et liquidorum.* Genuae: ex typographia Jo. Mariae Farroni, Nicolai Pesagni et Petri Francisci Barberii.

——. 1666. *Opere diverse.* Genova: per Pietro Giovanni Calenzani.

——. 1998. *De motu naturali gravium solidorum et liquidorum.* Introduzione e cura di Giovanna Baroncelli, Biblioteca della scienza italiana 19. Florence: Giunti.

Bloch, O. R. 1971. *La Philosophie de Gassendi. Nominalisme, matérialisme et métaphysique.* Le Haye: M. Nijhoff.

Borelli, G. A. 1667. *De vi percussionis liber.* Bologna: ex Typographia Jacobi Montij.

Cabeo, N. 1646. *Nicolai Cabei ... S.J. in quatuor libros meteorologicorum Aristotelis commentaria ... Tomus Primus-Secundus.* Romae: typis haeredum Francisci Corbelletti.

Caramuel, J. 1644. *Sublimium ingeniorum crux. Jam tandem aliquando deposita a Joanne Caramuel Lobkowits, Gravium lapsum cum tempore elapso componente concordiamque experimentis et demonstrationibus geometricis firmante.* Lovanii: apud Petrum van der Heyden.

Carranza, N. 1962. "Averani Niccolò." In *Dizionario Biografico degli Italiani,* 1960–4:659–660. Rome: Istituto della Enciclopedia Italiana.

Caruso, E. 1987. "Honoré Fabri gesuita e scienziato." In *Miscellanea Secentesca. Saggi su Descartes, Fabri, White,* 85–126. Department of Philosophy, University of Milan, Milan: Cisalpino-Goliardica.

Caverni, R. 1891–1900. *Storia del metodo sperimentale,* 6 Vols. Florence: G. Civelli.

Costabel, P. and M. P. Lerner. 1973. *Le nouvelles pensées de Galilée par Marin Mersenne.* 2 Vols. Paris: J. Vrin.

Costabel, P. 1974. "Morin Jean-Baptiste." *Dictionary of Scientific Biography,* edited by C. C. Gillispie, 9:527–528. New York: Charles Scribner's Sons.

Costantini, C. 1969. *Baliani e i Gesuiti. Annotazioni in margine alla corrispondenza del Baliani con Gio. Luigi Confalonieri e Orazio Grassi.* Florence: Giunti Barbera.

D'Elia, A. 1985. *Christiaan Huygens. Una biografia intellettuale.* Milan: F. Angeli.

Dati, C. 1663. *Lettera ai Filateti ... della vera storia della cicloide.* Florence: all'Insegna della Stella.

De Vregille, P. 1906. "Un enfant du Bugey. Le Père Honoré Fabri." *Bulletin de la Societé Gorini. Revue d'Histoire ecclesiastique et d'archéologie religieuse du Diocese de Belley* 3:5–15.

Dear, P. 1988. *Mersenne and the Learning of the Schools*. Ithaca and New York: Cornell University Press.

Drake, S. 1970a. "Baliani, Giovanni Battista." *Dictionary of Scientific Biography* 1:424–425.

——. 1970b. "Uniform Acceleration, Space and Time." In *British Journal for the History of Science* 5:21–43.

——. 1973. "Jacques-Alexandre Le Tenneur." *Dictionary of Scientific Biography* 8:267–279.

——. 1974. "Impetus theory and quanta of speed before and after Galileo." *Physis* 16:47–75.

——. 1975. "Free Fall from Albert of Saxony to Honoré Fabri." *Studies in History and Philosophy of Science* 5:347–66.

Fabri, H. 1659. *Brevis adnotatio in Systema Saturnium*. Roma.

Fellmann, E. A. 1959. "Die mathematische Werke von Honoratus Fabry." *Physis* 1:6–25.

——. 1971. "Fabri Honoré." *Dictionary of Scientific Biography* 4:505–507.

——. 1992. "Honoré Fabri (1607–1688) als Matematiker — eine Reprise." In *The investigation of difficult things. Essays on Newton and the history of exact sciences in honor of D. T. Whiteside*, edited by P. M. Harman and A. Shapiro, 97–112. Cambridge: Cambridge University Press.

Ferrone, E. 1982. *Scienza, natura, religione. Mondo newtoniano e cultura italiana nel primo Settecento*. Naples: Jovene editore.

Frajese, A. 1971. "Cabeo Niccolò." *Dictionary of Scientific Biography* 3:3.

Galilei, G. 1890–1909. *Le Opere di Galileo Galilei*. Edited by A. Favaro, 20 Vols. Florence: Barbera.

Galluzzi, P. and M. Torrini, eds. 1975. *Edizione Nazionale delle Opere dei Discepoli di Galileo. Carteggio 1642–8*. Vol. I. Florence: Giunti-Barbera.

Galluzzi, P. 1976. "Evangelista Torricelli. Concezione della matematica e segreto degli occhiali." *Annali dell'Istituto e Museo di Storia della Scienza di Firenze* 1(1):71–95.

——. 1977. "Galileo contro Copernico. Il dibattito sulla prova 'galileiana' di G.B. Riccioli contro il moto della Terra alla luce di nuovi documenti." *Annali dell'Istituto e Museo di Storia della Scienza di Firenze* 2(2):87–148.

——. 1993. "Gassendi e l'affaire Galilée delle leggi del moto." *Giornale critico della filosofia italiana* 72(74)1:86–119.

Gassendi, P. 1642. *Petri Gassendi de motu impresso a motore translato Epistolae duae, in quibus aliquot preacipuae tum de motu universe, tum speciatim de motu Terrae attributo difficultates explicantur*. Parisiis: Louis de Heucqueville. In Gassendi 1658.

——. 1646. *Petri Gassendi De proportione qua gravia decidentia accelerantur Epistolae tre, quibus ad totidem epistolas R. P. Petri Cazraei respondetur*. Parisiis: Apud Ludovicum de Heucqueville.

——. 1647. *Institutio astronomica iuxta hypothesis tam veterum quam Copernici et Tychonis.* Parisiis: apud Ludovicum de Heuqueville. In Gassendi 1658, 6:266ff.

——. 1649. *Petri Gassendi Apologia in Jo. Bapt. Morini Librum, cui titulus Alae Telluris fractae.* Lyon.

——. 1650. *Apologia: Recueil des lettres de Sieurs Morin, de la Roche, de Neuré et Gassendi en suite de l'apologie de Sieur Gassend touchant la question de motu impresso a motore translato, où par occasion il est traité de l'astrologie iudiciarie.* Paris: A. Courbé.

——. 1658. *Opera Omnia in sex tomos divisa, quorum seriem pagina prefationes proximè sequens continet / hactenus edita auctor ante obitum recensuit, auxit, illustrauit; posthuma verò totius naturae explicationem complectentia in lucem nunc primùm prodeunt, ex bibliotheca illustris uiri Henrici Ludovici Habertii Mon-Morii,* 6 Vols. Lugduni: sumptibus Laurentii Anisson et Ioan. Bapt. Devenet.

——. 1727. *Petri Gassendi Diniensis acclesiae praepositi... Opera omnia in sex tomos divisa... Curante Nicolao Averanio,* 6 Vols. Florentiae: apud Joannem Cajetanum Tartini et Sanctem Franchi.

——. 1994. Pierre Gassendi 1592–1655. *Actes du Colloque International,* Digne-les-Bains, 18–21 Mai 1992, Digne-les-Bains: Société Scientifique et Littéraire des Alpes de Haut Provence, 2 vols.

Grillo, E. 1963. "Baliani Giovanni Battista." *Dizionario Biografico degli Italiani,* 1960– 5:553–557.

Hine, W. L. 1973. "Mersenne and Copernicanism." *Isis* 64:18–32.

Huygens, C. 1646. *De motu naturaliter accelerato.* In *Oeuvres complètes par la Societé Hollandaise des sciences,* 22 Vols. 11:68–75. La Haye. 1888–1950.

Ingegno, A. 1972. "Cabeo Niccolò." *Dizionario Biografico degli Italiani,* 1960–, 15:686–690.

——. 1888–1950. *Oeuvres complètes par la Societé Hollandaise des sciences,* 22 Vols. La Haye.

Koyré, A. [1939] 1961. *Études galiléennes.* Translated in Italian by M. Torrini. 1976. *Studi galileiani.* Turin: Einaudi.

——. 1953. "An Experiment in Measurement." *Proceedings of the American Philosophical Society* IIIC 2:222–37.

——. 1955. "A Documentary History of the Problem of Fall from Kepler to Newton." *Transactions of the American Philosophical Society,* New Series 45, 4:55–66. Translated in French by J. Tallec. 1973. *Chute des corps et mouvement de la Terre de Kepler a Newton.* Paris.

——. 1965. *Newtonian Studies.* Cambridge Mass. Translated in Italian by P. Galluzzi. 1972. *Studi newtoniani.* Turin: Einaudi.

La Loubère, A. 1658. *Propositiones geometricae sex.* Toulouse.

Le Cazre, P. 1645a. *Physica demonstratio qua ratio, mensura, modus ac potentia accelerationis motus in naturali descensu gravium determinantur. Adversus nuper excogitatam a Galilaeo Galilaei Florentino Philosopho ac Mathematico de eodem motu pseudo-scientiam.* Parisiis: Apud Iacobum du Brueil.

——. 1645b. *Vindiciae demonstrationis physicae de proportione qua gravia decidentia accelerantur. Ad Clarissimum Petrum Gassendum.* Parisiis: G. Leblanc.

Le Tenneur, J.A. 1640. *Traité des quantitez incommensurables.* Parisiis: de l'Imprimerie de J. Dedin.

——. 1646. "Disputatio physico-mathematica de aequabili motus acceleratione et motu Telluris in qua Galilaei decerta de motu gravium perpendiculariter cadentium confirmantur. Et ostenditur vulgarem seu Ptolemaican de motu Solis opinionem non minus esse contrariam sacris litteris quam Copernicanam de motu Terrae ... Ad R. P. Petrum Cazreum."

——. 1649. *De motu naturaliter accelerato.* Parisiis: apud L. Boullengeri.

Lenoble, R. [1943] 1971. *Mersenne ou la naissance du mécanisme.* Paris: J. Vrin.

Lukens, D. 1979. "An Aristotelian Response to Galileo: Honoré Fabri, S. J., on the Causal Analysis of Motion." Ph.D. thesis, University of Toronto.

Mersenne, M. 1634a. *Le mechaniques de Galilée, mathematicien et ingénieur du Duc de Florence avec plusieurs additions rares et nouvelles, utiles aux architects, ingénieurs, fonteniers, philosophes et artisans, traduites de l'Italien par L.P.M.M.* À Paris: chez Henry Guenon.

——. 1634b. *Questions theologique, physiques, morales et mathématiques.* Parisiis.

——. 1637. *Harmonie universelle contenant le theorie et la pratique de la musique: où il est traité des consonances, des dissonances, des genres, des modes, de la composition, de la voix, des chats, & de toutes sortes d'instrumens harmoniques,* 2 Vols. À Paris: chez Sébastien Cramoisy.

——. 1639. *Le nouvelles pensées de Galilei, mathematicien et ingenieur du Duc de Florence. où par des inventions merveilleuses & des demonstrations incommues iusque ^ present il est traitté de la proportion des mouvements, tant naturels, que violents, & de tout ce qu'il y a de plus subtil dans les mechaniques & dans la phisique / traduit d'Italien en François.* À Paris: chez Pierre Rocolet.

——. 1644a. *Ballistica.* In *Cogitata physico mathematica in quibus tam naturae quam artis effectus admirandi certissimis demonstrationibus explicantur....* Parisiis: sumptibus Antonii Bertin.

——. 1644b. *Universae geometriae mixtaeque mathematicae Synopsis et bini refractionum demonstratarum tractatus.* Parisiis: apud Antonium Bertier.

——. 1647. *Novarum observationum physico-mathematicarum ...Tomus tertius. Quibus accessit Aristarchus Samius de mundi systemate.* Parisis: Sumptibus Antonii Bertier.

Middleton, W. E. K. 1964. *The History of the Barometer.* Baltimore: J. Hopkins Press.

Morin, J. B. 1631. *Famosi et antiqui problematis de Telluris motu vel quiete hactenus optata solutio....* Paris.

——. 1634. *Responsio pro Telluris quiete ad Jacobi Landsbergii doctoris medici Apologia pro telluris motu.* Paris.

——. 1643. *Alae Telluris fractae, cum physica demonstratione quod opinio Coper-nicana de Telluris motu sit falsa, et novo conceptum de Oceani fluxu atque refluxu....* Parisiis.

Moscovici, S. 1967. *L'expérience du mouvement. Jean-Baptiste Baliani disciple et critique de Galilée.* Paris: Hermann.

Mousnier, P. 1646. *Tractatus physicus de motu locali.* Lugduni: apud Ioannem Champion.

Mousnier, P. and H. Fabri. 1648. *Metaphysica demonstrativa.* Lyon.

Palmerino, C. R. 2000. "Infinite Degrees of Speed. Marin Mersenne and the Debate over Galileo's Law of Free Fall." *Early Science and Medicine* 4(4):29–328.

Riccioli, G. B. 1651. *Almagestum novum, Astronomiam veterum novamque complectens observationibus aliorum, et propriis novisque theorematibus, problematibus, ac tabulis promotam, in tres tomos distributam quorum argumentum sequens pagina explicabit.* Bononiae: ex haeredibus V. Benatii.

Sommervogel, C., ed. 1890–1932. *Bibliothèque de la Compagnie de Jésus*, 9 vols. Paris.

Sortais, G. 1920–1922. *La philosophie moderne depuis Bacon jusqu'a Leibniz. Études historiques*, 2 vols. Paris.

Tannery, P., C. De Waard, M. B. Rochot, A. Beaulieu, eds. 1945–1988. *Correspondance du P. Marin Mersenne religieux Minime*, 17 Vols. Paris.

Torricelli, E. 1644. *De motu gravium naturaliter descendentium, et proiectorum.* In *Opera Geometrica. De sphaera et solidis sphaeralibus*, 95–243 (irregular pagination). Florence: Typis Amatoris Massae e Laurentij de Landis.

——. 1919–1944. *Opere di Evangelista Torricelli*, edited by G. Loria, G. Vassura, 5 Vols. Faenza.

Istituto e Museo di Storia della Scienza
Florence

MARIO BIAGIOLI

Replication or Monopoly?
The Economies of Invention and
Discovery in Galileo's Observations
of 1610

The Argument

I propose a revisionist account of the production and reception of Galileo's telescopic observations of 1609–10, an account that focuses on the relationship between credit and disclosure. Galileo, I argue, acted as though the corroboration of his observations were easy, not difficult. His primary worry was not that some people might reject his claims, but rather that those able to replicate them could too easily proceed to make further discoveries on their own and deprive him of credit. Consequently, he tried to slow down potential replicators to prevent them from becoming competitors. He did so by not providing other practitioners access to high-power telescopes and by withholding information about how to build them. This essay looks at the development of Galileo's monopoly on early telescopic astronomy to understand how the relationship between disclosure and credit changed as he moved from being an instrument-maker to becoming a discoverer and, eventually, a court philosopher.

The following revisionist account of the production and reception of Galileo's telescopic observations of 1609–10 focuses on the relationship between credit and disclosure. Traditionally, the historiography on Galileo's discoveries has clustered around two very different views of evidence. Stillman Drake treated telescopic evidence as unproblematic, dismissing Galileo's critics as stubborn and obsurantist.[1] Others, instead, have argued that Galileo's discoveries were not self-evident

---

[1] Stillman Drake contended that "the arguments that were brought forward against the new discoveries were so silly that it is hard for the modern mind to take them seriously" (Galilei 1957, 73). In his other publications on the subject, he focused on Galileo's process of discovery (especially in Drake 1976a, 153–168) but did not discuss the difficulties others may have faced in trying to replicate them. He only remarked that "good" astronomers had problems corroborating his claims because suitable telescopes were hard to come by in 1610, and that "bad" philosophers were so committed to their bookish knowledge that they could not deal with Galileo's observations (Drake 1978, 159, 162, 165–6, 168). The telescope's epistemological status is treated as a non-problem, and perceptual issues are mentioned only in one case, to say that Galileo, because of an eye condition, had learned to peer

and that their making and acceptance depended on specific perceptual dispositions (possibly connected to his training in the visual arts), commitments to heliocentrism, or unique (and perhaps tacit) skills at telescope-making.[2]

By questioning the transparency of the process of observation and discovery, the perceptual relativists have produced interpretations far more thought-provoking than Drake's. And yet they do not seem able to account for the fact that, despite all the perceptual and cosmological implications they find in Galileo's discoveries (and the ambiguous epistemological status of the instrument that produced them), his claims were commonly accepted within nine months from their publication in March 1610.[3] This is all the more remarkable considering that the satellites of Jupiter were not visible for about two months during that summer, contemporary networks of philosophical communication were neither broad nor fast, and the corroboration of Galileo's claims required learning how to construct and use a brand new kind of instrument.[4]

A different picture emerges when we shift our view to focus on Galileo's own observational protocols and how he did (or rather did not) help others replicate his discoveries. Galileo, I argue, acted as though the corroboration of his observations

---

through his clenched fist or between his fingers to improve his sight and that this may have given him the idea to stop down the objective lens to improve its performance (ibid., 148).

[2] Feyerabend looked at how Galileo's telescopic evidence (mostly about the moon) could convince other observers and readers (or rather how they could not convince them without additional ad hoc hypotheses and "propaganda" tactics). However, unlike Drake, he did not analyze Galileo's process of discovery or his own reasons to believe in what he saw. Feyerabend saw Galileo's telescopic evidence as simultaneously problematic and productive. In his view, Galileo's evidence was deeply problematic but it was only by being so that it triggered conceptual change. It could become unproblematic only later, once it was framed within a new set of "natural interpretations" (Feyerabend 1978, 99–161). Like Feyerabend, Samuel Edgerton has studied Galileo's visual representations of the moon and has concluded that he was able to read the bright and dark patterns on its surface as pointing to physical irregularities (and to represent them in wash drawings that were then translated into engravings) because he had been trained in the artistic technique of chiaroscuro. Because of that training, Galileo saw the moon as a "landscape" and pictured it as such. Instead, astronomers like Harriot (who had observed the moon with a telescope a few months before Galileo) did not have the same artistic training, did not see what Galileo saw, and pictured the moon not as rugged but just as spotted (Edgerton 1984, 225–232). In part, Edgerton has relied on the work of Terrie Bloom who has argued that Harriot was able to "see" the spottedness of the moon as an index of its morphological irregularities only after he read Galileo's "Sidereus nuncius" and viewed its engravings. The *Nuncius* provided Harriot with the "theoretical framework" he needed to see what he couldn't see before (Bloom 1978, 117–22). Drawing a difference between encountering (or looking) and discovering, Bernard Cohen has argued that Galileo discovered what he did because of a theoretical mindset informed by a mix of anti-Aristoteleanism and incipient Copernicanism (Cohen 1993, 445–72). Van Helden, instead, has focused on the practical and perceptual challenges posed by early telescopes to argue that the making and replicating of Galileo's observations was a remarkable achievement, not a problem-free task. The conditions for such an achievement included suitable telescopes, considerable labor, appropriate observational setups, good eyesight, conceptual dispositions, and, ultimately, a tacit "gift" at observing.

[3] I take the Roman Jesuits' confirmation of Galileo's claims on December 17, 1610, as a conservative date for the closure of the debate. For a summary of the controversial nature of Galileo's discoveries and instrument, see Galilei 1989, 88–90. In a different text, van Helden has remarked: "Now much has been made of the conservative opposition to these discoveries, but I should like to suggest that in view of the circumstances, the time it took Galileo to convince all reasonable men was astonishingly short" (Van Helden 1974, 51).

[4] Galileo's manuscript log shows a gap in his observations of the satellites between May 21 and July 25 during Jupiter's conjunction with the Sun (Galilei 1890–1909, III:437–439).

were easy, not difficult. His primary worry was not that some people might reject his claims, but rather that those able to replicate them could too easily proceed to make further discoveries on their own and deprive him of future credit (Galilei 1989, 17). Consequently, he tried to slow down potential replicators to prevent them from becoming competitors. He did so by *not* providing other practitioners access to high-power telescopes and by withholding detailed information about how to build them.[5]

But as important as it was for Galileo to keep his fellow astronomers in the dark, such negative tactics alone would not have allowed him to gain credit from his discoveries and move from his post at the university of Padua to a position at the Medici court in Florence as mathematician and philosopher of the grand duke — goals clearly on his mind in 1610. He needed proactive tactics as well. First, he did his best to make sure the grand duke saw the satellites of Jupiter (which Galileo had named "Medicean Stars") by sending detailed instructions to Florence on how to conduct these observations, and then by going to court himself at Easter time (Galilei 1890–1909, X:281, 304). Second, through the prompt publication of the *Sidereus nuncius* in March of 1610 he tried to establish priority and international visibility — resources he needed to impress his prospective patron, not just the republic of letters.

The *Nuncius* was carefully crafted to maximize the credit Galileo could expect from readers while minimizing the information given out to potential competitors. Although it was researched, written, and printed in less than three months, it offered detailed, painstaking narratives of Galileo's observations and abundant pictorial evidence about his discoveries. It also said precious little about how to build a telescope suitable for replicating his claims.

Galileo gave a synthetic narrative (rich in dates and names but poor in technical details) of how he developed his instrument and remarked that one needed a telescope as good as his own to observe what he was describing in the book, but he did not tell his readers how he ground suitable lenses (which was the distinctive skill that gave him an edge over early telescope-makers), nor did he mention the dimension of his telescopes, the type of glass, the size and focal length of the lenses he used, and the diaphragm he had placed on the objective lens to improve its resolution (Galilei 1989, 37).[6] In the book, he provided only a bare diagram of the

---

[5] Galileo's concerns with priority and monopoly have been noticed before. Drake has remarked on Galileo's reluctance to give out information about the telescope as an "unwillingness to give away advantages" (Drake 1970, 155). Albert van Helden and Mary Winkler have argued that "[Galileo] was able to monopolize telescopic astronomy for the first several years and make almost all the important discoveries" (Van Helden and Winkler 1992, 214–6). In a different text, van Helden remarked that "because he won the instrument race, Galileo was able to monopolize the celestial discoveries" (Van Helden 1984, 155). However, they have not seen these monopolistic tendencies as central to the story of the making and acceptance of Galileo's discoveries.

[6] While other mathematicians besides Galileo were able to figure out the relationship between the focal length of the lenses and the enlarging power of the instrument, he was quickly able to develop remarkable skill at grinding lenses for telescopes — lenses that were outside of the standard repertoire of glass makers (van Helden 1984, 154–5).

instrument and mentioned that his optical scheme involved a plano-convex objective and a plano-concave eyepiece (fig. 1). He also told his readers that unless one had at least a good 20-power telescope, "one will try in vain to see all the things observed by us in the heavens" (ibid., 38). He then proceeded to tell how to measure the enlarging power of telescopes, allegedly to prevent his readers from wasting their precious time trying to observe what they could not possibly see (ibid.). While he promised his readers a forthcoming book on the workings of the telescope, he never published it, nor do we have any manuscript evidence of such a project (ibid., 39). He presented his instrument as the standard of reference while

RECENS HABITAE.   7

fpicillis ferantur fecundum lineas refractas E C H.
E D I. coarctantur enim , & qui prius liberi ad F G.
Obiectum dirigebantur, partem tantummodo HI. cõ-

præhendent: accepta deinde ratione diftantiæ E H. ad lineam H I. per tabulam finuum reperietur quantitas anguli in oculo ex obiecto H I. conftituti , quem minuta quædam tantum continere comperiemus. Quod fi Specilio C D. bracteas , aliás maioribus , aliás verò minoribus perforatas foraminibus aptauerimus, modo hanc modo illam prout opus fuerit fuperimponentes, angulos alios, atque alios pluribus, paucioribufquè minutis fubtendentes pro libito conftituemus, quorũ ope Stellarum intercapedines per aliquot minuta adinuicem diffitarum, çitra vnius, aut alterius minuti peccatum commodè dimetiri poterimus. Hæc tamen fic leuiter tetigiffe, & quafi primoribus libaffe labijs in præfentiarum fit fatis, per aliam enim occafio nem abfolutam huius Organi theoriam in medium proferemus. Nunc obferuationes à nobis duobus proximè elapfis menfibus habitas recenfeamus, ad magnarũ profectò contemplationum exordia omnes veræ Philofophiæ cupidos conuocantes.
    De facie autem Lunæ, quæ ad afpectum noftrum vergit

Figure 1. Schematic diagram of the telescope in Galileo's *Sidereus nuncius* (Venice: Baglioni, 1610), p. 7 (reproduction courtesy of Owen Gingerich).

withholding such a reference (but, as I will show in a moment, such a lack of disclosure did not necessarily destabilize his claims).[7]

This narrative seems to clash with the evidence that, shortly after the publication of the *Nuncius*, Galileo distributed several telescopes throughout Europe. The contradiction, however, is easily resolved. These instruments were sent to princes and cardinals, not to mathematicians. Princes and cardinals were not Galileo's peers but rather belonged to the social group of his prospective Medici patron. While their endorsements could strengthen Galileo's credibility with the grand duke, their social position prevented them from competing with him in the hunt for astronomical novelties. Furthermore, most princes and cardinals were already familiar with low-power telescopes because since 1609 glass-makers had been peddling these instruments to them, not to astronomers or philosophers.[8] Galileo himself had heard of the telescope from people connected to diplomatic networks of correspondence. The first two instruments to come to Italy in 1609 were owned by Count de Fuentes in Milan and Cardinal Borghese in Rome.[9] By the end of 1609, low-power telescopes went from being wondrous devices to cheap gadgets (by nobles standards) produced in several Italian cities by traveling foreign artisans and local spectacle-makers (Galilei 1890–1909, X:248, 264, 252, 267, 282, 306).[10] Princes sought and used telescopes on terrestrial and, more rarely, celestial objects well before rumors of Galileo's discoveries had began to circulate, and before most

---

[7] At first, Galileo's tactics seem to resemble those of Newton during the debate on his theory of light and colors. Like Galileo, Newton withheld a great deal of information about the instruments he used in his early experiments (Schaffer 1989, 67–104). But if Newton's actions (while not motivated by priority concerns) clashed with the philosophical sociabilities of that time and put him at risk of being "given the lie," I hope to show that Galileo's tactics (whilst clearly driven by self-interest) were socially acceptable in the field in which he operated and were epistemologically justified by the specific observational practices within which he used the telescope.

[8] Paolo Sarpi, the theological (and often technical) advisor to the Venetian Senate was Galileo's primary source of information about early telescopes, before one of them actually arrived in Venice in August 1609. Through his diplomatic connections, Sarpi had heard of the Dutch invention of the telescope in November 1608, and wrote about it to a number of correspondents in France. One of them, Jacques Badoer, wrote back from Paris in the Spring of 1609 with a more detailed description of the instrument which, by that time, was commonly sold by Parisian glass makers. Sarpi probably showed Galileo this letter in July 1609 (Galilei 1989, 37). On Sarpi's correspondence about early telescopes, see Drake 1970, 142–4.

[9] On August 31, 1609 (a few days after Galileo presented his telescope to the Venetians), Lorenzo Pignoria wrote from Padua to Paolo Gualdo in Rome that Galileo's telescope was "similar to the one that was sent to Cardinal Borghese from the Fiandres" (Galilei 1890–1909, X:234, 255). It appears that Pignoria received news of the arrival in Rome of Cardinal Borghese's telescope in July (Galilei 1890–1909, X:226, 250). Girolamo Sirtori reported that an instrument was delivered to Count de Fuentes in Milan on May 1609 by a Frenchman (Sirtori 1618, 24–5). The presence of telescopes in Naples was already mentioned in an August 28, 1609, letter from Giovanni Battista della Porta to Federico Cesi (Galilei 1890–1909, X:230, 252). Porta did not say he owned a telescope, but that he had seen one, probably an instrument owned by a noble.

[10] By March, telescopic observations of the moon were being conducted in Siena by Domenico Meschini, a gentleman who claimed to be in contact with other people in Rome who were also observing it with their own instruments. This is found in a postscript to a letter written to Galileo from Munich on April 14 paraphrasing a letter sent from Siena to Florence and then to Munich (Galilei 1890–1909, X:291, 341). In March 1610, Giovanni Battista Manso, a noble, wrote from Naples saying that, while the *Nuncius* had not yet arrived there, low power telescopes were available and were being used, with moderate success, to observe the irregularities of the moon (Galilei 1890–1909, X:274, 293).

astronomers had developed any serious interest in telescopes.[11] Emperor Rudolph II, for instance, observed the moon early in 1610, before the *Nuncius* was published (Kepler 1965, 13).[12] In Galileo's eyes, princes and cardinals constituted a low-risk, high-gain audience. Being more familiar with the telescope than Galileo's colleagues, they were likely to both appreciate the superior quality of his instruments and to corroborate his discoveries. At the same time, they were not going to compete with him and, having little professional and philosophical stake in his discoveries, they were less motivated to oppose them.

Galileo's differential treatment of his various audiences proved successful. He did take some short-term risk by relinquishing the credit he could have received from other mathematicians and astronomers through early widespread replications. But by the end of 1610 he had developed a monopoly on telescopic astronomy which he then maintained with the resources available to him as mathematician and philosopher of the grand duke of Tuscany.[13]

This essay looks at the development of Galileo's monopoly to understand how the relationship between disclosure and credit changed as he moved from being an instrument-maker to becoming a discoverer and, eventually, a court philosopher. My narrative does not follow a chronological order but is organized by a set of interrelated questions: How was Galileo enabled to make his observations? What kinds of textual information and skills were necessary to reproduce them? How could he justify his non-cooperative practices and yet have his findings accepted? What kinds of observational narratives could he develop to minimize disclosure and maximize credit? What was the relationship between the tactics of Galileo the inventor of the telescope and Galileo the author of the *Nuncius*? By following Galileo's trajectory from the development of the telescope in 1609 to the achievement of a Medici-based monopoly on telescopic astronomy in 1610, I show that he drew resources from various economies (of invention, of discovery, of artworks) without fitting completely into any one of them.[14] I argue that the categories "replication," "disclosure," and "evidence" need to be reframed within this hybrid economy — an economy that was different, in scale and structure, from that which emerged in late seventeenth century natural philosophy.

---

[11] On April 1610, a diplomat from Modena wrote Count Ruggeri that Prince Paolo Giordano Orsini was back from the Netherlands where he had purchased a number of telescopes, probably to give them as gifts to other Italian princes who did not have them yet (Galilei 1890–1909, X:304, 347).

[12] Since September 1609, the Emperor had been purchasing telescopes from Venice (Galilei 1890–1909, X:241, 259) and perhaps others from northern Europe.

[13] Van Helden has remarked on how quickly Galileo reacted to what he perceived as challenges to his status as the leading telescopic astronomer, and has explained that behavior as an expression of Galileo's concern with maintaining Medici patronage (van Helden 1984, 156–7). However, the relationship worked in the other direction as well, that is, Galileo's monopoly had been made possible by Medici patronage.

[14] I use the term "economy" to refer to systems of exchange, not just to monetary economies (capitalistic or otherwise).

Unaided Corroborations

None of the astronomers or savants who reproduced Galileo's observations of the satellites of Jupiter by the end of 1610 did so with his direct help. The first independent confirmation came in May from Antonio Santini. A Venetian merchant with no particular background in astronomy, optics, or instrument-making, Santini was able to build a high-power telescope and observe the satellites of Jupiter within two months of the publication of the *Nuncius*.[15] He conducted more successful observations in September, after the satellites had become visible again.[16] Although he knew Galileo and was probably among those who performed observations with him at Padua or Venice, Santini received neither telescopes nor instructions as to how to grind lenses from Galileo.[17] However, because he resided in a glass-making center and had probably seen the low-power telescopes that circulated in Venice since the summer of 1609, Santini could have been in a position to replicate Galileo's skills by himself.

Kepler was the second to see the satellites in late August and early September 1610 with one of Galileo's instruments (Kepler 1611). However, that was not according to Galileo's plans because, as I discuss later, the telescope used by Kepler was not intended to go to him.

The third replication came on December 17 from the Jesuit mathematicians at

---

[15] Santini was originally from Lucca. After his mercantile phase in Venice, he became a monk, and finally a mathematics professor in Rome (Galilei 1890–1909, XX:531–2). In a June 1610 letter to Galileo, Santini mentioned the observations of the satellites he had conducted some time before. The letter itself is a plea on behalf of Giovanni Magini to convince Galileo that Magini was not involved in Horky's printed attack on Galileo. The letter ends by saying that Magini had endorsed Santini's corroboration of Galileo's observations despite the fact that Magini, because of poor eyesight, had been unable to see them (Galilei 1890–1909, 10:337–378). By the end of May, Jupiter was too close to the Sun to be observed, which means that Santini's observations must have been carried out in the second half of May at the latest. On September 25, Santini confirmed to have clearly seen the satellites of Jupiter before conjunction *Giove vespertino* (Galilei 1890–1909, X:397, 435). That Santini's observations were not made through Galileo's telescope but through an instrument of his own production is supported by the fact that by June Santini was already a supplier of good lenses and telescopes to Magini (Galilei 1890–1909, X:338, 378–9) and that, in the several letters exchanged with Galileo during 1610 he never mentioned having observed with him. Santini's corroboration was made public in Roffeni's *Epistola apologetica contra caecam peregrinationem cuiusdam furiosi Martini* (Bologna: Rossi 1611, reproduced in Galilei 1890–1909, III, Part I:198).

[16] Santini's observations are reported several months after they took place (or perhaps earlier letters mentioning them are lost). In a September 25 letter to Galileo, Santini wrote: "Finalmente mi risolsi di rivedere Giove mattutin, se bene, per quello aspetta a me, haveo tanta confermassione dall'averlo visto vespertino, che non dubitavo se li pianeti intorno a esso da lei scoperti vi fossero o no (se però non si desse là sopra qualche alterassione). Lo rivedetti lunedimattina, alle ore 10, giorno che fu de' 20 stante, e trovai li 4 pianeti tutti orientali. Alli 23 poi li riveddi del modo che notirò' da basso [one to the left and three to the right of Jupiter]" (Galilei 1890–1909, X:397, 435). That Santini's September observations were cast as a belated re-checking "I finally decided to see Jupiter again," not as an urgent matter, confirms Santini's confidence in his earlier corroborations.

[17] It is possible Santini inspected one of the many low-power telescopes available in Venice since the autumn of 1609. However, he didn't necessarily need to have access to that information. It appears that several of the mathematicians and glass makers who produced early low-power telescopes (Harriot, Marius, Galileo, Lipperhey, Janssen, and Metius) did so after receiving only a verbal description of them (van Helden 1974, 39 n. 3).

the Collegio Romano.[18] They too had not received telescopes or instructions from Galileo, but used instruments sent them by Santini or produced locally by one of Clavius' students, Paolo Lembo, and perfected by Grienberger, a fellow-Jesuit (Galilei 1890–1909, XI:466, 33–4). Neither Lembo nor Grienberger had previous experience making optical instruments. Two other successful observations of the satellites were achieved in 1610 in France (Peiresc and Gaultier, November 1610) and England (Harriot, October 1610)—both of them without any direct help from Galileo or his telescopes.[19]

Galileo's tendency not to share telescopes or information about their construction was most striking in Kepler's case. Just a month after the publication of the *Nuncius*, Kepler had endorsed Galileo's discovery of the satellites of Jupiter in a long letter which was immediately published as the *Dissertatio cum nuncio sidereo*.[20] Kepler publicly endorsed Galileo's discoveries despite the fact that he was not able to replicate them because, at that time, he had access only to low-power instruments owned by his patron, Rudolph II (Kepler 1965, 13).[21] These instruments were powerful enough to observe the irregularities of the lunar surface, but their magnification and clarity was not sufficient to detect the satellites of Jupiter. In August, Kepler pleaded with Galileo to send him a telescope saying "You have aroused in me a passionate desire to see your instruments, so that I at last, like you, might enjoy the great spectacle in the sky" (Galilei 1890–1909, X:374, 413–4).[22]

Although by this time Galileo had already sent instruments to princes and cardinals (and was in the process of sending more), he did not oblige Kepler. He excused himself by suggesting that Kepler deserved only the best of telescopes, which unfortunately Galileo no longer owned because it had been placed "among

---

[18] Clavius confirmed the existence of the satellites in a December 17 letter to Galileo (Galilei 1890–1909, X:484–5). The Jesuits had been recording their sightings of the satellites since November 28 (Galilei 1890–1909, 3, Part 2:863), but had observed them also on November 22, 23, 26, and 27 — as reported by Santini to Galileo in a December 4 letter in which he also included diagrams of the Jesuits' observations (Galilei 1890–1909, X:479–80). The Jesuits seemed particularly cautious. Clavius had written Santini that, even after the November 22–27 observations, "we are not sure whether they are planets or not" (ibid., X:480). Others seem to have observed the satellites in Rome before the Jesuits. In a November 13 letter to Galileo, Ludovico Cigoli stated that Michelangelo Buonarroti, a friend of Galileo and Cigoli's, had been an eyewitness *testimonio oculato* of the satellites on several occasions, and that his testimonials had been able to convince a few skeptics (ibid., X:428, 475). Moreover, in a June 7 letter to Galileo, Martin Hasdale wrote from Prague that he had received a letter from Cardinal Capponi in Rome saying that Roman mathematicians approved of Galileo's discoveries, though he did not mention names (ibid., X:328, 370).

[19] These replications had no historical role in the story I am telling here. On these observations, see Roche 1982, 9–51; Humbert 1948, 316. In 1614, Simon Marius, a German mathematician claimed to have discovered Jupiter's satellites earlier than Galileo, but his claims have been disputed since (Galilei 1989, 105 n. 61). Marius' priority claims are in his *Mundus Jovialis*, translated in Prickard 1916, 367–503.

[20] Kepler's letter to Galileo was dated April 19, 1610 (Galilei 1890–1909, X:297, 319–40). The dedication of the printed version is dated May 3, 1610.

[21] For a discussion of some of Kepler's reasons for endorsing Galileo's claims without being able to replicate them, see Biagioli forthcoming.

[22] Kepler continued: "Of the spyglasses we have here, the best ones are ten-power, others three-power. The only twenty-power one I have has poor resolution and luminosity. The reason for this does not escape me and I see how I could make it clearer, but we don't want to pay the high cost."

more precious things" in the grand duke's gallery to memorialize the discovery of the Medicean Stars (Galilei 1890–1909, X:379, 421). Galileo also intimated that, at the moment, he was temporarily unable to produce more instruments because, being in the process of moving from Padua to Florence, he had disassembled the machine he had constructed to grind and polish lenses (Galilei 1890–1909, X:379, 421).[23] The Emperor too had requested (with some insistence) an instrument through the Medici ambassador in Prague, and had vented his frustration at not being given priority over cardinals whom he knew Galileo had provided with telescopes.[24] However, the imperial pleas, like Kepler's, went unanswered. When Kepler finally observed the satellites of Jupiter in August and September and published his findings in another short text, the *Narratio de observatis a se Quattuor Iovis satellitibus erronibus*, he did so with a telescope Galileo had sent to the Elector of Cologne — not to him or to the Emperor.[25]

Galileo's behavior may seem particularly ungrateful as Kepler's *Dissertatio* was the first and only strong endorsement he had received from a well-known astronomer before he obtained his position at the Medici court in the summer of 1610. Thanks to Kepler, Galileo was able to confront his critics with a powerful testimonial and to quench some of the Grand Duke's anxieties about having his family name attached to artifacts. Perhaps Galileo thought that, given Kepler's commitment to the Copernican cause, he would have supported his discoveries anyway (which he did) and that, therefore, he did not need any further help or sign of gratitude. Moreover, in the *Dissertatio*, an enthusiastic Kepler exclaimed: "I should rather wish that I now had a telescope at hand, with which I might anticipate you in discovering two satellites of Saturn" (Kepler 1964, 14). The use of

---

[23] On October 1, Galileo's lens-grinding machines (which he said had to be set in place with mortar) were still inoperative (ibid., X:402, 440).

[24] Giuliano de' Medici to Galileo, April 19, 1610 (Galilei 1890–1909, X:.296, 319). In July, better telescopes reached Prague from Venice, but none of them were made by Galileo (ibid., X: 360, 401–2). On July 19, Giuliano de' Medici acknowledges the arrival of additional ephemerides of the satellites (not telescopes) Galileo had sent to Kepler (ibid., X: 362, 403). In the same letter, he urges Galileo to send an instrument to the Emperor (ibid., X:404). On August 9, Galileo is told that the Emperor has received a better telescope from Venice, but that Galileo's instrument (that some thought had been received by the Medici ambassador) had not yet been seen (ibid., X:375, 418). It does not appear it was ever there, as on August 17 Galileo is told of the Emperor's aggravation (ibid., X:378, 420). Interestingly, the imperial court at Prague was not on the first list of potential recipients of telescopes Galileo submitted to the Medici on March 19 (ibid., X:277, 298, 301), but was added only in May (ibid., X:311, 356). While much of the evidence points to the fact that Galileo did not wish Kepler to have a telescope, on May 7 he asked the Medici for permission to send one in the diplomatic pouch from Venice to Prague. He also remarked that he did not have any good telescopes ready (ibid., X:307, 349–50). The Medici authorized the shipment on May 22 (ibid., X:311, 356), but on May 29, the Medici resident in Venice expresses worries that the telescope could get damaged during shipping (ibid., X:323, 364). That does not seem to have been a problem as often telescopes were shipped disassembled. It could be that Galileo thought it would be useless to send a telescope to Prague at the end of May as Jupiter was no longer observable, and that he could have taken the two months before it became visible again to produce a better instrument. However, he never sent such an instrument, probably because, by the time Jupiter was visible again, he had already received a contract from the Medici.

[25] In September 1610, Giuliano de' Medici informed Galileo of Kepler's observations and of his decision to publish the Narratio (Galilei 1890–1909, X:329).

the verb "to anticipate" may have drastically decreased Kepler's chances of receiving the instrument he sought.

Galileo displayed a similarly uncooperative attitude toward other potential allies.[26] On April 17, Ilario Altobelli asked him for lenses or a telescope so that, he claimed, he could help Galileo with testimonials against his critics. But he seemed a bit too eager to determine the periods of the Stars and received nothing in the end.[27] Magini, who had been one of Galileo's early detractors but had slowly changed his mind, asked him for a eyepiece on October 15, but was not gratified.[28]

Galileo did not help the Jesuit mathematicians of the Collegio Romano either. Right after moving back to Florence from Padua in September 1610, he wrote Clavius that he had heard the Jesuits were having problems seeing the satellites of Jupiter. That did not surprise him, Galileo continued, as he knew all too well that one needed an "exquisite instrument" to replicate his observations (Galilei 1890–1909, X:391, 431). However, he did not volunteer to send Clavius such an exquisite telescope but simply advised him to build a sturdy mount for whatever instrument he had because even the small shaking caused by the observer's pulse and breathing was enough to disrupt the observations (ibid., 431). He concluded that, in any case, he would show Clavius the "truth of the facts" during his forthcoming visit to Rome (ibid., 432). On October 9, Santini wrote Galileo that the Jesuits had not yet seen the satellites and added that "I think that these big shots, I mean in term of reputation, are playing hard to get so that Your Lordship may feel obliged to send them an instrument."[29] Even then, however, Galileo did not send the Jesuits a telescope.

He was much more forthcoming with patrons and courtiers. In a January 7 letter written from Padua (either to Antonio de' Medici or Enea Piccolomini) Galileo gave more useful tips about the telescope than in the *Nuncius* or in anything else he wrote that year. In it, he told the Florentine courtier (whom he was probably instructing how to show the satellites to the grand duke) how to minimize the shaking of the telescope caused by the observer's heartbeat and breathing, how to maintain the lenses, and how much excursion one should allow the tubes carrying the two lenses so as to achieve proper focusing (as he probably had sent or was planning to send him two lenses but no casing) (Galilei 1890–1909, X:259, 277–8).[30] More importantly, he stressed that the objective lens needed to be

---

[26] Raffaello Gualterotti requested lenses on March 6, 1610, (Galilei 1890–1909, X:268, 287) and Alessandro Sertini on March 27, 1610 (ibid., X:282, 306).

[27] "*et m'ingegnero"d'adattare il tubo in forma della fiducia nel dorso dell'astrolabio per osservare anco i periodi; e scrivero' a V.S. il tutto in lingua latina, accio' lo possi poi annettere nelle sue osservationi*" (Altobelli to Galileo, April 17, 1610 [Galilei 1890-1909, X:294, 317]).

[28] Magini said to have received three large lenses from Santini, and that he thought to have a very good one among them. But he lacked good eyepieces, and asked Galileo to send some (Galilei 1890–1909, X:408, 446).

[29] "*Io dubito che alcuni di questi pezzi più grossi, voglio dire di più riputassione, non stiano duri, accio' V.S. si metta di necessità di mandargli lei uno instrumento*" (Galilei 1890–1909, X:407, 445).

[30] However, we do not have any evidence that Galileo actually sent the lenses or a telescope to Florence before his visit during the Easter vacation. The only mention of a telescope in Florence is from April 20, 1610 (Galilei 1890–1909, X:299, 341). The letter mentions a telescope kept in the

stopped down with a diaphragm. As the lenses' shape was particularly irregular toward the edges, covering that part would significantly reduce aberrations. Clavius was told by Galileo of this significant tip almost a year later (after he confirmed the satellites' observation in December 1610) and only because the Jesuit had asked Galileo why the telescopes he had sent to Rome (to cardinals) had stopped-down objectives.[31]

Predictably, Galileo did not loan his own instrument. He organized or participated in public observational seances in Venice, Padua, Bologna, Pisa, Florence, and later in Rome, but it appears that he never left the telescope in alien hands, not even for a few hours.[32] These meetings were meant to provide demonstrations rather than to foster independent replications. Galileo would arrive, demonstrate, and depart. While people could look *through* the telescope, it appears that they did not have much of a chance to look *at* or *into* it. During his visit to Bologna in April 1610, Martinus Horky had to sneak around Galileo's guard (probably while he was asleep) to make a cast of the telescope's objective lens (Galilei 1890–1909, X:301, 343).

Galileo's fears about the consequences of giving good telescopes to mathematicians or helping them construct their own were not unjustified. He knew from personal experience that, after receiving an approximate verbal description of a telescope, one could build a prototype in a single day, move from 3-power instruments to 9-power telescopes in a few weeks, and that it was possible to develop a 20-power instrument in about four months and a 30-power one in less than seven months.[33] Soon after, he witnessed a merchant like Santini build telescopes good enough to observe the satellites of Jupiter within two months from the publication of the *Nuncius* in March, and then supply lenses and entire telescopes to both Magini and the Roman Jesuits.[34] He also knew that increasingly powerful instruments were being constructed in Venice and elsewhere.[35] For

---

Medici storage rooms, but does not say it is by Galileo. Also, because Galileo was in town at that time (and is actually mentioned in the letter), this could be an instrument he had brought with him from Padua.

[31] Clavius's query is in (Galilei 1890–1909, X:437, 485). Galileo's response is in (Galilei 1890–1909, X:446, 501): "*Hora, per rispondere interamente alla sua lettera, restami di dirgli come ho fatto alcuni vetri assai grandi, benchè poi ne ricuopra gran parte, et questo per 2 ragioni: l'una, per potergli lavorare più giusti, essendo che una superficie spaziosa si mantiene meglio nella debita figura che una piccola; l'altra è che volendo veder più grande spazio in un'occhiata, si può scoprire il vetro: ma bisogna presso l'occhio mettere un vetro meno acuto et scorciare il cannone, altramente si vedrebbero gli oggetti assai annebbiati. Che poi tale strumento sia incomodo da usarsi, un poco di pratica leva ogni incomodità; et io gli mostrerò come lo uso facilissimamente.*"

[32] Galileo to Kepler, August 19, 1610 (Galilei1890, X:379, 422). On meetings in Venice and Padua see Galileo to Vinta, March 19, 1610 (Galilei 1890-1909, X:277, 301).

[33] On Galileo's quick progress, see van Helden 1984, 150–5. On the earlier developments of the telescope, see van Helden 1977, Part 4:1–67.

[34] Santini's lenses and telescopes to Magini are mentioned in Galilei 1890–1909, X:338, 378; 356, 398; 408, 446; 400, 437; 414, 451. On Santini's gifts of telescopes to the Jesuits, see ibid., XI:466, 33–4.

[35] Galileo's correspondence indicates that by mid-1610, low-power telescopes were common, their price had dropped, and the market was so saturated that some telescope-makers were moving on to other (probably more provincial) cities. Then, Magini wrote him in October that Cardinal Giustiniani had managed to attract to Bologna a skilled glassmaker from Venice ("Bortolo," the son of the

instance, his friend Castelli reported that at the beginning of February a friar, Don Serafino da Quinzano, had shown him the moon through a 9-power telescope that he had built on his own (Galilei 1890–1909, X:287, 310–1).

His worries would have only been increased had he known of the Jesuits' quick progress.[36] By the end of 1610 Grienberger had successfully modified the eyepiece of a second good telescope Clavius had received from Santini thereby turning it into a 34-power instrument (ibid., XI:466, 34)[37] According to Grienberger, this telescope was better than those produced by Galileo which the Jesuits had been able to test in Rome (telescopes Galileo had sent to cardinals, not to astronomers). According to information provided by Grienberger, the power of this instrument was somewhat superior to anything Galileo had produced at that time. The fact that making telescopes was a quickly-spreading skill was evident not only to Galileo. At the end of September 1610, Santini wrote him: "I do not understand, now that the telescope has become so common and easy, how come the practitioners of the speculative sciences have not managed to clarify this matter [the existence of the satellites] and express their consensus."[38] The issue, then, was not whether people could develop powerful telescopes, but only how many weeks or months it would take them to move from 3-power to 20-power instruments.[39]

That Galileo worried about priority disputes rather than about the difficulties others might face in replicating his discoveries is confirmed by his statement that the *Nuncius* had been "written for the most part as the earlier sections were being printed" for fear that by delaying publication he would have "run the risk that someone else might make the same discovery and preceded me [to print]" (Galilei 1890–1909, X:277, 300). The way Galileo behaved in 1610 suggests he thought he had only a limited amount of time to discover whatever there was to be discovered with telescopes of that power range.[40] As he put it in the *Nuncius*, "Perhaps more

---

Emperor's glassmaker) who was quite good at grinding lenses for long (that is, high-power) telescopes and that Magini planned to use his services (Galilei 1890–1909, X:408, 446). On October 15, 1609, Lorenzo Pignoria wrote from Padua that there were "most excellent telescopes," adding that they were produced by a few artisans, that is, not just by Galileo (ibid., 243, 260). Hasdale wrote to Galileo from Prague that the Emperor was getting increasingly better telescopes from Venice, one of which, apparently, had been produced by an artisan who worked for Galileo (ibid., 360, 401–2). On April 24, Gualterotti mentions a good telescope made by "Messer Giovambattista da Milano" whose quality Galileo appears to have praised (ibid., 300, 341). Santini sent a new telescope to Florence (to the Venetian ambassador) on November 6, 1610, and asked Galileo to take a look at it (ibid., 423, 464–5). Galileo liked it (ibid., 433, 479).

[36] Since the summer of 1610, the Roman Jesuit Paolo Lembo had been producing increasingly good telescopes with which the mathematicians of the Collegio Romano were eventually able to see the satellites — though only when the sky was very clear. According to Grienberger, Lembo had developed his first telescopes on his own, without information or examples from the outside (Galilei 1890–1909, XI:466, 33–4).

[37] I wish to thank Albert van Helden for decoding this figure from Grienberger's letter.

[38] "*Io non so come, essendosi fatto tanto comune e facile questo uso del cannone, non sia da quelli che attendono alle specolative chiarito questa partita e dato l'assento*" (Galilei 1890–1909, X:397, 435).

[39] Harriot, for instance, had a 10 power telescope by July 1610, a 20-power by August, and a 32-power by April 1611 (Roche 1982,17).

[40] The first phase of the race for telescopic discoveries was effectively over by 1612 with the

excellent things will be discovered in time, either by me or by others, with the help of a similar instrument" (Galilei 1989, 36). Even the first available report about the use of the telescope indicates that those who managed to construct or have access to an instrument quickly pointed it to whatever celestial body they could spot.[41] How close the race must have been can be gathered from a January 1611 letter from Grienberger to Galileo in which he mentioned that even before the Jesuits had heard about his discovery of the phases of Venus at the end of December, they had independently observed them (Galilei 1890–1909, XI:466, 34).

While cosmological commitments may have played a role in setting the direction of further observations (as in the case of the discovery of the phases of Venus), this astronomical hunt seemed primarily propelled either by plain curiosity or by the desire to discover more novelties and get credit for them. Considerations of the possible pro-Copernican or anti-Ptolemaic significance of these discoveries were not a common concern in the first half of 1610, but emerged *after* Galileo's claims had been accepted (Biagioli 1993, 94–6).

Because of the speed with which others were learning how to build telescopes suitable for astronomical use, Galileo's uncooperative stance may have been the determining factor in achieving a monopoly over that first astronomical crop. He was first to discover the unusual appearance of Saturn (in the summer of 1610) and the phases of Venus (in the fall), and to determine the periods of the satellites — a result that both reinforced the epistemic status of the Medicean Stars and brought him more visibility.[42] His monopoly became almost self-sustaining. He managed to reclaim credit for the discovery of the sunspots from the Jesuits (although they had been first to publish that discovery late in 1611) and, years later, he succeeded in defending the referential status of his telescopes when other instrument-makers, like Fontana in Naples, had produced more powerful ones.[43]

---

discovery of sunspots. The second wave of discoveries started only in 1655 with Huygens (Van Helden 1984, 155).

[41] *Ambassades du Roy de Siam envoyé à l'Excellence du Prince Maurice, arrivé à la Haye le 10. Septemb. 1608* reports that one of the very early telescopes had been aimed at the stars in the Netherlands as early as fall 1608 (The Hague 1608, 11).

[42] In the *Nuncius*, he exhorted other astronomers to find the satellites' periods (Galilei 1989, 64). By this time he had only a figure for the outer satellite, which he put at about fifteen days (Galilei 1890–1909, X:271, 289). That figure was corrected to more than sixteen days in the spring 1611 (ibid., XI:532, 114). There he also gave estimates for the period of the innermost at less than two days. He published his first full description of the satellites periods in 1612 in his Discourse on Bodies in Water. Those values were very close to modern ones. On Galileo's investigation of these periods, see Stillman Drake 1979, 75–95.

[43] "From an early point, then, the authority of instruments was intertwined with personal authority. A strong argument can be made that after about 1612, Galileo's lead in telescope making had disappeared and that others had instruments of comparable quality. Yet Galileo ruled until his death as the undisputed master of telescopic astronomy (van Helden 1994, 19). On related issues, see also van Helden and Winkler 1992, 214–6.

A Field, not a Community

Galileo's non-cooperative attitude and his focus on developing a Medici-based monopoly of telescopic astronomy reflected more than just his fears about being deprived of credit for future discoveries. Galileo and his readers did not belong to a professional community that could provide the kind of credit and rewards he sought. Furthermore, the lack of consensus about style of argumentation and standards of evidence as well as the scant interdependence among the members of this field hindered closure of the debate.[44]

Galileo's correspondence shows that his discoveries were discussed in a field geographically dispersed over several courts and universities or punctuated by isolated individuals linked only through selective correspondence networks. It included few professional astronomers but many physicians, men of letters, diplomats, students, polymaths, and variously educated gentlemen. Political and religious boundaries mattered. A French or German mathematician did not have much incentive to engage, or even less to agree with the claims put forward by someone who operated on the other side of the Alps — unless, as in Kepler's case, the legitimation of those claims could provide him with further resources for his own Copernican program.

Of the several critiques of Galileo's findings that circulated in 1610, only one made it to print (Horky 1610, 129–145).[45] Critiques were more commonly presented first in private conversational settings and then communicated, often anonymously, through networks of scholarly and courtly correspondence and gossip. The remarkable metamorphoses that affected what went in and out of these channels did little to stabilize the debate. The proliferation of opinions was also fostered by the courtly format in which many of these views were presented and developed (Biagioli 1993, 72–83). Upon receiving a copy of Galileo's *Nuncius*, a prince or his courtiers could ask court mathematicians and physicians for an opinion about the book. Critical responses were almost *de riguer* in these contexts as they could generate lively and entertaining debates, but, by the same token, they did not tend to facilitate closure.[46]

---

[44] My analysis is broadly informed by Pierre Bourdieu's notion of "field," and especially by his discussion of how fields are established (Bourdieu 1985, 723–44; idem 1999, 31–50).

[45] A second critique (Sizi 1611, 203–250) was written in 1610, but was published only in 1611, after Galileo's claims had been widely accepted. It had little or no impact on the debate.

[46] These people were given little time to formulate their views and often were expected to respond on the spot, sometimes without having seen the book or tried a telescope. Because of this conversational format, commentators were only moderately accountable for their views and could modify or even reverse them at a moment's notice without much embarrassment or professional liability. For instance, a major astronomer like Magini could support (and largely share) Martinus Horky's vehement critique of Galileo (Galilei 1890–1909, X:303, 345; 324, 365), but then turn around and write (or have others write) that he was a Bohemian madman as soon as Horky's attack on Galileo seemed to backfire (ibid., X:334, 376; 335, 377; 337, 378; 338, 379; 344, 384–5). In a differently structured republic of letters, the remarkable contradiction between Magini's public and private stances may have not been without liabilities. Moreover, some of these critiques seemed to be aimed not so much at

To modern ears, the tone of several of these early critiques may appear harsh, even libelous. However, it would be wrong to take this simply as a sign of strong emotions stirred by cosmological and philosophical incommensurabilities. More mundanely, such a tone reflected the kind of discourse generated by controversial novelties (and by the sudden stardom of their producer) in a dispersed and marginally interdependent field.[47] The same field that allowed Galileo to adopt an uncooperative stance toward other astronomers did not compel his critics to treat him respectfully either (Biagioli 1993, 60–73). Although the *Nuncius* triggered many conversations, their end point was conversation, not closure. The *Nuncius* was like a message in a bottle, a carefully packaged message let to drift, hoping for the best.

Periodic Evidence vs. Instantaneous Perception

At first glance, the *Nuncius* appears to present a straightforward account of sequential observations. But the narrative structure that wove these observations into physical claims displayed a kind of demonstrative logic — a logic that well matched Galileo's epistemological and social predicament at the beginning of 1610.

Galileo's first goal was to gain assent for his claims, minimize the risk of losing priority over future discoveries, and cast his reluctance to provide information about the telescope as inconsequential to the acceptance of his discoveries. Second, he could not present himself as someone whose claims could be accepted on the grounds of his personal credibility. By the time the *Nuncius* was published, few readers knew of its author. Narratives that de-emphasized the author's personal qualities while stressing their internal logic helped Galileo bypass the problems posed by his modest professional and social status. Third, narratives whose acceptance did not appear to hinge on their author's adherence to specific disciplinary "forms of life" had a better chance to be understood and accepted by Galileo's diverse audiences.

The logic of Galileo's narratives rested on the specificity of his observational protocols.[48] The production and reproduction of his observations was a time-consuming process not only in the sense that much labor and effort went into it,

Galileo's stars but rather at his sudden stardom. In September 1610, Magini remarked to Monsignor Benci that, "in some universities, other mathematicians are paid better." For instance, recently Mr. Galilei has received 1000 florins from the Venetians, and is currently retained by the Grand Duke with 1200 scudi for life, although I know in my conscience that I am not at all inferior to but rather superior to him" (ibid., X:388, 429).

[47] On the transition from this kind of sociabilities to more interdependent ones, see Biagioli 1996a, 193–238.

[48] These protocols have been discussed, in various degrees of depth, in Drake 1979, 75–95; van Helden 1989, 10–6; Chalmers 1990, 54–5; and Dear 1995, 107–11. What, in my view, has not been previously addressed is how Galileo's observational practices dovetailed with his concerns about minimizing disclosure and maximizing credit.

but, more importantly, in the sense that the evidence behind those discoveries was inherently *historical*. Like other astronomical phenomena, the satellites of Jupiter were observed as a *process* (and, I argue, were probably observable only as a process).

One does not see the precession of the equinoxes by looking in the direction of the celestial pole for a few hours but detects it by comparing and interpolating the observations of the motion of the celestial pole through the stars over centuries. Similarly, one did not "see" the satellites of Jupiter just by pointing the telescope toward that planet for a few minutes. That would have shown, at best, a few bright dots. What enabled their discovery was not a specific gestalt that immediately turned those dots into satellites, but a commitment to produce the suitable apparatus and conduct observations over several days so as to detect the *periodic motions* of the satellites and differentiate them from other visual patterns (be they fixed stars or optical artifacts produced by the instrument). Because of the features of early telescopes (narrow field of vision, double images, color fringes, and blurred images especially toward the periphery), people who looked through a telescope for only a few minutes could legitimately believe that Galileo's claims were artifactual, as numerous spurious objects could be seen through a telescope's eyepiece at any given time.

This view of Galileo's process of discovery is no *a posteriori* reconstruction but conforms to his log entries, to the *Nuncius*, and to a letter written immediately after his first observation of Jupiter. When he observed Jupiter for the first time on January 7, 1610, Galileo wrote to a friend that he had seen three fixed stars near the planet, two to the east and one to the west (Galilei, 1890–1909, X:259, 277). In the *Nuncius* he added that these stars seemed "brighter than others of equal size" and "appeared to be arranged exactly along a straight line and parallel to the ecliptic" but, in and of itself, their peculiar appearance and arrangement did not cause him to doubt that they were fixed stars (Galilei 1989, 64).[49] At first, he "was not in the least concerned with their distances from Jupiter," but on the following night he noticed that while the three stars had remained close to the planet, they had all moved to the west (ibid., 65). Even then, Galileo did not think that the stars had shifted.[50] He assumed, instead, that Jupiter must have moved (though he was puzzled by the fact that, according to his tables, it should have gone in the opposite direction) (ibid., 65). Clouds prevented him from observing on the following night. On January 10, however, he was surprised to see that only two stars were visible and that they had again switched sides, this time from the west to the east (ibid., 65–6). He could make sense of the missing star by thinking that it must have been

---

[49] In the January letter he had already remarked that planets appeared well demarcated ("like small full moons") when observed through the telescope, but that fixed stars remained so shimmering that their shape could not be detected. It seems, therefore, that the three "fixed stars" around Jupiter had struck him as being of the size of stars while looking more like planets.

[50] "at this point I had by no means turned my thought to the mutual motions of these stars" (Galilei 1989, 65).

hidden by Jupiter, but could not believe that Jupiter had moved around again (ibid., 66). On January 11, there were still only two stars to the east of Jupiter, but they had moved much further to the east of the planet, were closer to each other, and one of them appeared much larger (though on the previous night they had appeared to be of equal size). Only at that point did he conclude that what he had observed were not fixed stars but planets (*stelle erranti*) and that "they had been invisible to everyone until now" (Galilei 1890–1909, III, Part 2:427; Galilei 1989, 66). Both the *Nuncius* and his log show that from that night on Galileo began to record the changing distances between them and Jupiter, having probably decided that the robustness of his claims rested on the determination of their motions (Galilei 1890–1909, III, Part 2:427).[51] On January 13, after having sturdied the telescope's mount, he observed a fourth satellite.[52] Since January 15, all his log entries were made in Latin, suggesting that on that date he decided to publish the *Nuncius* and to include the daily positions of the satellites in it (ibid.).

To Galileo, then, the evidence that counted was not a snapshot of individual luminous dots around Jupiter, but the "movie" of their motions (Galilei 1989, 67–83). It was "historical logic" that linked his string of observations and turned the luminous bodies near Jupiter into satellites, not fixed stars. Since the title page of the *Nuncius*, in fact, Galileo identified the satellites with their motions — "four planets flying around the star of Jupiter at unequal intervals and periods with wonderful swiftness" — a characterization that was then repeated in the text (Galileo 1989, 26, 36).

The *Nuncius*' mapping of the satellites' motions did not stop on January 13, but continued with painstaking descriptions of more than sixty configurations (which he also represented as diagrams) of the four satellites over forty-four almost consecutive nights (fig. 2). The textual and diagrammatic description of their movements occupies a large portion (about 40 per cent) of Galileo's text. Taken at face value, this section may appear tedious (Drake's first English translation edited out most of it) as it does not present complex arguments or exciting evidence.[53] And yet, Galileo included it and continued to observe the satellites for several more weeks despite being already certain of his claims and despite his fear that any delay in publication could deprive him of priority. His actions clearly indicate the importance he placed on this section — a section he then planned to expand in a revised edition of the *Nuncius* (Galilei 1890–1909, X:332, 373).[54] Although after

---

[51] In the *Nuncius* he added that the procedure he followed to measure these distances was the one described at the beginning of the book (Galilei 1989, 66). On Galileo's shift from fixed stars to planets, see also Drake 1976a, 153–168. For a reconstruction of the visibility of Jupiter's satellites in the period Galileo first observed them (and what he may have missed because of the quality of his telescope), see Meeus 1964, 105–6.

[52] "Havendo benissimo fermato lo strumento" (Galilei 1890–1909, III, Part 2:427; Galilei 1989, 67).

[53] In Galilei 1957, Drake edited out all the observations from January 14 to February 25.

[54] Such an edition, however, never materialized. The last observation of the satellites reported in the *Nuncius* is from March 2, and the book was off the press on March 13. The lunar observations included in the *Nuncius*, instead, dated from much earlier. According to Ewan Whitaker's reconstruc-

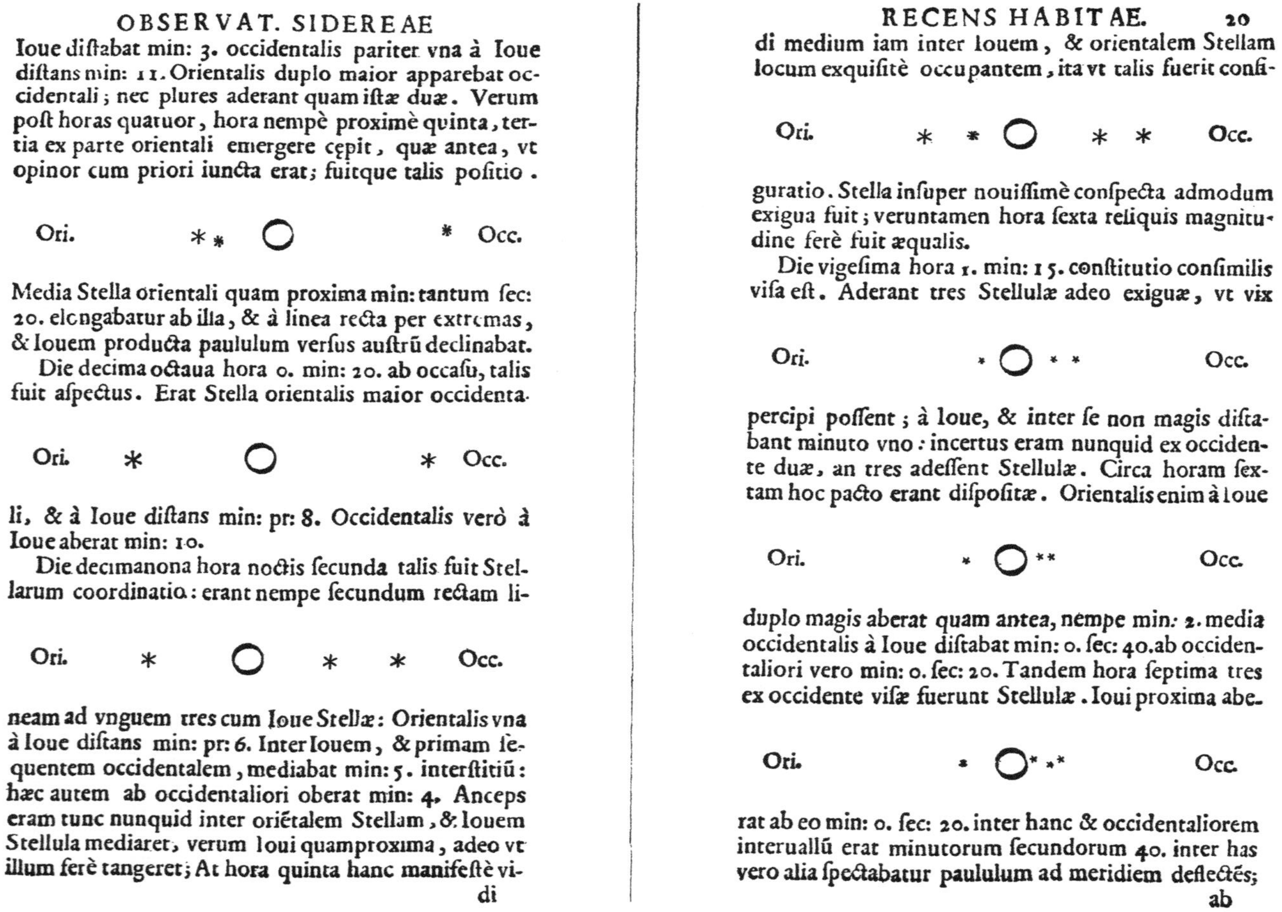

OBSERVAT. SIDEREAE

Ioue diſtabat min: 3. occidentalis pariter vna à Ioue diſtans min: 11. Orientalis duplo maior apparebat occidentali; nec plures aderant quam iſtæ duæ. Verum poſt horas quatuor, hora nempè proximè quinta, tertia ex parte orientali emergere cœpit, quæ antea, vt opinor cum priori iuncta erat; fuitque talis poſitio.

Ori.   * *   O   *   Occ.

Media Stella orientali quam proxima min: tantum ſec: 20. elongabatur ab illa, & à linea recta per extremas, & Iouem producta paululum verſus auſtrū declinabat.
Die decima octaua hora o. min: 20. ab occaſu, talis fuit aſpectus. Erat Stella orientalis maior occidenta-

Ori.   *   O   *   Occ.

li, & à Ioue diſtans min: pr: 8. Occidentalis verò à Ioue aberat min: 10.
Die decimanona hora noctis ſecunda talis fuit Stellarum coordinatio: erant nempe ſecundum rectam li-

Ori.   *   O   *   *   Occ.

neam ad vnguem tres cum Ioue Stellæ: Orientalis vna à Ioue diſtans min: pr: 6. Inter Iouem, & primam ſequentem occidentalem, mediabat min: 5. interſtitiū: hæc autem ab occidentaliori oberat min: 4. Anceps eram tunc nunquid inter orietalem Stellam, & Iouem Stellula mediaret; verum Ioui quamproxima, adeo vt illum ferè tangeret; At hora quinta hanc manifeſtè vidi

RECENS HABITAE.   20

di medium iam inter Iouem, & orientalem Stellam locum exquiſitè occupantem, ita vt talis fuerit configuratio.

Ori.   * *   O   *   *   Occ.

Stella inſuper nouiſſimè conſpecta admodum exigua fuit; veruntamen hora ſexta reliquis magnitudine ferè fuit æqualis.
Die vigeſima hora 1. min: 15. conſtitutio conſimilis viſa eſt. Aderant tres Stellulæ adeo exiguæ, vt vix

Ori.   * * *   O   Occ.

percipi poſſent; à Ioue, & inter ſe non magis diſtabant minuto vno: incertus eram nunquid ex occidente duæ, an tres adeſſent Stellulæ. Circa horam ſextam hoc pacto erant diſpoſitæ. Orientalis enim à Ioue

Ori.   * *   O   Occ.

duplo magis aberat quam antea, nempe min: 2. media occidentalis à Ioue diſtabat min: o. ſec: 40. ab occidentaliori vero min: o. ſec: 20. Tandem hora ſeptima tres ex occidente viſæ fuerunt Stellulæ. Ioui proxima aberat ab eo min: o. ſec: 20. inter hanc & occidentaliorem interuallū erat minutorum ſecundorum 40. inter has vero alia ſpectabatur paululum ad meridiem deflectés; ab

Figure 2. Example of Galileo's diagrams of the motions of the Medicean Stars in *Sidereus nuncius* (Venice: Baglioni, 1610), p. 20 (reproduction courtesy of Owen Gingerich).

the publication of the *Nuncius* he refrained from giving telescopes to other astronomers, he sent them records of his continued observations (Galilei 1890–1909, X:362, 403). In fact, if one trusted Galileo's description of their rapid movements (and that such movements seemed to lie within a specific plane), it would have been difficult to claim that the satellites were optical artifacts produced by the telescope. The determination of the satellites' periods would have provided even stronger evidence for the Stars' existence, but a preliminary mapping of the luminous dots' regular motions along a plane already cast them as strong candidates for physical phenomena.

Interesting, the only feature of the telescope Galileo discussed at some length at the beginning of the *Nuncius* was not its construction and optical principles, but its use for measuring angular distances, that is, for tracking the movements of the satellites and detecting their periods (Galilei 1989, 38–9). And Galileo's exhortation to his colleagues to go beyond what he had done only concerned the periods of objects he had already detected, not new discoveries. Read in the context of his monopolistic ambitions, his saying: "I call on all astronomers to devote themselves to investigating and determining their periods" does not sound like an altruistic tip, but an attempt to channel his competitors' drive in directions useful to him. Even if they preceded him at determining the periods, their confirmation of the physical reality of the satellites would have still helped him. Similarly, the "virtual witnessing" which the *Nuncius* offered to those readers who did not have telescopes was part of a strategy of control, not of collaboration or community-building (Shapin 1984, 81–520; Shapin and Schaffer 1985, ch. 2). His reports were not aimed at facilitating independent replications, but at satisfying his readers with narrative simulations of his own experience so that they would not feel the need to pursue it on their own. He cast his readers not as colleagues in an emerging philosophical community, but as remote, credit-giving consumers.

Galileo used the same "historical" logic of observation to argue that, contrary to received views, the lunar surface was not smooth but rugged like the earth's (Galilei 1989, 40).[55] He also adopted such an approach a few years later in his book on sunspots which was suggestively titled "*History* and Demonstrations Concerning Sunspots" (Galilei 1613; emphasis mine).

In the *Nuncius*, he opened his discussion of the moon's appearance with the observation that "when the Moon displays herself to us with brilliant horns, the boundary dividing the bright from the dark part does not form a uniformly oval line, as it would happen in a perfectly spherical solid, but is marked by an uneven, rough, and sinuous line" (Galilei 1989, 40). As with the satellites of Jupiter,

---

tion of the dating of Galileo's lunar observations and drawing, all but one were done by December 18 (Whitaker 1978, 155–69). His essay also recapitulates the previous debate about the dating of such observations). This shows that from January 7 to March 2 Galileo dedicated himself almost exclusively to observing the satellites to substantiate a claim he was already sure of by January 11.

[55] See also his description of the lunar surface in the January 7 letter (Galilei 1890–1909, X:259, 273–7).

Galileo's problem was to show that physical objects (valleys and ridges) were behind the irregular visual appearance of the terminator. And, as with the satellites of Jupiter, his argument did not stop at one snapshot of the irregular pattern of bright and dark spots on the lunar surface but continued with a discussion of how that visual pattern changed in time:

> Not only are the boundaries between light and dark on the Moon perceived to be uneven and sinuous, but, what causes even greater wonder is that very many bright points appear within the dark part of the Moon, entirely separated and removed from the illuminated region and located no small distance from it. Gradually, after a small period of time, these are increased in size and brightness. Indeed after 2 or 3 hours they are joined with the rest of the bright part, which has now become larger. In the meantime, more and more bright points light up, as if they were sprouting, in the dark part, grow, and are connected at length with that bright surface as it extends farther in this direction. (Galilei 1989, 42)

He then repeated this same kind of "historical" analysis for specific and particularly conspicuous dark and bright spots, showing how their changing appearances were consistently connected to the changing angle at which sunlight struck the lunar surface as the Moon went through its phases. He also set up a sort of "crucial experiment" by showing the moon at first and second quadrature, that is, when the moon is half full but its bright and dark sides are switched around (fig. 3). By doing so, he tried to show how the irregular patterns of lights and shadows are inverted in the two cases and that, therefore, they constituted the negative and positive picture of the same physical features of the lunar surface.

As in his discussion of the movements of the satellites of Jupiter, Galileo used pictorial representations to guide his readers through the changing patterns of lunar lights and shadows, though in this case the movie was quite "jumpy" and most of the narrative burden was put on the text (fig. 4).[56] The argument's logic, however, was the same in both cases. The existence of lunar valleys and mountains did not hinge on a few disjointed observations, but on the pattern traced by dark and bright spots as they changed through several interrelated observations (Van Helden 1989, 21–2). Being consistently connected to the phases of the Moon, these changing visual patterns could not be easily dismissed as optical artifacts produced by the telescope (Galilei 1989, 44–5).[57] Therefore, while having the appearance of

---

[56] For a discussion of the relationship between Galileo's somewhat crude pictures of the moon and his more accurate narrative, see van Helden and Winkler 1992, 207–9. Galileo was aware that the copper plates used in the *Nuncius* were not as good as they could be and planned to include better illustrations of the moon in a revised edition that, however, never appeared (Galilei 1890–1909, X:332, 373).

[57] A few years later, during the debate with the Jesuit mathematician Christoph Scheiner on the discovery and nature of sunspots, Galileo resorted again to periodic evidence, not snapshots. His claims about the status of sunspots as objects were inseparable from the description of their periodical movements and of how their shape changed in time. Tellingly, the title of his book was *History and Demonstrations about Sunspots*.

RECENS HABITAE. 10

Hæc eadem macula ante fecundam quadraturam

nigrioribus quibufdam terminis circumuallata confpi-citur; qui tanquam altiffima montium iuga ex parte Soli auerfa obfcuriores apparent, quà verò Solem re-fpiciunt lucidiores extant; cuius oppofitum in cauita-tibus accidit, quarum pars Soli auerfa fplendens ap-paret, obfcura verò, ac vmbrofa, quæ .x parte Solis fita eft. Imminuta deinde luminofæ fuperficie, cum primum tota fermè dicta macula te..ebris eft obducta, clariora mòtium dorfa eminenter tenebras fcandunt. Hunc duplicem apparentiam fequentes figuræ com-moftrant.

OBSERVAT. SIDEREAE

&um daturam. Depreffiores infuper in Luna cernun-tur magnæ maculæ, quàm clariores plagæ; in illa enim tam crefcente, quàm decrefcente femper in lucis tene-brarumque confinio, prominente hincindè circa ipfas magnas maculas contermini partis lucidioris; veluti in defcribendis figuris obferuauimus; neque depreffiores tantummodo funt dictarum macularum termini, fed æquabiliores, nec rugis, aut afperitatibus interrupti. Lucidior verò pars maximè propè maculas eminet; a-deò vt, & ante quadraturam primam, & in ipfa fermè fecunda circa maculam quandam, fuperiorem, borea-lem nempè Lunæ plagam occupantem valdè attollan-tur tam fupra illam, quàm infra ingentes quædam emi-nentiæ, veluti appofitæ præferunt delineationes.

Hæc

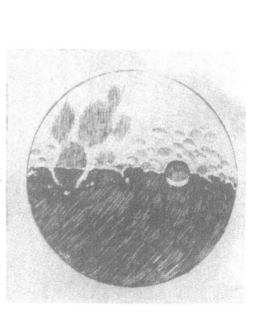

Figure 3. Two of Galileo's illustrations of the lunar surface in *Sidereus nuncius* (Venice: Baglioni, 1610), p. 10 (reproduction courtesy of Owen Gingerich).

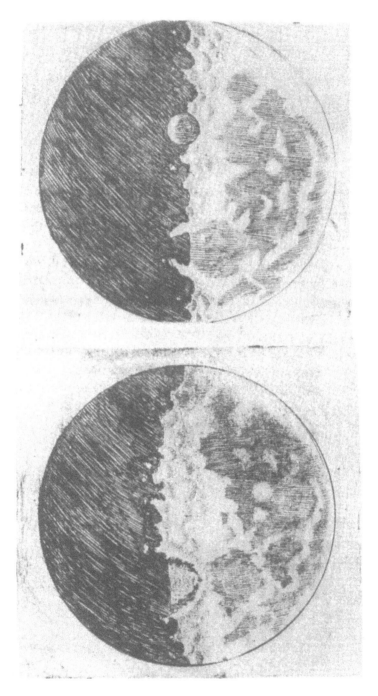

Figure 4. Additional illustrations of the lunar surface in *Sidereus nuncius* (Venice: Baglioni, 1610), p. 11 (reproduction courtesy of Owen Gingerich).

"natural histories" of satellites or lunar peaks and valleys, the structure of Galileo's arguments resembled that of demonstrations. They were not syllogistic demonstrations (and they used qualitative representations, not numerical entities or logical categories) and yet they explained effects from physical causes.

However, even those willing to accept that such visual patterns were not artifactual did not need to agree that they were about ridges and valleys. They may have been a movie, but a movie about what? Throughout the discussion of the changing visual appearance of the moon during its phases, Galileo made repeated analogies to how terrestrial mountains and valleys are variously illuminated and cast shadows during the day. The analogy may be read as an anti-Aristotelian argument because it simultaneously undermined the unique status of the earth while claiming that the Moon was not as pristine as the philosophers expected it to be. I believe Galileo would have not opposed such a reading of his argument. However, there was a more specific, local role for the earth-moon analogy in the *Nuncius*.

In the case of the satellites of Jupiter, Galileo argued that they were real because they had periodical motions, but did not need to convince anyone that planets (the category in which he placed the Stars) had periods. The case of the lunar valleys and mountains, however, was different. While he needed to hinge their physical status on the periodicity of their appearances, here he did not have an astronomical exemplar for that kind of "movie." Terrestrial mountains and valleys provided him with that exemplar. Once Galileo's claims about the ruggedness of the Moon had been accepted, their implications reverberated back on the cosmological status of the Earth, but in the writing of the *Nuncius* Galileo needed to turn the inferential arrow in the other direction: he needed a messy Earth to show that he was telling the truth about the Moon.

While I do not argue that everyone should have felt compelled to accept Galileo's logic, there was nothing revolutionary about the protocols and inferences he asked his readers to follow. The way he processed telescopic evidence to argue for the existence of the satellites of Jupiter or for the irregularity of the lunar surface was the same used by traditional astronomers to detect the precession of the equinoxes or other time-based phenomena (with the important difference that, in this case, the periods involved were in the order of days, not centuries). The *Nuncius*' crucial novelty as a narrative was that it translated these practices from a series of numerical observations into a form that could be appealing to the philosophically curious, not only to professional astronomers. Although Galileo made his inferences from geometrical entities (the angular distances of the satellites from Jupiter, the relation between lunar shadows and the height of lunar mountains, etc.) he presented his claims in visual terms — as movies about satellites and shadows. Judging from how few people rejected the *Nuncius*, it appears that its narratives succeeded at least in casting Galileo's claims as plausible.

Time and Its Markets

The few practitioners who, after reading the *Nuncius*, went on to observe the satellites of Jupiter did adopt the observational practices Galileo had laid out in his text. For example, Kepler and the Roman Jesuits confirmed Galileo's claims after conducting a series of interrelated observations of the Medicean Stars.[58] The Jesuits remained doubtful about the reality of the satellites after observing them for a few nights, but their skepticism gave way after conducting daily observations over two weeks and noticing their revolutions.[59] As Clavius wrote on December 17: "Here in Rome we have seen them [the Medicean Stars]. I will attach some diagrams at the end of this letter from which one can see most clearly that they are not fixed stars, but errant ones, as they change their position in relation to Jupiter" (Galilei 1890–1909, X:391, 484).

Analogously, on October 9, Santini wrote Galileo that he had seen the satellites again, "several times, in different positions, so that I have no doubt [about their existence]" (Galilei 1890–1909, X:407, 445). In May 1611, Luca Valerio, a Roman mathematician, added a more explicit methodological spin to these remarks:

> It has never crossed my mind that the same glass [always] aimed in the same fashion toward the same star [Jupiter] could make it appear in the same place, surrounded by fours stars which always accompany it ... in a fashion that one evening they might appear, as I have seen them, three to the west and one to the east [of Jupiter], and other times in very different positions, because the principles of logic do not allow for a specific, finite cause [the telescope] to produce different effects when [the cause] does not change but remains the same and maintains the same location and orientation. (Favaro 1983, 1:573)[60]

While we have much evidence about the importance of time in the corroboration of Galileo's observations, we have no indications that tacit, cosmologically-informed perceptual gestalts played a role in that process. Both Galileo and Kepler were Copernicans, but the Jesuits were not (though they were growing increasingly skeptical about the Ptolemaic system). There is no clear evidence about Santini's cosmological beliefs, but none of his letters addressed those issues thus suggesting that he was not particularly concerned with the discoveries' cosmological implications. Cosmological beliefs, it seems, motivated the observers' behavior but did

---

[58] The structure of Kepler's *Narratio* resembled that of the *Nuncius*. In it, Kepler listed daily observations of the satellites of Jupiter from August 30 to September 9 (Kepler 1606, 319–322). Right after the last entry Kepler simply wrote that these observations confirmed Galileo's claims and that he returned the telescope to the Elector of Cologne.

[59] The Jesuits' observational log shows that they had been recording the daily positions of the satellites since November 28 (Galilei 1890–1909, III, Part 2:863).

[60] A similar point is made, in a more humorous fashion, by Galileo in a May 21, 1611, letter to Piero Dini in which he promises 10,000 scudi to whomever can construct a telescope that shows satellites around one planet but not others (Galilei 1890–1909, XI:532, 107).

not frame their perceptions.[61] Those who observed the satellites had to invest weeks and months in the project, and did so because they had something to gain (or at least nothing to lose) from corroborating Galileo's claims.

Symmetrically, the rejection of these discoveries did not result from cosmological or perceptual incommensurabilities or from the lack of a satisfactory description of the telescope's workings. Simply, those who opposed Galileo's claims did not take sufficient time to conduct long-term observations. By observing for only a short time, they could plausibly argue that the evidence available to them was, at best, insufficient. And because of the structure of the field, there were no shared professional norms that compelled Galileo's opponents to abide by his rules and invest time and resources to engage in the long-term observations needed to test his assertions, or to require Galileo to give them telescope-time, telescopes, or instructions about how to build them. Galileo's monopolistic attitudes were as ethical or unethical as his critics' allegedly stubborn or obscurantist dismissals.

In the case of the philosopher Cremonini, geocentric beliefs translated into an absolute refusal to observe. It was reported that he did not want to look through the telescope fearing it would give him a headache.[62] However, it is not that Cremonini was unable to see the satellites of Jupiter because he was an Aristotelian, but simply that such an observation would have been a very unwise investment of time and resources for someone of his disciplinary affiliation and professional identity. Cremonini was not being unreasonable; he simply did not wish to commit professional suicide.

Unlike Cremonini, other critics did look through the telescope, though only for a short time. Because of the brevity of their observations, they remained vocally skeptical about Galileo's reading of those changing patterns of bright spots as satellites. In a letter sent to Kepler right after Galileo's visit to Bologna, Horky wrote that the instrument worked wonderfully when aimed at terrestrial objects but performed poorly when pointed at the sky. Horky was probably correct saying that fixed stars appeared double, a fact that may have made him justifiably skeptical about Galileo's other claims (Galilei 1890–1909, X:301, 343). But while he did not share Galileo's perception of the significance of those spots, he did see them nevertheless. A few weeks later, in the *Peregrinatio contra nuncium sidereum*, Horky added that when he tried to observe Jupiter he saw "two globes or rather two very minute spots" near Jupiter on April 24, and detected "all four very small spots" on April 25.[63] He did not believe that those spots were satellites and yet the

---

[61] Valerio's case is more ambiguous because a few years after he wrote the letter I cited, his membership in the Accademia dei Lincei was suspended when he declined to endorse the Academy's full support of Galileo's pro-Copernican position in the "Letter to the Grand Duchess." However, Valerio's stance in 1615 was not informed by direct geocentric commitments, but rather from the desire to stay out of dangerous cosmological debates. On this dispute, see Biagioli 1995, 139–66.

[62] On Cremonini's refusal to confront Galileo's discoveries see Galilei 1890–1909, XI:526, 100 and especially XI:564, 165.

[63] "24 Aprilis nocte sequente vidi duos solummodo globulos aut potius maculas minutissimas". When Horky asked Galileo why the two other two stars were not visible despite the fact that the night

fact that Galileo's own records for those two nights report exactly the same configurations shows that Horky's cosmological beliefs did not prevent him from registering the phenomena as Galileo saw them (fig. 5).[64]

Giovanni Magini was another of Galileo's early opponents. A professor of mathematics at Bologna, a supporter of geocentric astronomy, and Horky's employer, Magini was among those who spent two nights observing with Galileo. He did not publish a critique of his claims but worked hard at undermining Galileo's credibility through letters describing his fiasco (Galilei 1890–1909, X:303, 345; 324, 365). However, when he described those events to Kepler a few weeks later, Magini adopted a much more accommodating stance, simply saying that those who observed with Galileo at Bologna were unable to see the satellites *perfectly* (ibid. X:315, 359).

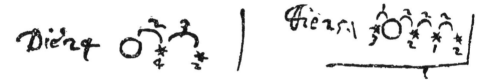

Figure 5. Entries for April 24 and 25 in Galileo's observational log (from Galilei 1890–1909, III, Pt.2, p. 436).

Magini had a point. He and Horky had reasonable grounds for skepticism (such as the double images produced by the telescope) and little incentive to take time to observe. Furthermore, Galileo did little to change their minds. First of all, he did not visit Bologna on the way to Florence (although it was there that a Medici carriage picked him up), but only on the way back to Padua, after he had shown the Stars to the grand duke and his family.[65] Eager to reach Florence as soon as possible, he actually changed his travel plans and skipped the stopover in Bologna altogether.[66] Furthermore, his one visit was very short and yielded only two

---

was clear, he allegedly received no answer (Galilei 1890–1909, III, Part 1:140). The next night, "Iupiter occidentalem exhibuerat, cum omnibus suis novis quator famulis supra nostrum Bononiensem Horizontem apparuit. Vidi omnes quator maculas minutissimas a Iove presilientes cum ipsius Galileo perspicillo, cum quo illas se invenisse gloriatur" (ibid., 141). Interestingly, Horky did not admit to having seen any of these "spots" in the April 27 letter to Kepler.

[64] Galileo's manuscript log in (Galilei 1890–1909, III, part 1:436), last line. Kepler too spotted the congruence between Horky's report and the configurations of the satellites he had received from Galileo (Galilei 1890 X:374, 416).

[65] That Galileo stopped to observe in Bologna on his way to Florence is a claim commonly found in the secondary literature, but is not supported by any evidence contained in Galileo's correspondence or observational log. On March 13, Galileo asked Vinta to send a carriage to Bologna "on the Monday of the week of Passion," that is, the week leading to Easter (Galilei 1890–1909, X:271, 289). Other letters confirm the appointment (Galilei 1890–1909, X:278, 303; 284, 307). In the Gregorian calendar, Easter fell on April 11, 1610. This means that the previous Friday was April 2, and that Monday of Easter week was April 5. I thank Owen Gingerich (and his remarkable collection of historical ephemerides) for providing the date of Easter 1610.

[66] Galileo's observational log shows that on the night of April 2 he was already close to Bologna (he

observational sessions. A few more nights could have made the satellites' periodic behavior more evident. But on April 26, a few hours after the end of the second session, Galileo left. Horky assumed that Galileo, demoralized by his failure, had left early in the morning to avoid further confrontations with his critics.[67] Instead, he simply needed to rush back to Padua to teach. In a March 13 letter to the Medici secretary, Galileo stated that the Easter recess at Padua lasted about 23 or 24 days and that he could leave only on April 2 (probably at the very beginning of the vacation) (Galilei 1890–1909, X:271, 289). This suggests that the recess ended on Monday April 26, the day Galileo left Bologna for Padua. He was cutting it quite close. But if he had strong reasons not to delay the departure any further, there is no evidence that he could not have arrived in Bologna a few days earlier.[68] He spent almost three weeks in Tuscany, but dedicated only two days to Bologna. Although he may have regretted his rush later, after realizing the harm done by Horky's and Magini's opposition, at that point Galileo seemed content with having shown the satellites to the grand duke and treated the assent of his "colleagues" in Bologna as a side dish that was not worth shortening his Tuscan stay by a few days.

Black-Boxes and Wrapping

If Galileo's observations did not require revolutionary gestalt switches, neither did their stabilization hinge on blackboxing his tacit instrument-making skills. Others were able to develop those skills in a matter of months without Galileo's instructions, thus suggesting that the "secret" of his telescope was only a trade secret. If one shared Valerio's conclusion (as Clavius, Santini, and Kepler did) that the satellites of Jupiter could not be dismissed as optical artifacts (because, under *ceteris paribus* conditions, one would expect telescopic artifacts to have a fixed appearance, not orderly motions), then one could consider the telescope's status as unproblematic despite the fact that no one, including Galileo, seemed able to provide a comprehensive explanation of how it worked.

Many studies of experimental replications use the notion of "blackbox." As a topos, the blackbox is usually conceived of as a container filled with some

---

observed in Firenzuola), suggesting he may have left Padua on April 1. On the 3rd he was already observing in Florence, indicating that he did not stop in Bologna and did not catch the Medici carriage that was supposed to pick him up on the 5th. On the 5th, in fact, he was already at San Romano, on the way to Pisa to meet the grand duke (Galilei 1890–1909, III, Part 1:436).

[67] "Galileo became silent, and on the twenty-sixth, a Monday, dejected, he took his leave from Mr. Magini very early in the morning. And he gave us no thanks for the favors and the many thoughts, because, full of himself, he hawked a fable. Mr. Magini provided Galileo with distinguished company, both splendid and delightful. Thus the wretched Galileo left Bologna with his spyglass on the twenty-sixth" (Horky to Kepler, April 27, 1610, [Galilei 1890–1909, X:301, 343]). English translation by Albert van Helden in Galilei 1989, 93.

[68] He probably arrived in Bologna on either the 23rd or the 24th, as he was still in Florence on April 20 (Galilei 1890–1909, X:299, 341), but was gone by April 24 (ibid., 300, 341).

knowledge that was initially tacit, private, and body-bound but was later rendered public, standardized, and (temporarily) unquestioned. The notion of blackbox, however, does not capture the process through which telescopic evidence was accepted. The telescope was neither inherently transparent (as Drake seemed to think) nor was it blackboxed at a later time (as the constructivists would assume). While philosophically opposite, these two positions share the assumption that legitimation is always needed and that it comes from some kind of knowledge that answers possible de-legitimizing doubts about the instrument's epistemological status (though the two camps would disagree on whether such knowledge is explicit or tacit, in the mind or in the body, about instruments or about the social qualifications of the people who use them). Instead, I argue that the telescope could be treated (or rather ignored) as a non-problem. What mattered was not the contents of the blackbox but its wrapping, that is, the narratives within which Galileo structured his reports.

Similar considerations apply to the status of Galileo the observer. The narrative logic of the *Nuncius* not only reduced the pressure on Galileo to disclose the workings and manufacture of the telescope, but it also cast his personal trust-worthiness as something of a non-question. Although readers of the *Nuncius* were asked to believe Galileo's claims about spending several nights on the roof of his house observing the changing positions of the satellites of Jupiter or the changing appearances of the moon, they were not required to trust the accuracy of all the specific observations he reported. Because Galileo's claims were about the recursiveness of certain patterns, their robustness did not rely on one crucial observation or experiment, nor did it hinge completely on his personal qualifications as a trustworthy observer.[69]

Inventions and Disclosure

Before 1610, Galileo participated in various professional and social groups which, in different ways, accustomed him to the value of limited disclosure and to the appreciation of economies of reward based on local patronage. He was by no means a reclusive scholar and yet he seemed quite content to limit his audience to the circles of Paduan academics, Venetian patricians, and Florentine courtiers with whom he discussed philosophy, music, mathematics, and literature.[70] He also

---

[69] Also, unlike later seminal texts of experimental philosophy, the *Nuncius* was not cast as the exemplar of a philosophical "form of life." Galileo tried to blackbox neither the telescope nor the community of its users. He adopted the customary protocols of long-term observational astronomy, but did not treat other astronomers as colleagues. The *Nuncius* tried to get credit from whatever constituency it could reach, and it did so by minimizing (not maximizing) the role of social conventions and values of any given community.

[70] Although he was forty-six by the time he wrote the *Nuncius*, Galileo had made no prior attempts to reach broader readerships through his publications, and his correspondence had been modest in volume and geographically limited to Italy. Even Kepler's 1597 invitation to engage in an epistolary dialogue about Copernicanism did not move him (Galilei 1890–1909, X:59, 69–71).

interacted actively with "low-culture" practitioners: artists, artisans, and engineers. Until 1610, the only publication under his name was a short instruction manual of a military compass — a device he developed and then sold privately to his students.[71]

Placed in this context, the monopolistic tactics Galileo displayed in the *Nuncius* and his carefully controlled distribution of telescopes were not just the actions of an author who "held back"; they could be seen also as the behavior of someone who had not "gone out" before and knew little about what to expect from larger audiences. Being new to the business of writing something like the *Nuncius*, he framed his tactics within the local and non-cooperative credit systems he was familiar with and extended them to cover much wider audiences, but without fully recasting them into a cooperative framework — a framework he had few exemplars for and from which, in any case, he had little to gain.

Some of Galileo's tactics came from his astronomical background, but others came from the world of inventors and instrument makers. He had been designing and producing instruments and machines prior to 1609, and his career as an inventor peaked precisely with the development of the telescope in the nine months leading to the publication of the *Nuncius*. Before he realized he could gain more credit for his discoveries than for his instrument, Galileo focused on the telescope as his ticket to success. By March 1610 he had fashioned himself as the discoverer of the Medicean Stars, but just a few months before he was still casting himself as the inventor of the first high-power telescope — an instrument he marketed for its military (not astronomical) applications. Galileo the inventor turned into Galileo the discoverer, but the metamorphosis was never complete.

Many readers seemed to recognize the inventor's "voice" in the *Nuncius*, as neither supporters nor critics questioned his secretive attitudes on ethical grounds. Some wished he had given out telescopes or information how to build them, but did not expect him to do so — at least not for free. The Elector of Fraising, for instance, read the *Nuncius* and, disappointed with how little Galileo had shared with his readers about the construction of high-power telescopes, offered him a reward if he communicated his secret to him and promised not to divulge it to others.[72] Galileo, then, was treated as an artisan entitled to have proprietary attitudes about the "secret" of the device he had developed — an artisan who had met the very low disclosure requirements typical of early modern inventions.[73] The

---

[71] He was also the probable author of a short 1605 pseudonymous satirical publication that, written in Paduan dialect, was meant for local consumption.

[72] On April 14, 1610, Galileo's brother, Michelangelo Galilei, reported that according to the Elector: "non havendo voi, in questo vostro primo libro, insegnato chiaramente tal fabbrica, li pare che sia di mancamento; et dice, se metterete innesecuzione quello che scrivete, che vi farete immortale; et vi prega, non volendo voi insegnare a altri detta fabbrica, al manco siate contento di volerne compiacere S.A., che vi si dimostrerà quel principe che egli e" (Galilei 1890–1909, X:290, 313).

[73] Later, on January 7, 1611, Mark Welser wrote Galileo that "I can tell you that information about how to build [telescopes] is much desired here [in Germany]," but did not intimate that Galileo's secrecy was seen as unethical (Galilei 1890–1909, XI:452, 14).

word "secreto" appeared often in correspondence discussing early telescopes, thus confirming that most of Galileo's contemporaries had a clear sense of the economy in which these instruments circulated.

There was no international patent law in Galileo's time. The protection of inventions depended on local legal and administrative practices, and was necessarily limited to the state that issued it.[74] Typically, the inventor was expected to show the appropriate officials a working example of the device for which he sought a temporary monopoly within that state's jurisdiction, but did not need to provide a full description of that device.[75] For instance, a north-European artisan approached the Venetian Senate in August 1609 asking for one thousand ducats for a low-power telescope, but did not want the Venetian authorities to examine the instrument, but only to look through it.[76] Paolo Sarpi, acting as the Senate's advisor, opposed the offer but not because he thought that the inventor's position was unethical.[77] What he objected to was the high price the artisan demanded for an instrument whose "secret" was proving to have a remarkable short half-life. Similarly, disclosure was not mentioned in any of the documents related to Hans Lipperhey's October 1608 application for a patent for the telescope he filed in the Netherlands (Van Helden 1977, 36–44).

If disclosure requirements were a local matter, so was the definition of inventor.[78] If local authorities deemed a certain device useful or protectable (or both), they

---

[74] According to Christine McLeod, Italian states had been at the forefront of the development of property rights for technical achievements, and these legal and administrative models were then transferred to northern Europe and England. In particular, Venice "was the first to regularize in law the award of monopoly patents, the Senate ruling of 1474 that inventions should be registered when perfected: the inventor thereby secured sole benefit for ten years, with a penalty of 100 ducats for infringement, while the government reserved the right to appropriate registered inventions." In order to expand the geographical coverage of their patents, inventors registered them in other states, provided they were deemed interesting enough to deserve that treatment (McLeod 1988, 11).

[75] In England, for instance, the legal demand for written specifications emerged only in the early eighteenth century, and such specifications were made public only towards the end of the century (MacLeod 1988, 11–2). Before then, "It was rare to demand anything of the patentee" (ibid., 13).

[76] On August 22, 1609, Giovanni Bartoli, the Medici agent in Venice, wrote to Florence about a foreign artisan's offer of a telescope to the Senate, and that the instrument was tried out from St. Mark's bell-tower (like Galileo's a few days later), but that many thought that its "secret" was well known in France and elsewhere, and that similar instruments were quite cheap outside Venice (Galilei 1890–1909, X:227, 250). The reference to the foreign artisan not allowing any internal inspection of the telescope is found in Drake 1978, 140, and in Drake 1970, 147–8. Drake cites no sources, but his description of the glassmaker's actions matches standard artisanal practices. MacLeod argues that in sixteenth-century England, an inventor was not required to share his secret if the technology he was bringing into the country helped the "furtherance of trade" (MacLeod 1988, 13).

[77] Sarpi may have done some "technology transfer" here. On August 29, 1609, Bartoli wrote to the Medici secretary that Sarpi told Galileo about "the secret he had seen [the foreigner's telescope]" and that Galileo, moving from that tip was able to produce a better instrument (Galilei 1890–1909, X:233, 255). Given Sarpi's role in the Venetian government, it would have been quite ethical for him to to facilitate Galileo's successful development of the telescope by feeding him information that could lead to a better instrument (and then to see it accepted and rewarded by the Senate).

[78] Such a definition of inventor makes sense in a context in which many inventors were itinerant artisans making a living out of spreading a country's technology into another (very much like the foreigner who first brought the telescope to Venice). In sixteenth-century England, "the rights of the first inventor were understood to derive from those of the first importer of the invention" (MacLeod 1988, 13). Such a definition was still held in the early seventeenth century (ibid., 18).

might issue a privilege (through a "letter patent") to a person who was not necessarily the original inventor but just the one who made available or perfected that technology within the jurisdiction of the privilege-granting authorities. One could obtain a privilege for the exclusive use of the printing press in Venice for a certain amount of time despite the fact that his name was not Gutenberg.[79] This, I think, casts interesting light on the debate about whether or not Galileo was the inventor of the telescope (Rosen 1954, 304–12).

Galileo's gift of the telescope to the Venetian Senate in 1609 did not amount to a proper patent application (most likely because he knew that such a privilege would have been unenforceable given how widespread telescope-making skills had become).[80] However, some of his interactions with the Venetian Senate (such as the monopoly he offered them for the use and production of the instrument, and the higher salary and tenure at Padua he received as a counter-gift from the Senate) conformed to artisanal and legal practices according to which he was the inventor of that kind of telescope within the jurisdiction of the Republic of Venice.[81] Galileo's tactics appear even less remarkable when we realize that some inventors did not request patents but donated their devices to their rulers in exchange for a job or a pension.[82] Even the patent application for the telescope filed by Hans Lipperhey with the States-General at the Hague on October 4, 1608, asked for an annual pension in case the patent itself were to be denied.[83] Galileo's gift of the

---

[79] On September 18, 1469, Johannes of Speyer received a privilege from the Collegio of the Signoria for printing in Venice and his dominion for five years (Gerulaitis 1976, 21).

[80] This problem had already caused Lipperhey to have his patent application denied in the Netherlands. On October 14, 1608, the States-General commented that "we believe there are others [other inventors] as well and that the art cannot remain secret at any rate, because after it is known that the art exists, attempts will be made to duplicate it, especially after the shape of the tube has been seen, and from it has been surmised to some extent how to go about finding the art with the use of lenses"(van Helden 1977, 38–9). Lipperhey, like Galileo after him, stressed the military application of the telescope.

[81] Galileo's dedication of the telescope to the Venetian Doge is not a simple letter, but a formal document that was officially debated and discussed by the Senate. If it did not ask for specific qui pro quo, that was for politeness' sake. The Senate did understand that Galileo was offering them a device (whose military applications were clearly laid out in the letter of presentation) in exchange for a better salary at Padua (which they did give him, together with tenure). Galileo's letter of presentation is in Galilei 1890–1909, X:228, 250–1.

[82] Inventors who had developed something directly useful to the state itself (as distinct from a technology that could foster a state's industry and trade) would not usually apply for a patent, especially knowing that states could take over their patents if they wished to do so. On the issue of state appropriation of patents, see the Venetian Senate ruling of 1474 in McLeod 1988, 11.

[83] "On the request of Hans Lipperhey, born in Wesel, living in Middelburg, spectacle-maker, having discovered a certain instrument for seeing far, as has been shown to the Gentlemen of the States, requesting that, since the instrument ought not to be made generally known, he be granted a patent for thirty years under which everyone would be forbidden to imitate the instrument, or otherwise, that he be granted a yearly pension for making the said instrument solely for the use of the land, without being allowed to sell it to any foreign kings, monarchs, or potentates; it has been approved that a committee consisting of several men of this assembly will be appointed in order to communicate with the petitioner about his invention, and to ascertain from the same whether he could improve it so that one could look through it with both eyes, and to ascertain from the same with what he will be content, and, upon having heard the answers to these questions, to advise [this body], at which time it will be decided whether the petitioner will be granted a salary or the requested patent" (Van Helden 1977, 36). Notice that the section about the pension matches quite closely what Galileo requested two years later in Venice.

telescope to the Senate in exchange for tenure and a salary raise fits squarely in this tradition.[84]

The workings of the early privilege system explain not only Galileo's secrecy but also his sense of to whom disclosure was due. As the Venetian Senate was the institutional patron to reward Galileo for his gift of the telescope, it was the Senate (not the readers of the *Nuncius* or fellow-astronomers) that Galileo felt obliged to share the secret of the telescope with. In fact, the only substantial description of Galileo's instrument (including the focal length of the objective lens, angle of view, and overall size of the instrument) is found in one of Sarpi's private letters.[85] Sarpi had access to this information because he was the Senate's advisor on telescopes, not only Galileo's close friend. Galileo had the same sense of obligation toward his next patron, the grand duke of Tuscany. In 1610, two months after the publication of the *Nuncius*, Galileo told Belisario Vinta, the grand duke's secretary: "I do not wish to be forced to show to others the true process for producing [telescopes], except to some Granducal artisan" (Galilei 1890–1909, X:307, 350). Disclosure was given to the source of credit, and in Galileo's social field credit came from patrons, not "colleagues."

There is some irony here. Much of Galileo's career ambitions after 1610 focused on gaining recognition as a philosopher, not a "mere" mathematician or instrument-maker. However, his career as a philosopher hinged on the fact that at that time he was not perceived as a philosopher but as a remarkable instrument-maker and that, as such, he was entitled to keep his secrets. Such a socially-sanctioned right to secrecy allowed him to develop a monopoly on observational astronomy and obtain the title of "philosopher" he desired so much, despite the fact that secrecy was not exactly a customary value among philosophers.

One could even say that Galileo's secrecy was not a right but a duty. Having been rewarded by the Senate for his telescope, he was obliged not to divulge its secret or

---

[84] Additionally, offering a device to a prince in exchange for a job or a pension made particular sense when such a device had no great commercial potential. In fact, in the summer of 1609, the telescope had only two financially rewarding applications: gadgets for rich gentleman, or military intelligence. Galileo already knew that the market for "play" telescopes was becoming quickly saturated and that prices were dropping fast. Also, the more telescopes one produced and sold, the more likely it was that his "secret" would be copied. The military market was more appealing. The telescope did have military applications, but one can also speculate that Galileo had plenty of reasons to amplify the range and importance of such applications, as he did in the formal presentation of the instrument to the Venetians. Selling the telescope to a prince was the best deal he could think of under those circumstances. After buying Galileo's device, the Venetians would have had all the interest to keep its secret for as long as possible, thereby lengthening Galileo's leverage. And Galileo could still enjoy tenure and a higher salary after the secret was gone.

[85] "Constat, ut scis, instrumentum illud duobus perspicillis (*lunettes vos vocatis*), sphaeric ambobus, altero superficiei convexae, altero concavae. Convexus accepimus ex sphaera, cuis diameter 6 pedum; concavum, ex alia, cuius diameter latitudine digiti minor. Ex his componitur insrumentum circiter 4 pedum longitudinis, per quod videtur tanta pars objecti, quae, si recta visione inspiceretur, subtenderet scrupula 1.a 6; applicato vero instrumento, videtur sub angulo maiori quam 3 graduum" (Sarpi to Leschassier, March 16, 1610 [Galilei 1890–1909, X:272, 290]). Why Sarpi felt free to share this information with his Parisian friend remains an open question.

to sell his instruments to anyone other than his employers.[86] During his flirtation with the Medici, Galileo was treading on delicate grounds as he was enticing a new patron with an instrument for which he had already been rewarded by another patron.[87] Because the Senate had rewarded the telescope for its military applications, sending an instrument to the Medici might have amounted to treason.[88] That Galileo kept promising to send the grand duke a good telescope but only took one to Florence himself at Easter time suggests that he probably felt he could not send an instrument to the Medici without violating his contract with the Venetians.[89] Had there been an international patent law, Galileo could have been in serious trouble. Instead, he could simply cross the river Po and start a new professional life.[90]

Finally, the conventions of early patents may explain, from another angle, why Galileo never provided a description of the optical processes of image-formation through a telescope. Unlike the historians and philosophers of science who have seen this alleged failure as potentially damaging of the epistemological status of the telescope, Galileo seemed unfazed by it (see for example Machamer 1973, 1–46, esp. 13–27). He did not seem to be familiar with the most relevant optical

---

[86] Unlike what I have written elsewhere (Biagioli 1993, 45–7) the fact that Galileo never sold his instruments was not just a matter of social self-fashioning but also of legal obligations. On October 3, 1609, Giovanni Bartoli, the Medici representative in Venice, wrote to Florence that Galileo's instruments were the best and that he was building 12 of them for the Senate. However, he continued, Galileo could not teach anyone how to build them because he had been ordered by the Senate not to divulge the secret (Galilei X:241, 260). Bartoli repeated the point a few days later (ibid., X:242, 260). And if Galileo could not divulge the secret, one can assume he could not sell his instruments either.

[87] On June 5, 1610, the Medici promised Galileo that: "in the meantime [your appointment] will be kept as secret as possible" (Galilei 1890–1909, X:327, 369). The Medici resident in Venice wrote to Florence on June 26 that "I have been asked if it is true that Dr. Galilei is going to serve the grand duke with a great salary. I answered I didn't know anything about it. If what they say is true, and is found out, it could give him trouble here" (ibid. X:343, 384).

[88] One way to interpret Galileo's behavior is that by the beginning of 1610 everyone understood that the "secret" of the telescope was hopelessly public, and that his contract with the Venetians was therefore more nominal than actual. In any case, it is interesting that Galileo's disclosure of the telescope to the Medici, his proposal to send several of them to European princes, and his acceptance of a contribution toward the cost of producing those instruments came only after both Galileo and the Medici understood that a position for him at the Florentine court was a serious possibility. And the distribution of instruments and the payment of the Medici's contribution took place only after he was told, during his Easter visit to Florence, that his employment in Florence was almost a fait accompli.

[89] The Medici seemed to understand that they were in a peculiar position. Around April 1610, Alfonso Fontanelli, a diplomat from Modena who had observed with Galileo's telescopes (probably at Pisa) told jokingly to the grand duchess that as soon as other nobles heard of the quality of the Medici's instruments, they would flood them with requests. To this, the grand duke and the grand duchess replied that the telescopes they had did in fact belong to the Venetians and that the Medici could not give them to anyone else (Galilei 1890–1909, X:304, 347). However, in that same period, the Medici were also evaluating Galileo's request to distribute telescopes to European princes through their diplomatic networks.

[90] We know from Sagredo that Galileo's departure from Venice (and especially the modalities of such departure) had upset several people there. I had previously thought that the Venetians' indignation had to do with what they must have perceived as Galileo's ingratitude (Biagioli 1993, 44–5). In light of this new evidence, the Venetians probably thought that Galileo had behaved unethically, perhaps even illegally. As a thought experiment, it may be interesting to consider what could have happened to Galileo had his new patron been a prince who, unlike the Medici, did not have friendly relations with the republic of Venice.

literature (Kepler's 1604 *Ad Vitellionem paralipomena*) nor did he seem anxious to fill his knowledge gap. Years later, he still did not think that Kepler's *Dioptrice* (a text in which the German mathematician discussed the process of telescopic image-formation) actually shed much light on the workings of the telescope and told Jean Tarde that Kepler's book was so obscure that maybe not even its author had understood it.[91] But no matter what Galileo's knowledge of optics might have been, he did not need to produce such an explanation.

Inventions, Discoveries, and Natural Monuments

The *Nuncius*, however, was not a private letter requesting a privilege from a local political body. Written in Latin and printed in 550 copies, it was aimed at a European audience. It did open with a brief and vague discussion of the telescope, but its stated purpose was to report discoveries. These discoveries, however, were still dedicated to and tailored to the patron, the Medici, who was as local as the patron to whom Galileo had previously offered the telescope. And while the *Nuncius* made his discoveries public, it did not break Galileo's artisanal secrecy about the telescope (because neither the Medici nor the Venetians wished him to divulge the secret of the instrument).

So, what kind of genre, what kind of economy did the *Nuncius* belong to? Into what "market category" did the Medicean Stars fall? For one, they were not discoveries in the modern sense of the term. Galileo did not operate in the kind of economy of discoveries that began to develop in the late seventeenth century: the placing of discoveries in the public domain in exchange for professional, non-monetary credit that accrued on the author's name. Instead, like any other inventor, Galileo received financial rewards from the Medici. At the same time, the Medicean Stars were not inventions to be used locally by a patron and kept secret within his jurisdiction. Their value was predicated on being as widely visible as possible. And while authenticity was not an issue in the economy of early inventions, the Medici did care about the fact that Galileo was the original discoverer of the Stars, not just someone who had "brought them" to Florence.

Indeed, what is worth noting is that the Stars were simultaneously private and public objects. The Medici did not own the Stars the way they could have owned an invention, and yet the Stars did belong to them in some ways. Galileo presented them simultaneously as natural entities and as monuments to the Medici. They were objects he had carved out of the state of nature for his patrons, and yet they were not like a statue chiseled out of a block of marble. They were not cast as artifacts Galileo had shaped with his hands and then sold to his patron. Nor were

---

[91] Galileo requested Kepler's *Ad Vitellionem paralipomena* from Giuliano de' Medici on October 1, 1610 (Galilei 1890–1909, X:402, 441). By December 1612 he had also a copy of Kepler's *Dioptrice* (ibid. XI:813, 448). Tarde's remarks are in "Dal Diario del Viaggio di Giovanni Tarde in Italia" (ibid. XIX:590).

they a work of art the Medici held and controlled as their property — objects they could keep in their galleries. They were tied symbolically to Florence and the grand duke only through their name — a name that had been legitimately given them by their original discoverer. The Medicean Stars were a "natural monument," something that was simultaneously natural and artifactual, global and local. One could think of them as a peculiar artwork — an artwork that was globally visible because the Medici had "loaned" it to all viewers at once.

The hybrid economy of the Medicean Stars matched the hybrid kind of credit Galileo received from them. He was neither a modern scientific author who put his work in the public domain in exchange for professional credit, nor an inventor who fully relinquished his rights by selling his work to a specific buyer in a specific place. He dedicated (but did not really sell) his discoveries to the Medici, because they were not something he could truly sell. The Medici gave him both financial rewards (of the kind given to artists or inventors who sell their work) and more symbolic recognitions (such as the title of philosopher) because what he gave them was not a piece of property exchangeable through monetary transactions. And because the Medicean Stars were not really sold to the Medici, Galileo could also put them in the "public domain" by communicating their finding to non-Florentine audiences (and receive non-financial, philosophical credit from them).[92] His function as author was equally hybrid: he was the discoverer of his patron's Stars. Like a court artist who could be famous and yet remain someone's artist, Galileo could only be the personal philosopher of the grand duke of Tuscany.

The Importance of Being Foreign

The tension between the local and global scale of Galileo's credit and the market of his discoveries can be traced to the hybrid kind of expertise and disclosure necessary to secure their acceptance and reward.

Inventions were evaluated locally, their reward was financial, and the process required little or no disclosure. Instead, the evaluation of later scientific discoveries depended on the judgment of a non-local community (a judgment that was based on the information disclosed by the author) and the discoveries were rewarded with philosophical credit, not money. The Medicean Stars fell in between. They were not assessed solely by a local patron nor by a global community, but through a hybrid arrangement in which a few external evaluations were brought to the local patron who integrated them with his own judgment. Similarly, their reward was neither fully local and financial nor completely global and philosophical. Disclo-

---

[92] As Merton has discussed in the case of modern science, eponymy is made possible by the fact that scientists do not receive direct monetary credit from their discoveries as they could, instead, from their inventions (Merton 1973, 286–324). Because they cannot claim real property on their findings, they may attach their names to them as a gesture of symbolic ownership for their work. In this case, however, eponymy was tied to the patron's (not the discoverer's) name.

sure requirements were selective too. A textual account of the discovery was made public, but telescopes were made available only to the patron and his peers (not to other astronomers), and the specifications of the instrument that made the Stars visible remained undisclosed and proprietary (as in the economy of invention).

The protocols of their evaluation were hybrid. Practices of philosophical legitimation and peer-evaluation were largely alien to the economy of inventions, but became increasingly important in the economy of discoveries. Inventions were rewarded because of their local usefulness, not because of their non-local truth status. Instead, the Stars' acceptance and reward did not result exclusively from a transaction in which the Medici "bought" the Stars because of their material usefulness to their house. Philosophical considerations about the truth of Galileo's findings (not just their usefulness for his local patron) did matter, as did the views of his peers (not just of his patrons). The process through which the Stars were evaluated and rewarded was a process that combined elements of the economies of invention and discovery. It was predicated on the Medici's perception that the Stars had a degree of usefulness (usefulness that offset the risks they would have taken by "buying" them), but the meaning of "usefulness," "risk," and "buying" was much less financial, material, and local than it was in the economy of invention.

Again, a comparison between Venice and Florence, between local and non-local markets, may shed some light on these differences. In Venice Galileo needed only to convince elderly senators to climb up the many steps to St. Mark's tower. Once up there, the breathless elders were as qualified as anyone else to realize that Galileo's telescope made distant ships visible. They didn't need to consult experts from outside Venice, nor did they want to do that because it was in their interest to keep that invention as secret as possible. This, however, does not mean that the senators were "experts" in telescopic matters, but only that they saw enough military potential in that instrument to be willing to invest in it. And, in any case, all they were risking was a limited amount of money — Galileo's salary raise at Padua.

Instead, Galileo's success at showing Cosimo II the Medicean Stars on a number of different occasions in April 1610 was a necessary (but not sufficient) step toward sealing his court appointment.[93] Although the grand duke appeared to be sufficiently impressed by what he was shown and let Galileo know that a position for him was in the making, he waited until July to make his offer official (Galilei 1890–1909, X:359, 400–1).[94] Kepler's endorsement of Galileo's claims in the *Dissertatio* and the positive reception of the *Nuncius* played a crucial role in

---

[93] In a June 25 letter, Galileo remarks that the grand duke "col proprio senso ha piu' volte veduto" (Galilei 1890–1909, X:339, 382). This claim was repeated in August. There Galileo added that Giuliano de' Medici and many other courtiers had witnessed the satellites as well (ibid., X:379, 422). In a May letter to Vinta, Galileo reminds him of what he had told him during his visit to Pisa regarding the possibility of a position at the Florentine court (ibid. X:307, 350).

[94] For a discussion of the grand duke's protracted hesitation, see Biagioli 1993, 133–9.

moving the Medici toward the final contract. But this does not mean that without Kepler's endorsement the Medici could not have trusted their own eyes or could not have found professionally qualified local talent to assess Galileo's claims. Because of the scale of the market the Stars were supposed to reach, the question faced by the grand duke was not primarily *who* had the necessary expertise to certify the Stars, but *where* that person needed to be.

The telescope was to be used locally in Venice, and its buyers had a clear idea of the material advantages it could provide.[95] Instead, the Medicean Stars' market was global, the grand duke did not have a clear sense of their potential usefulness for the house of Medici, and such usefulness was symbolic, not material (and therefore difficult to assess). Similarly, the risk the Medici would be taking by accepting the Stars was more symbolic than financial (but serious nevertheless). Had the Stars proved artifactual, the laughter of other princes would have hurt much more than wasting some money on Galileo's salary. Galileo's framing his discoveries in the Medici's dynastic mythologies did help the grand duke realize what he could gain from his Stars (Biagioli 1993, 103–57).[96] But the Medici were not the sole prospective users of the Medicean Stars — objects whose appreciation needed to exceed the Florentine market. Consequently, they could no longer do a "consumer test" on themselves like the Venetian senators had done with the telescope. They needed a third opinion.

At the same time, the Medici saw the Stars more as a monument (with peculiar visibility requirements) than as an object of knowledge. The Stars were not just an item of private knowledge that needed to be rendered public so that it could travel reliably outside of Florentine settings. Given the economy in which the Medici operated, the issue was not how to "blackbox" the Stars, but how to "sell" them as broadly as possible. Like the Venetian senators, the Medici thought primarily in terms of usefulness, investment, and risk, and less in terms of philosophical legitimation and peer-evaluation. Therefore, when the grand duke looked for external endorsements of the Stars, he acted like a company that seeks consumer opinions from its products' projected market niches, not opinions of experts qualified to certify the truth of Galileo's claims. Had he needed trustworthy people who could confirm Galileo's observations, the grand duke could have found them in Florence (and he was one of them).[97] What mattered the most, instead, were the

---

[95] Furthermore, the senators did not have to assess the *absolute* quality of the instrument, but whether it was good enough (i.e. worth the money), and how it performed in comparison to other instruments they had seen before. An August 1609 letter from Venice reports that the senators took the low-power telescope offered by a foreign artisan to the top of St. Mark's tower (Galilei 1890–1909, X:227, 250). This is exactly what they did with Galileo's instrument a few weeks later. While the senators had to *compare* two instruments and assess their *relative* usefulness, the Medici were asked to endorse the existence of the Stars, that is, to make an *absolute* claim.

[96] Some of my claims have been disputed by Michael Shank (Shank 1994, 236–43). My response is in Biagioli 1996b, 70–105.

[97] Not only had the grand duke seen the satellites of Jupiter several times in April with his courtiers, but he also observed the moon together with Galileo with one of his very early telescopes in the autumn of 1609 ("Che la luna sia un corpo similissimo alla terra, già me n'ero accertato, et in parte

views of *foreign* experts, not because they were more qualified but because they were foreign (because they spoke from where the Stars were supposed to go).

Kepler's early endorsement was crucial despite the fact that he was unable to replicate Galileo's observations at that time. It was crucial because Kepler was a well-qualified practitioner and bore the prestigious title of "Imperial Mathematician," but even more because Kepler was in Prague, not Florence. His "expertise" was a philosophic-economic hybrid: it was partly rooted in his professional credentials and partly in his physical location in an external market niche. And Prague was a key market because it was in places like the Imperial court that the glory of the Medici was supposed to be appreciated, not because it constituted an important node in the philosophical community. Kepler could have been wrong, and yet his endorsement told the grand duke that the Medicean Stars were likely to be celebrated in Prague (or that the Emperor would have shared in the flop).

The workings of this hybrid economy help to explain Galileo's aggressive distribution of telescopes to princes and cardinals. On March 19, between the publication of the *Nuncius* and his trip to Florence, Galileo wrote the Medici secretary that to celebrate the Grand Duke's glory in an appropriate fashion,

> it would be necessary to send to many princes, not only the book, but also the instrument, so that they will be able to verify the truth. And, regarding this, I still have ten spyglasses which alone among hundred and more that I have built with great toil and expense are good enough to detect the new planets and the new fixed stars. I thought to send these to relatives and friends of the Most Serene Grand Duke, and I have already received requests from the Most Serene Duke of Bavaria, the Most Serene Elector of Cologne, and the Illustrious Cardinal del Monte. ... I would like to send the other five to Spain, France, Poland, Austria, and Urbino, when, with the permission of the Grand Duke, I would receive some introduction to these princes so that I could hope that my devotion would be appreciated and well received. (Galilei 1890–1909, X:277, 298)

The secretary agreed. He replied: "Our Most Serene Lord agrees that the news [of the discovery] should spread and that telescopes should be sent to princes. He will make sure that they will be delivered and received with the appropriate dignity and magnificence" (ibid., X:285, 308).

By giving telescopes to princes and cardinals, Galileo was trying to get credit from important people who were not going to become his competitors.[98] But he

---

fatto vedere al Serenissimo Nostro Signore, ma però imperfettamente, non avendo ancora occhiale della eccellenza che ho adesso" (Galilei 1890–1909, X:62, 280). Since then, the grand duke showed himself extremely interested in the telescope and its development (well before Galileo's discoveries) and even helped Galileo's work by sending him in Padua glass blanks made to his specifications by Medici artisans in Florence ("gli si mandano i cristalli conforme all'avviso suo" (ibid. X:240, 259). Galileo's visit to Florence in the fall of 1609 is not mentioned explicitly in any of his letters, but hints can be found in ibid. X:247, 262; 250, 265; 254, 268.

[98] Courtly audiences were not risk-free. Princes tended to ask their mathematicians and philo-

did not treat them as Collins-style "core sets" — colleagues whose authority could certify new knowledge (Collins 1985, 142–7). Rather, he treated princes the way he had treated Venetian senators when, as an inventor, he had approached them a few months before. While in 1610 he marketed the Medicean Stars and the telescope (not the telescope alone), he repeated on a much larger scale what he had done locally in Venice in 1609. Also, at this time he acted as the Medici's "sales representative," not as an individual inventor. He hoped that, at each site, princes would "buy" (from the Medici and him) a package including the telescope, the *Nuncius*, and the Medicean Stars the way the senators had bought the telescope (from him) before. The process was analogous: it did not involve disclosure, blackboxing, or optical explanations but just the purchase (symbolic, not monetary) of what, for lack of a better term, I would call a "discovention."

After bringing Kepler's endorsement to Florence, Galileo was told that his position at the Florentine court was almost a fait accompli (Galilei 1890–1909, X:307, 350). At this point he started acting like an inventor who, having been rewarded by his prince for his work, did not care very much if other artisans could not reproduce his invention elsewhere. The endorsement from Prague seemed to be all he had needed from the republic of letters. He did not treat Kepler as a colleague, but as someone who just read the *Nuncius* competently and did not deserve any special sign of reciprocity (such as, for example, a good telescope).

With the grand duke on board, Galileo did not seem to worry about finding further witnesses to his discoveries. In August 1610, Kepler wrote Galileo asking for names of people who had seen the satellites (as well as for the usual telescope), so that he could use them to silence the critics who were still active in Prague (ibid. X:374, 416). Galileo replied:

> You, dear Kepler, ask me for other witnesses. I will offer [the testimony of] the Grand Duke of Tuscany who, a few months ago, observed the Medicean Stars with me at Pisa, and generously rewarded me. ... I have been called back to my fatherland, with a stipend of one-thousand scudi a year, with the title of Philosopher and Mathematician of His Highness, with no duties but plenty of free time. (Ibid. X:379, 422)

While Galileo may not have been able to mention Cosimo II as a witness in a written rebuttal to the mathematicians and philosophers who had questioned his claims (as that would have not been appropriate to the prince's honor), the Grand Duke gave him a position and title that, in that economy, was much more effective than a testimonial from a philosophically qualified practitioner.

Galileo was able to achieve a striking reversal: he acted as if his colleagues'

---

sophers for their opinions about Galileo's claims, and there was no guarantee that these would be positive ones. But if Galileo could not control these debates, he could expect the princes to have a relatively benign disposition toward his discoveries because of their familiarity with telescopes and of the respect they were expected to demonstrate towards the Medici (Biagioli 1993, 96–8).

difficulties to replicate his discoveries were to be read as signs of his own authority, not of the possibly problematic status of his claims. He almost turned his telescope into a "mystery," a sign of his "uniqueness." If people still had difficulties replicating Galileo's discoveries, it was their problem — a problem which confirmed that Galileo's telescopes were better than theirs. Now he could wait comfortably and work at producing more discoveries (as he did). The Jesuits' endorsement was still very important, but now Galileo could wait for them to corroborate his findings and go to Rome (as he did in the Spring of 1611) to be celebrated, not to help them with observations.

Conclusion

One always discloses in order to gain credit. Today, those who publish a discovery or invent a device always give something away in the process of getting their claims recognized or their devices patented. Competitors may be able to use the publication of a patent to circumvent it (or use it for free once it expires), and the publication of a scientific claim can allow other scientists to make further related discoveries and take credit for them. At the same time, discoverers or inventors receive something in return for disclosure. A scientist receives professional credit from publications, and a patent-holder is granted a temporary monopoly of the invention he/she has registered. Galileo operated in a different economy, one in which the checks and balances between credit and disclosure were drawn and managed differently.

The analysis of Galileo's monopolistic tactics has provided a window on the dynamics of the field in which he operated. Along the way, I have tried to show that these tactics were part of economies which constituted the objects they rewarded, not just the modalities of their crediting. Depending on the economies in which it circulated, Galileo's work could be put in different boxes (invention, discovery, artwork), each of them attached to different standards of visibility, disclosure, and secrecy — practices that framed the conditions of acceptance of that work. The boundaries between these economies, however, were far from natural. Galileo's tactics changed between 1609 and 1610, but not because his work had neatly evolved from "invention" to "discovery." The telescope and the Medicean Stars were different kinds of objects not by virtue of some essentialist taxonomy but because their features made them potentially suitable for different markets — markets of different scales, notions of usefulness and value, and kinds of reward. While the line between invention and discovery has come to demarcate two different regimes ("economic" and "scientific"), this example indicates that what we are looking at are, in fact, two economies (not an economy opposed to a non-economy).

By reframing the context in which Galileo operated, I have argued that the making and reception of his findings were both simpler and more complicated than previously thought. Simpler because, in my view, their observation did not depend on opaque perceptual dispositions or on tacit instrument-making skills. More complicated because the process was time-consuming and took place in a field whose members operated in different social and disciplinary economies, and because there was no ready-made "market category" in which Galileo's discoveries could be placed, assessed, and then rewarded in the way he wished them to be. I do not claim that Galileo's discoveries were facts that could be recognized as such once the "accidental" obstacles produced by his uncooperativeness could be removed, or that the debate could have been much shorter had his critics not attacked him with "obscurantist" arguments or had they not refused to take the time to observe. Once we understand these actions in the economies in which they took place, they no longer appear to be passive obstacles. Galileo's tactics (as well as those of his competitors and critics) were not accidental constraints on the path to truth, but constitutive elements of the production of the objects he called "Medicean Stars."

Acknowledgments

An early version of this paper was delivered at Harvard in July 1994. Since then, I have accumulated many debts of friendship with those who commented on this essay in its several oral and written embodiments: Jim Bennett, Allan Brandt, Bernard Cohen, Sande Cohen, Lorraine Daston, Peter Dear, Yehuda Elkana, Rivka Feldhay, Maurice Finocchiaro, Alan Franklin, Peter Galison, Owen Gingerich, Philippe Hamou, Roger Hart, Steve Harris, Matt Jones, Nick Jardine, Rob Kohler, Peter Machamer, Everett Mendelshon, Katy Park, Jürgen Renn, Simon Schaffer, Randy Starn, Noel Swerdlow, Sherry Turkle, Marga Vicedo-Castello, Norton Wise, and many others whose names I'll suppress in an attempt to maintain a semblance of authorship in this text. Special thanks to Al Van Helden, my advisor in *res telescopica*, with whom I have discussed this essay since its very inception (but who is not responsible for any mistakes I may have made).

References

Biagioli, Mario. 1993. *Galileo Courtier*. Chicago: University of Chicago Press.
——. 1995. "Knowledge, Freedom, and Brotherly Love: Homosociality and the Accademia dei Lincei, 1603–1630." *Configurations* 3:139–66.
——. 1996a. "Etiquette, Interdependence, and Sociability in Seventeenth-Century Science." *Critical Inquiry* 22:193–238.
——. 1996b. "Playing with the Evidence." *Early Science and Medicine* 1:70–105.
——. Forthcoming. *The Check is in the Mail*.
Bloom, T. 1978. "Borrowed Perceptions: Harriot's Maps of the Moon." *Journal for the History of Astronomy* 9:127–22.
Bourdieu, Pierre. 1985. "Social Space and the Genesis of Groups." *Theory and Society* 14:723–44.
——. 1999. "The Specificity of the Scientific Field and the Social Conditions for the Progress of Reason." In *The Science Studies Reader*, edited by Mario Biagioli, 31–50. New York: Routledge.
Chalmers, Alan. 1990. *Science and Its Fabrication*. Minneapolis: University of Minnesota Press.
Cohen, Bernard. 1993. "What Galileo Saw: The Experience of Looking Through a Telescope." In *From Galileo's "Occhialino" to Optoelectronics*, edited by P. Mazzoldi, 445–472. Padova: Cleup Editrice.
Collins, H.M. 1985. *Changing Order*. London: Sage.
Dear, Peter. 1995. *Discipline and Experience*. Chicago: University of Chicago Press.
Drake, Stillman. 1970. "Galileo and the Telescope." In *Galileo Studies*, 142–4. Ann Arbor: University of Michigan Press.
——. 1976a. "Galileo's First Telescopic Observations." *Journal for the History of Astronomy* 7:153–168.
——. 1976b. *The Unsung Journalist and the Origin of the Telescope*. Los Angeles: Zeitlin & ver Brugge.
——. 1978. *Galileo at Work: His Scientific Biography*. Chicago: University of Chicago Press.
——. 1979. "Galileo and Satellites Prediction." *Journal for the History of Astronomy* 10:75–95.
Edgerton, S. 1984. "Galileo, Florentine 'Disegno' and the 'Strange Spottedness' of the Moon." *Art Journal* Fall:225–232.
Favaro, Antonio. 1983. *Amici e corrispondenti di Galileo*. Edited by Paolo Galluzzi. Florence: Salimbeni.
Feyerabend, P. 1978. *Against Method: Outline of an Anarchistic Theory of Knowledge*. London: Verso.
Galilei, Galileo. 1613. *Istoria e dimostrazioni intorno alle macchie solari e loro accidenti: comprese in tre lettere scritte all'illustrissimo signor Marco Velseri*. Roma: Mascardi.

——. 1890–1909. *Le opere di Galileo Galilei.* Edited by Antonio Favaro. Florence: Barbera.

——. 1957. *Discoveries and Opinions of Galileo.* Translated by Stillman Drake. New York: Doubleday.

——. 1989. *Sidereus nuncius: or the Sidereal Messenger.* Translated with introduction, conclusion, and notes by Albert Van Helden. Chicago: University of Chicago Press.

Horky, Martinus. 1610. *Brevissima peregrination contra nuncium sidereum.* Modena: Cassiani. (Reprinted in Galilei 1890–1909, III, Part 1:129–45).

Gerulaitis, Leonardas Vytautas. 1976. *Printing and Publishing in Fifteenth-Century Venice.* Mansell: London.

Humbert, Pierre. 1948. "Joseph Gaultier de la Vallette, astronome provencal (1564–1647)." *Revue d'histoire des sciences et de leurs applications* 1:316.

Kepler, Johannes. 1606. *Narratio.* Pragae: Ex Officina calcographica Pauli Sessii.

——. 1611. *Narratio de observatis a se quatuor Iovis satellibus erronibus.* Frankfurt: Palthenius. (Reproduced in Kepler 1941, Band IV:317–25).

——. 1941. *Gesammelte Werke.* Munchen: Beck'sche Verlagsbuchhandlung.

——. 1965. *Conversation with the Sidereal Messenger.* Translated by Edward Rosen. New York: Johnson.

Machamer, Peter K. 1973. "Feyerabend and Galileo: The Interaction of Theories, and the Reinterpretation of Experience." *Studies in History and Philosophy of Science* 4:1–46.

McLeod, Christine. 1988. *Inventing the Industrial Revolution.* Cambridge: Cambridge University Press.

Meeus, Jean. 1964. "Galileo's First Records of Jupiter's Satellites." *Sky and Telescope* 27(2):105–6.

Merton, Robert. 1973. "Priorities in Scientific Discoveries." *The Sociology of Science: Theoretical and Empirical Investigations.* Chicago: University of Chicago Press.

Prickard, A.O. 1916. "The 'Mundus Jovialis' of Simon Marius." *The Observatory* 39:367–503.

Roche, John. 1982. "Harriot, Galileo, and Jupiter's Satellites." In *Archives internationales d'histoire des sciences* 32:9–51.

Rosen, Edward. 1954. "Did Galileo Claim He Invented the Telescope?" *Proceedings of the American Philosophical Society* 98:304–12.

Schaffer, Simon. 1989. "Glass Works: Newton's Prism and the Uses of Experiment." In *The Uses of Experiment*, edited by David Gooding, Trevor Pinch, and Simon Schaffer, 67–104. Cambridge: Cambridge University Press.

Shank, Michael. 1994. "Galileo's Day in Court." *Journal for the History of Astronomy* 25:236–43.

Shapin, Steven. 1984. "Pump and Circumstance: Robert Boyle's Literary Technology." *Social Studies of Science* 14:481–520.

Shapin, Steven and Simon Schaffer. 1985. *Leviathan and the Air Pump*. Princeton: Princeton University Press.

Sirtori, Girolamo. 1618. *Telescopium: sive ars perficiendi*. Frankurt: Iacobi.

Sizi, Francesco. 1611. *Dianoia astronomica, optica, physica*. Venice: Bertani. (Reprinted in Galilei 1890–1909, III, Part 1:203–50).

Van Helden, Albert. 1974. "The Telescope in the Seventeenth Century." *Isis* 65:38–58.

——. 1977. "The Invention of the Telescope." *Transactions of the American Philosophical Society* 67(4):1–67.

——. 1984. "Galileo and the Telescope." In *Novità celesti e crisi del sapere*. Edited by Paolo Galluzzi, 150–7. Florence: Giunti.

——. 1989. "Introduction." In *Sidereus nuncius: or the Sidereal Messenger*. Translated by Albert Van Helden, 10–6. Chicago: University of Chicago Press.

Van Helden, Albert, and Mary Winkler. 1992. "Representing the Heavens: Galileo and Visual Astronomy." *Isis* 83:207–6.

——. 1994. "Telescopes and Authority from Galileo to Cassini." *Osiris* 9:19.

Whitaker, Ewan. 1978. "Galileo's Lunar Observations and the Dating of the Composition of the 'Sidereus nuncius'." *Journal for the History of Astronomy* 9:155–69.

Department of History of Science
Harvard University

Appendix:
A Forgotten Controversy

A Forgotten Controversy
Introductory Note to the Appendix

More than a century ago three eminent Galileo scholars, Raffaello Caverni, Antonio Favaro, and Emil Wohlwill, discussed the emergence of Galileo's science of motion and the documentary evidence pertaining to it. Among the works of these scholars, only Favaro's *Edizione Nazionale* of Galileo's works is still widely used, while the contents of their other writings only play a minor if any role in the current English-speaking literature. The disappearance from historical memory of many of the substantial contributions by these authors is closely associated with a narrowing of the perspective under which Galileo's science and its context have been discussed in more recent scholarship.

This Appendix comprises three short contributions by Caverni, Favaro, and Wohlwill respectively, each prefaced by an essay that discusses their work in the context of the period. Although these contributions by three turn-of-the-century masters of Galileo scholarship certainly do not represent the pinnacle of their respective intellectual achievements, the writings here presented for the first time in English translation have the advantage of being closely related to each other, forming part of a controversy about the origin of Galileo's discoveries of the law of fall and of the parabolic shape of the projectile trajectory, as well as about the authenticity of certain unpublished texts by Galileo.

From the perspective of currently accepted views on Galileo's discoveries, the exchange, partially represented by the three contributions reproduced here, may appear to be rather odd: In the first volume of his monumental *History of the Experimental Method in Italy*, Caverni claims that Galileo had stolen the discovery of the parabolic shape of the trajectory from a disciple. In the fifth volume, an excerpt of which is translated here, Caverni published a hitherto unknown writing of Galileo on the hanging chain, crediting him with the dubious "discovery" that the hanging chain allegedly has, just as the projectile trajectory, a parabolic shape. Wohlwill, in his paper reproduced here, disputes Caverni's claim that Galileo committed intellectual theft and argues that Galileo in fact discovered the parabolic shape of the trajectory early in his career, after having first found the law of fall. In his later biography of Galileo, Wohlwill eventually develops the unorthodox and henceforth neglected view that Galileo possibly discovered first the parabolic shape of the trajectory and only then inferred from this discovery the law of fall. Reacting to the publication of Caverni's last work, Favaro, in a short paper also included in this Appendix, refutes the authenticity of the supposedly Galilean text published by Caverni but abstains from commenting on its content. In summary, it seems that the texts of Caverni, Favaro, and Wohlwill were rightly neglected in the past hundred years, since they merely deal with marginal subjects in Galileo's work such as the hanging chain, unorthodox opinions about his achievements, such as the claim that Galileo stole the discovery of the parabolic trajectory or made it even earlier than that of the law of fall, and even with forgeries of historical sources, justly forgotten after they had been revealed as such.

Revisiting this controversy from a broader perspective, however, it appears in a different light. Seen against the background of the intellectual context of Galileo's work as it becomes evident from his unpublished manuscripts, from the writings of his contemporaries, and from the practical challenges of the time, the three texts by Caverni, Favaro, and Wohlwill contribute to a new account of the establishment of the law of fall and of the parabolic shape of the projectile trajectory. According to this account, Galileo's achievements emerge not as isolated discoveries but as part of the development of the shared knowledge of the time. (For an extensive discussion, see Renn, Damerow, and Rieger, "Hunting the White Elephant," in this volume.) It furthermore turns out that, from this perspective, Favaro, Wohlwill, and even Caverni are each correct in their specific claims, in spite of their diverging views.

First of all, a close investigation of Galileo's unpublished manuscripts, including an examination of the inks used by Galileo in some of them, has made it possible to add further evidence to Favaro's assertion that the supposedly Galilean text published by Caverni must be a fake. In fact, the text published by Caverni mentions certain manuscripts by Galileo which can actually be identified among his papers (Ms. Gal. 72, folios 41/42 and 113). According to the supposedly Galilean text by Caverni, these manuscripts document the way in which Galileo produced, by means of a hanging chain, curves of projectile trajectories for the practical purposes of artillery, a description which turns out to be a plausible interpretation of these manuscripts. But the text also claims that these curves had been generated with the help of charcoal powder. A closer inspection of the drawings in these manuscripts, as well as their analysis by means of a PIXE (*Particle Induced X*-ray *Emission*) analysis, performed at the Istituto Nazionale di Fisica Nucleare in Florence, has, however, shown that these curves must have been produced by letting ink seep through a perforated sheet rather than sweeping it with charcoal powder from a feather duster, as claimed by Caverni. Thus Favaro's arguments in favor of forgery could be confirmed.

Second, a reconstruction of Galileo's investigative pathway in the context of his cooperation and exchange with contemporaries such as Paolo Sarpi has vindicated also Wohlwill's unorthodox suggestion, made in his biography of Galileo, that the search for the shape of the projectile trajectory, a question of practical interest at the time, was indeed at the root of his later science of motion, in contrast to the now generally held view.

Finally, it has turned out that even Caverni was right after all, despite having fabricated a Galilean text. On closer inspection, Caverni's text represents not only a shrewdly produced forgery but an ingenious reconstruction of Galileo's preoccupation with a challenging object of study neglected by most historians of Galileo's science, the curve of a hanging chain. In this odd way, Caverni points to the fact that the hanging chain was a subject crucial to Galileo's thinking on projectile motion. A careful analysis of the extant manuscript material makes it evident that Caverni was justified in claiming that Galileo, towards the end of his life, planned to complete the *Discorsi* with a proof of the alleged parabolic shape of the hanging chain, although, for all we know, this plan was never actually realized by Galileo himself.

The texts of this appendix have been edited by Giuseppe Castagnetti and rendered into English by a team of translators: Giuseppe Castagnetti, Lindy Divarci, Susan Kutcher, Leigh Rogers, and Fiorenza Zanoni-Renn.

Jürgen Renn
Max-Planck-Institut für Wissenschaftsgeschichte, Berlin

GIUSEPPE CASTAGNETTI AND MICHELE CAMEROTA

Raffaello Caverni and his *History of the Experimental Method in Italy*

Raffaello Caverni is a controversial figure among the scholars of Galileo's work. Almost entirely ignored for half a century, his research has been more attentively considered only after his major work (Caverni 1891–1900) was reprinted in 1970.[1]

Caverni was born in Montelupo, not far from Florence, on 12 March 1837, into a modest family of brick manufacturers and carters.[2] At the age of thirteen he began preparation for the Catholic priesthood in Florence, where he attended the secondary schools. He was ordained in June 1860. During his vocational studies, Caverni showed particular interest in the natural sciences and deepened his scientific knowledge by attending classes in astronomy, mathematics, mechanics, and hydraulics at the Istituto Ximeniano in Florence. Since the autumn of 1859 to the end of 1870 he taught philosophy and mathematics at the seminary for priests in Firenzuola, a small village in the high Appennines. During this long period of retired life Caverni considerably, even if disorderly, extended his reading in the humanities as well as in the sciences. Naturalist observation in the field was also one of his favorite occupations. A number of notebooks and journals with book extracts, philosophical and religious reflections, literary attempts, and intimate thoughts still testify to Caverni's autodidactical work.[3]

At the end of December 1870 Caverni finally took over the little parish of Quarate in Bagno a Ripoli, in the outskirts of Florence, where he lived until his death on 30 January 1900. The new appointment allowed him to pursue his scientific interests along with his pastoral duties for the following 29 years. In 1872 Caverni started to publish in periodicals for the family loose articles on curious mathematical problems, explanations of physical phenomena of everyday life, and also historical notes on scientific discoveries and inventions of instruments (Martini 1901, 296; Giovannozzi 1910, 265). Some of those unsystematic articles were later collected in a volume (Caverni 1882).

---

[1] The first Italian reprint was published by Forni, Bologna, in 1970. We refer to the second reprint, Caverni 1972. All references are concentrated in a bibliographic section at the end of the Appendix.

[2] The biographical data are drawn from Procacci 1900, Martini 1901, Giovannozzi 1910 (also reproduced in Betti and Pagnini 1991, 17–34), Tabarroni 1972, Cappelletti and Di Trocchio 1979, Maffioli 1985, Betti and Pagnini 1991. For Caverni's bibliography, see Mieli 1919–20, Cappelletti and Di Trocchio 1979, Betti and Pagnini 1991.

[3] The papers are preserved by the Caverni family in Prato, Italy (Betti and Pagnini 1991, 175).

The same lack of any apparent system characterizes Caverni's first book (Caverni 1874): a collection of essays on the most disparate subjects of mechanics, optics, astronomy, meteorology, and so on, that had been dealt with by Galileo, Borelli, Torricelli, and other seventeenth-century Italian savants. Clearly, the conceptual aspects of the "Galilean problems" rather than their historic authenticity were the focus of Caverni's treatment. As he explained in the foreword,[4] his intention was not to faithfully reproduce old texts or reconstruct historical development, but rather to make understandable for a broader public of young readers what seemed to him of particular scientific interest. Thus, he had no problem with rewriting passages without marking his own changes (Betti and Pagnini 1991, 105).

The book attracted the attention of another beginner in the history of science, the young university professor Antonio Favaro, who wrote to Caverni praising his work and asking for quotations about the causes of earthquakes in the works of Galileo and his followers.[5] This was not to be the last time that Favaro asked for Caverni's help in matters concerning Galileo.

In the years 1875–76 Caverni published a series of articles "on the philosophy of natural sciences," later also collected in a book (Caverni 1877a), dealing with problems posed to the traditional conceptions in ethics, theology, and metaphysics by the rapid growth of scientific knowledge. Essentially, he held a conciliatory position based on the orthodox dualistic view of the world recognizing, on the one side, the cognitive contribution of sciences in their own specific fields, the material world, while maintaining, on the other side, that sciences can say nothing about the spiritual world. The new and subversive idea was that even Darwin's theory of evolution should not be seen as contradicting religious beliefs, being only a hypothesis on the origin of animal species and their morphologic changes, and therefore not denying divine creation in principle. Of course, according to Caverni the biblical account was not to be taken literally (Giovannozzi 1910, 267–68); on the contrary, Galileo's hermeneutics was declared to be an "example to be imitated by future exegetes" (quoted in Maffioli 1985, 27). This show of intellectual boldness was not appreciated by the Catholic Church and the book was listed on the Index in 1878 (Giovannozzi 1910, 268; Tabarroni 1972, xi-xii), whereas it received warm appreciation by Favaro (Maffioli 1985, 56).

Caverni's intention, however, had not been to claim unquestioned validity of the Darwinian theory but to appeal to his fellow churchmen to carry on the dispute on a pertinent ground with better scientific arguments. For his part, as he made clear in a successive book (Caverni 1881), he was to a certain degree sceptical as to the reliability of paleontology in general and the theories on the origin and prehistory of man in particular, which were, in his opinion, extremely conjectural, with weak

---

[4] The foreword is reprinted in Betti and Pagnini 1991, 104–07. This book also includes a selection of texts from Caverni's published works as well as manuscripts.

[5] The relations between Favaro and Caverni are narrated and documented by Maffioli 1985, which we follow in our account; for Favaro's letter of 20 March 1875 see ibid., 56. On Favaro and his editions of Galileo's works, see the biographical note in this Appendix.

empirical foundation (Giovannozzi 1910, 268; Cappelletti and Di Trocchio 1979, 86).

As a true polymath with a paramount educational commitment, Caverni also wrote books on the language of Dante (Caverni 1877b) and on Italian literary style (Caverni 1879). Some years later followed popularizations of physics (Caverni 1884), of botanics (Caverni 1886b) and of mineralogical observations (Caverni 1888). In all the books of scientific popularization, he occasionally made historical digressions with the intent to arouse in the young readers esteem for the deeds of Italian savants and a desire to emulate them (Caverni 1972, 1:260–61). The wish to glorify the fatherland and to praise the Italian achievements in every field and especially in science can, indeed, be detected as a constant guiding motive in all the works of Caverni, whose "soul [was] woven with faith and nationality" (Giovannozzi 1910, 267).

Since the middle of the seventies, stimulated by Favaro who had become a personal friend and repeatedly asked for help in documentary research (Maffioli 1985, 31, 56–8), Caverni concentrated his studies on the history of Italian science. He got in the habit of going every Wednesday to the Biblioteca Nazionale Centrale of Florence, which had recently opened in premises of the Palazzo degli Uffizi adjacent to the Galleria (Tabarroni 1969, 564), and must have begun very soon to study the so-called Galilean Collection gathered by the Grand Dukes of Tuscany and now kept at the Library. This collection of more than 300 volumes of documents includes not only many manuscripts of Galileo, but also of his contemporaries as well as pupils and followers, documenting also the activity of the Accademia del Cimento and covering the period from the end of the sixteenth to the end of the seventeenth century and further. Caverni became one of the best and perhaps as yet unequaled experts of the entire collection, as shown by the material examined in his later publications.

Caverni's first scholarly publication in the history of science was on the invention and development of the thermometer (Caverni 1878). The subject was dealt with in the very traditional manner of priority disputes, aiming to state who had been the first to invent the instrument, who had improved it, and so on. His approach to the subject, however, was new compared with the works of contemporary Italian historians. The novelty was not so much that Caverni in his treatment kept to the sources and refused to follow established accounts on the principle of authority (Caverni 1878, 531), since these methodological rules were already accepted by the historians, even if, maybe, not always thoroughly respected. Caverni's new contribution consisted instead in the effort to reconstruct the process of invention by embedding it in a conceptual world. The story of the invention, with its personae and circumstances, was accompanied by a reconstruction of what was going on in "Galileo's mind" (Caverni 1878, 547). Addressing his attention to the philosophical frame in which the comprehension of heat phenomena could take place, Caverni showed that Galileo followed Platonian conceptions explaining heat as due to the major or minor quantity of heat corpuscles present in the body and argued that only on the basis of this theory was Galileo able to conceive of an instrument measuring the quantity of heat (ibid., 534–35, 543–44).

Of course, this kind of reconstruction was not possible without a certain degree of speculation with which not every historian would necessarily agree. Nevertheless some years later, in a chapter of his book on Galileo's Paduan period, Favaro explicitly followed Caverni in reconstructing Galileo's intellectual path (Favaro [1883] 1966, 197, 202, 204) adding only some new considerations on historical circumstances.

Notwithstanding Favaro's expressions of esteem, Caverni, in reviewing the book of his friend, made some critical remarks showing the differences of approach between the two historians (Caverni 1883; Maffioli 1985, 30–1). In Caverni's opinion, Favaro had given more room to the description of Galileo's "external life" than to the explanation of "the intimate life of the mind," meaning Galileo's intellectual work, of which Favaro had given only a resumptive exposition based on other authors instead of analyzing it. According to Caverni, as a consequence of this lack of penetrating study Favaro had accepted the account given by Vincenzio Viviani on the discovery of the isochronism of the oscillations of the pendulum on the basis of considerations concerning Viviani's trustworthiness, but he had not realized that some of Galileo's assertions on the phenomenon were experimentally untenable. "The fact is," Caverni objected, "that the discovery was not occasioned by an observation, but was a corollary of geometrical mechanics." Viviani's account should be considered as a tale (Caverni 1883, 478–79).

Furthermore, Caverni reproached Favaro for having accepted Galileo's own claim of priority in the discovery that the trajectory of a projectile is a parabolic curve, without trying to explain why, then, Galileo never made use of or even mentioned the discovery until the end of his life (ibid., 479). Even in appreciating Favaro's work Caverni had a polemic undertone, since he pointed out that Favaro's studies on details of Galileo's life and world did not carry on the hagiographic tradition but contributed to "humanizing" Galileo (ibid., 477). We have to bear in mind that Favaro did not question the accepted stereotype of Galileo as the lonely and heroic founder of modern science but, on the contrary, he contributed to the fostering of the myth.[6]

In the review of Favaro's book there appears, although only indirectly and not thematized, another important change in the interpretation of Galileo's science that was to be developed in Caverni's later writings. The fact that Caverni explained the discovery of isochronism as a deduction from a more general theory and not as a result of occasional observation seems to indicate that he considered the conceptual system and not the experiment as the true determining factor in Galileo's scientific procedure.

Also in another review of a book written by Favaro (Favaro 1886a) Caverni

---

[6] On the Italian historiography on Galileo and Caverni's new approach, see Micheli 1988 and Landucci 1996.

praised Favaro's erudition but in fact showed incomprehension for his work and at the same time made inadvertently clear how his own approach to the history of science was not thoroughly reflected (Caverni 1886a). Caverni did not appreciate the extensive exposition of the astrological works of Magini given by Favaro in order to reconstruct the cultural context of the cosmological controversies at Galileo's time. In Caverni's opinion, astrology was only superstition without connection to astronomy. Not aware of contradicting himself as to the role of the conceptual system in the process of scientific discovery, Caverni maintained that pure empirical evidence as given by the use of the telescope was enough to ruin astrology. According to him, Magini did not want to use it and stuck to astrology only because he was envious of Galileo (Caverni 1886a, 569). Clearly, Caverni did not yet have a coherent approach and jumped, so to say, from an explanation of a scientific tenet based on alleged personal factors to another explanation that took into account the conceptual reasons behind what the later story demonstrated to be an "error."

In spite of the already evident differences, Favaro involved Caverni in his project of a new edition of Galileo's works — which was to become the so-called Edizione Nazionale (Galilei 1890–1909) — since the early stages of its preparation in January 1883. He even proposed that they share the scientific responsibility of the edition and that Caverni take on the task of commenting on the scientific texts. In particular, Caverni was asked to work on the arrangement of the so-called *De motu antiquiora*, a set of Galileo's writings belonging to the Pisan period which were then published by Favaro in the first volume of the Edizione Nazionale, following in part the order suggested by Caverni (Maffioli 1985, 72–9). Yet, contrary to the original project, in 1887 it was finally decided that Galileo's texts should be accompanied only by historical notes and textual criticism, without the scientific explanations that Caverni had been envisaged for. As a consequence, Caverni was not appointed to the editorial board. This exclusion had an obvious negative effect on Caverni's relations with Favaro, but the collaboration between the two continued for a couple more years.

Meanwhile, since January 1886, Favaro was encouraging Caverni to take part in the contest for the "Tomasoni Prize," offered by the Venitian learned society "Reale Istituto Veneto di Scienze, Lettere ed Arti" for a work on the "history of the experimental method in Italy," especially "as applied to the physical sciences" from the beginning ot the fifteenth to the end of the seventeenth century (Favaro et al. [1889–90] 1972, 5).[7] Favaro was not only a member of the institute, but also of the jury, and his later behavior suggests that he intended to favor Caverni. Three years later, in March 1889, Caverni sent to the institute a manuscript of more than 3,000 pages, making roughly 2,000 printed pages, treating almost exclusively Galileo and his disciples, with scarcely any attention given to the time before and

---

[7] On Caverni's participation to the Tomasoni Contest and the coming into being of his *Storia del metodo sperimentale* we follow Tabarroni 1972 and Maffioli 1985.

after them (Maffioli 1985, 40). Nevertheless, the prize was given to Caverni but the tangled and occasionally farcical story of the adjudication and payment of the sum led to the definitive breakdown of personal relations between Favaro and Caverni. The reciprocal embitterment left a mark in their successive publications.

Caverni, for his part, taking notice of some of the critiques expressed by the judges (Favaro et al. [1889–90] 1972) — actually by Favaro — as to the scope of the treatment, started to enlarge the original manuscript to an encyclopedic work, the *Storia del metodo sperimentale in Italia* (Caverni 1891–1900), that unfortunately remained unfinished because of the author's sudden death. In spite of its title, Caverni's *Storia* is not limited to the Italian scene but encompasses the history of science from ancient Greece to the end of the eighteenth century and further and deals, at least briefly, with all the major figures of the ancient world and of the European sciences, sweeping from Heron of Alexandria to Newton, from Archimedes to Volta.

Caverni intended to publish at least seven volumes but only six appeared, the last one posthumous and unfinished.[8] The first volume begins with a 240-page introductory essay expounding the author's methodological and philosophical guidelines interwoven with a compendium of the history of science with particular attention to Italy. Following to it are some chapters concerning the invention of instruments and technical devices. Caverni then deals in greater detail with the progress of optics, theory of heat, magnetism and electricity, metereology, and astronomy in the second volume, and with anatomy, physiology, zoology, botany, and mineralogy in the third. The fourth and fifth volumes are dedicated to mechanics and dynamics, focusing essentially on Galileo, so that some chapters can be read as a close commentary to Galileo's *Discorsi e dimostrazioni matematiche intorno a due nuove scienze* and the related manuscripts in the Florentine Library. Finally, the sixth volume deals with the progress of hydraulics up to Torricelli, a great part dedicated to the then unknown work of Viviani. According to Caverni's own plan, the seventh volume should also deal with hydraulics and even "other volumes" on the history of "moral sciences" were envisaged in case the author had enough time and energy (Caverni 1972, 1:260).

It has been observed that the work of Caverni represents the first systematic history of science in Italy conducted over a large documentary base and in accordance with a general, although rather simplistic and "naïve," theory concerning the acquisition and progress of knowledge (Micheli 1980, 619; idem 1988, 184). Caverni's philosophical system certainly deserves deeper study. Here, however, we give only a succinct exposition of it.

---

[8] The volumes were published in 1891 (vol. 1), 1892 (vol. 2), 1893 (vol. 3), 1895 (vol. 4), 1898 (vol. 5). The last one bears the date 1900, but it was probably published after February 1905 or even after February 1910, since the old friend of Caverni, Father Giovannozzi, did not know of the existence of a sixth volume when he delivered his memorial speech on 6 February 1905 and the speech was published without correction exactly five years later (Giovannozzi 1910, 257, 272). The first notice of the existence of a sixth volume dates from summer 1916 (Favaro [1919–20a] 1992a, 141). A further chapter belonging to the sixth volume has been published in Giovannozzi 1928.

Refusing straight-away but without argumentation "ontologist" systems and sensualism, Caverni takes as the starting point of his theory the conceptions of the common sense philosopher Thomas Reid, allegedly confirmed by pedagogists and novelists. According to Reid — in Caverni's interpretation — the first knowledge follows from the first human experience of maternal love, that is from an emotional experience between two persons. According to a "law of love and understanding" that is not specified further, the growth of knowledge is not due to spontaneous development but to the transmission of knowledge among people. From this Caverni infers "the necessity of traditions," in the sense that the transmission and growth of knowledge can take place only within cultural traditions (Caverni 1972, 1:26–7).

Successively, Caverni elaborates a system whereby the development of mankind parallels the development of a human being from infancy to adulthood, a system in three phases reminiscent of Vico and Comte, although the phases are characterized in a different way. At each phase a new step in the growth of knowledge takes place. Following the theories of his contemporary pedagogist Raffaele Lambruschini regarding the psychological evolution of the child (Landucci 1996, 197), Caverni explains that at first the child is in "contemplation" of the world, it perceives the "superficial space," the "forms" of objects, and therefore "the first science learned by man is geometry." Subsequently, the child realizes that the world is something other than oneself, but "before it learns that the world rules itself with its own laws, it would like to be the lawgiver." Since the child's process of learning is the same as the process of acquisition of knowledge which a "civilized nation" goes through, it follows that "the law governing the intellectual development of an individual is the same one governing the intellectual development of a whole civilized nation." That means that in the history of every culture the sciences concerning "forms," such as geometry and mathematical astronomy, are the first to establish themselves, followed by a period in which general explanatory systems are conceived and imposed upon the world before the true laws of nature are discovered by the physical sciences. Among all nations, knowledge acquired in a particular phase is condensed in the work of a school founder and transmitted through it to posterity (Caverni 1972, 1:27–9).

According to this "historical law," the Platonian doctrines are seen as the analogue of the contemplative phase, during which mankind, and respectively the child, grasps the "forms," whereas Aristotelian philosophy corresponds to the "delirious" phase, during which the child imposes its own laws on the world. It is true, Caverni writes, that Platonian philosophy does not lead to experimental science, and moreover rejects it; but, on the other side, it contributes to the establishment of the theoretic preconditions for experimental science. Archimedes and Galileo are considered followers of Plato because they developed a geometrical science. Nor was Aristotle the initiator of experimental science; on the contrary, he provoked a regression because he attributed to reason the faculty of conceiving the general concepts from empirical particulars and devised artificial explanations for phenomena (ibid., 1:30–7).

Also the later Peripatetics as well as the Italian rationalists of the sixteenth century conceived dogmatic and sterile systems that did not promote scientific progress (ibid., 1:60). Scientific progress since the later Middle Ages was not due to the followers of the two rival schools but to the non-academic practitioners of the arts, meaning painters as well as navigators, architects, and poets, epitomized by the figures of Leonardo, Columbus, Leon Battista Alberti, and Dante. According to Caverni, their merit was that they had directly observed, "interrogated" nature, in contrast to the bookish science of the Peripatetics of the Middle Ages. They represented the third phase of human development, after the "Platonian illusions" and the "Aristotelian deliriums," corresponding to the phase of individual development during which "man begins, through the candid use of senses, to acquire stable possession of the world" (ibid., 1:67–8).

Generally, however, Caverni tended to deny to the followers of Aristotle, or to those he considered to be speculative Aristotelians like, for example, Descartes and Bacon, any merit in the progress of science, while stressing the alleged Platonism of those who contributed to that progress, or at least the Platonic elements in their theories, so that in the end it appears evident that Caverni considered Platonism to be the only philosophical system that truly promotes science. In order to carry through his dichotomic schematization of the entire history of science,[9] Caverni was forced to oversimplify and even misinterpret the thought of many authors.[10]

In this scheme, Galileo was listed among the Platonists because he considered geometry to be the instrument indispensable to discovering the laws of nature and because in some statements he allegedly denied validity to Aristotelian logic (ibid., 1:143). More generally, the Platonism of Galileo's "philosophia naturalis" seemed to consist for Caverni in the conception that the laws of nature as abstractions from empirical contingencies are "forms" with "geometrical regularity" (ibid., 1:241). Actually, Caverni did not support his interpretation with an analysis of Galileo's writings in connection with the culture of the time, so that the assertion remained unsatisfactorily corroborated and was questioned (Schiaparelli [1892] 1930, 20 n. 1).[11]

Even this short account makes clear that Caverni's theory concerning the development of science had at least two aspects that gave him a noted place in the history of scientific historiography. First of all, Caverni, far from being an erudite collector of sheer facts, was clearly aware of the debate of his time concerning the laws of scientific progress. In our opinion, he tried to syncretize the suggestions

---

[9] "All the varieties of doctrinal systems appeared in the course of the centuries can be easily reduced to two types, the Aristotelian and the Platonian one" (Caverni 1972, 1:38–9).

[10] In a review of the first volume the astronomer and historian of science Giovanni Schiaparelli blamed Caverni for having dealt with the history of philosophy following his feelings of likes and dislikes for the different thinkers (Schiaparelli [1892] 1930, 19). Micheli writes "Caverni's admiration for Plato is as excessive as his denigration of Aristotle" (Micheli 1988, 184).

[11] On the feebleness of the treatment of Galileo's philosophy by Caverni, see Landucci 1996, 197–99.

resulting from the then recent application of theories concerning biological evolution to the development of thought, as exemplified by his so to speak "recapitulation theory" of the growth of knowledge into a system based on the idealistic philosophy prevailing in the Italian Catholic tradition.[12] Furthermore, Caverni approached the history of science from a challenging psychological perspective that, unfortunately, went unnoticed at the time. Without pretending to establish any kind of genealogical relation, we wish only to remark on the analogy between Caverni's and Piaget's approach to cognitive development.

However, as shown by the way he contrasted Platonism with Aristotelism, certain tenets and aspects of Caverni's system were very idiosyncratic, being more the expression of a passionate personality with strong preferences for some philosophical traditions, than the result of systematic reflection. Caverni also did not always stick to his theory of cognitive development, in particular concerning the problematic relation between experience and conceptual understanding, and often contradicted himself.

Of the entire philosophical scheme only the guiding concept of tradition found a fruitful application in Caverni's historiography of science. Caverni left undefined what he meant by the term "tradition," as well as how, in his opinion, tradition operates in history; but we can broadly understand the alleged necessity of tradition as the assertion that there is no solution of continuity or revolution in the transmission and growth of scientific knowledge. In Caverni's words, from the necessity of tradition it follows that "it is a philosophical error to admit the existence of truly creative geniuses" (Caverni 1972, 1:26–7).[13] Accordingly, Caverni tried to show all along his work that new ideas and discoveries were the result of resumption and development of old ones. Caverni did not live long enough to take part in the disputes about continuity and revolution in the history of science but he was recognized as a kind of predecessor by Duhem who, some years later, explicitly agreed with Caverni's historiographic program although he rejected some of his specific reconstructions (Duhem 1909).

Actually, the results of Caverni's program are of a very different value. In some cases his search led simply to the redefinition of priority claims concerning invention of instruments or discoveries. In other cases Caverni detected in the words or concepts of older authors the precursors of later developments without consideration of their meanings in the original context. By emphasizing continuity on the basis of analogic linkages rather than of scientific meanings he tended to overlook the differences and thus misinterpreted the real progress. So, for example, he interpreted Newton's force of attraction merely as a more precise determination of Gilbert's magnetic forces (Caverni 1972, 2:356), or he saw in Newton's *vis*

---

[12] Caverni's philosophy has been connected to that of the Italian catholic priest Antonio Rosmini, who lived in the first half of the nineteenth century (Giovannozzi 1910, 265; Micheli 1988, 184).

[13] Similar statements can be found, e.g., in Caverni 1972, 4: 31, 69, 100, 328.

inertiae only a new name for an older concept already used by Giovanni Battista Benedetti and Galileo (ibid., 4:302–03).

However, an important merit of Caverni was the light he shed on the scientific situation before Galileo, so that he could not be seen any more as an isolated genius, but as a one among others who continued an already existent trend of studies (ibid. 4:374). In particular, Caverni showed Galileo's dependency upon a tradition reaching from the pseudo-Aristotelian *Quaestiones Mechanicae* and Archimedes through Jordanus Nemorarius and Leonardo, till Niccolò Tartaglia, Benedetti, Guidobaldo del Monte and many other scholars who had not been studied before (ibid., 4:579–80; 5:169–70; 6:48–52, 75, 109–11; Segre 1991, 44–5).

More important, in his history of mechanics and hydraulics around Galileo, Caverni did not limit himself to superficial analogies but tried to reconstruct conceptual developments in connection with broader theoretical or experimental contexts. Following step by step the arguments of different authors, Caverni explained particular turns in the solution of problems or showed how new perspectives were opened up as a result of the intrinsic constraints of the conceptual and material instruments employed. Exemplary in this sense are the treatments of fall and resistance of bodies in the fourth volume, where alongside with the major authors a number of lesser known scientists are also considered in pages and pages of commented theorems and propositions.

Of particular interest is the reconstruction of the discovery of the parabolic shape of the trajectory of a projectile following the writings of Guidobaldo del Monte, Galileo, and others (Caverni 1972, 4:506–78). Caverni documented the fact that Guidobaldo as well as Galileo based their studies on this subject on a conceptual analogy established between the trajectory of a projectile and the catenary, that is the curve described by hanging chains, in so far as they both were considered to be the result of the "composition of natural and violent motion." As a consequence it was possible to suppose that the trajectory would make a curve like the catenary, possibly a parabola, and to perform experiments in order to prove it (ibid., 4:515–17). Furthermore, although he claimed that Galileo recognized the parabolic shape of the trajectory only after Cavalieri in 1632 derived it from the law of fall, Caverni suggested that the connection between the parabolic trajectory of a projectile and the law of fall could have easily been noticed already at an earlier stage and in this way he traced a link between the discovery of the law of fall and the studies on ballistics (ibid., 4:517–19). The suggested link was worthy of consideration and was resumed by Wohlwill (Wohlwill 1969, 1:145–46) but generally overlooked by the historiography for almost a century.

Caverni's study of conceptual developments was a great novelty in Italian history of science, which was then predominantly concerned with narration of deeds and summarization of books. Caverni instead maintained that the main subject of scientific historiography should be the "storia intima del pensiero," the development of thoughts in the scientist's mind. The difference of approach that had already come to light in the reviews of Favaro's books was now underscored in polemic references to "certain new editors" who transcribe manuscripts without understanding them (Caverni 1972, 5:262; see also 1:144).[14] Caverni's focus on conceptual developments seems to originate from a specific comprehension of some intellectual features of scientific activity and from his intention to make this activity understandable in its own terms to a public of scientifically interested readers. For that reason he often insisted on the necessity of explanatory comments to the scientific texts (Caverni 1878, 535 n. 1; Caverni 1972, 1:83).

For a long time, the Italian historians of science were not able to appreciate and profit from Caverni's enormous pioneering work, partly because their approach was essentially erudite and not concerned with the genesis of concepts. Partly, also, because Caverni's re-evaluation of traditions and continuity in the transmission of knowledge was at odds with the predominant tendency to stress the radicality of change during the so-called scientific revolution in the seventeenth century, especially concerning the work of Galileo.[15]

Unfortunately, Caverni himself was partly responsible for the fact that his valuable insights were not appreciated and fell into oblivion after his death. One of the reasons was the anti-Galileian bias of his *Storia*. Introducing a disrupting factor in his scheme, Caverni argued that science from time to time becomes like an old kingdom without inner power and energy. In such phases according to Machiavelli only a tyrant that imposes his absolute will would be able to renew the state. Employing the metaphor of science as a kingdom, Caverni pictured Galileo as a repugnant personality who gave a decisive impulse to scientific progress albeit by imposing a tyrannical power upon the scientific community, appropriating the results of traditions while occulting them, and plagiarizing the discoveries of others (Caverni 1972, 1:127–28, 136–37, 140–43, passim; Schiaparelli [1892] 1930, 21). In order to belittle Galileo's contributions Caverni occasionally even concocted very implausible stories that would not stand historical analysis (e.g., Caverni 1972, 2:357–61; Favaro 1919–20b).[16]

This somewhat gothic image could not be accepted at a time when the predominant myth of Galileo was that he was alone in vindicating scientific rationality against obscurantism, a myth that was intermingled with the political struggle fought by the young Italian State against the Catholic Church. The reaction was correspondingly harsh, culminating in a kind of a "trial" against Caverni before the Italian historians of science nearly 20 years after the defendant's death, on the occasion of a congress of Italian scientists in April 1919 (Micheli 1987, 296; Del

---

[14] On Caverni's conception of history of science as history of conceptual development, see, e.g., ibid., 4:100; 5: 34, 644–45. A perspicuous evaluation of Caverni's approach is given by Maffioli 1985, 27, 49–52.

[15] On the reception of Caverni's contributions to history of science, see Tabarroni 1969, 566–67: Tabarroni 1972, vii–viii; Baldini 1980, 384–85; Maffioli 1985, 27, 30; Micheli 1987, 296–98; Micheli 1988, 184.

[16] Wohlwill called Caverni's account of Galileo's part in the discovery of parabolic trajectory a "fiction" (Wohlwill 1899, 607, 612). Wohlwill's article is republished in this Appendix. The judgement was shared by Tannery 1900.

Lungo 1919). The critics, while recognizing the originality and richness of Caverni's research, blamed him as one who denigrated Galileo, but they did not discuss the general principles inspiring the *Storia del metodo*. In particular, Wohlwill and Del Lungo gave alternative accounts of discoveries of which Caverni had disputed Galilean paternity (Wohlwill 1899, Del Lungo 1921–22), while the implacable rival Favaro missed no opportunity to point out alleged factual errors and distorted interpretations in the work of the old friend (Favaro 1907a, 1919–20a and 1919–20b; Maffioli 1985, 30 n. 14). Today we can say that Caverni was right in bringing to light the scientific setting that preceded and surrounded Galileo, thus reducing his achievements to their more realistic dimensions. But in times of ideological conflict only a sober presentation would have had any chance of being fairly considered.

It was not only his exaggerations against Galileo that made Caverni unreliable as a historian for his contemporary colleagues. Caverni had not the habit of a meticulous scholar even by the standards of his time. Although he pretended to base his account exclusively on primary sources (Caverni 1972, 1:261–62; 5:645), and indeed mastered an impressive amount of new materials even making use of Galilean scratches that were ignored ever after, in fact Caverni showed an inexcusable disregard for the rules of accuracy and adherence to the sources, following which, during the nineteenth century, historiography was establishing itself as an accountable discipline. Caverni did not really acknowledge that textual criticism, which he openly despised (ibid., 4:341; Maffioli 1985, 35 n. 34), as well as correct bibliographical data and reconstruction of historical circumstances, are indispensable for any kind of historiography. Generally, he tended to base his account on the alleged intrinsic reasons of a development, but he did not care to deliver an adequate theory of conceptual production that would support his interpretation. On the other hand, he sometimes took into consideration for his reconstruction extrinsic facts like evolution of handwriting or differences in inks, but he did not follow clearly stated criteria in examining and interpreting the data and often fitted them arbitrarily into a preconceived story (Caverni 1972, 4:341–73). Caverni did not even shrink from forging texts, as Favaro documented in an article listing at least six passages of the *Storia del metodo*, sometimes many pages long, expressly quoted by Caverni as authored by Galileo or Viviani but in fact invented by Caverni himself (Favaro 1919–20a).[17]

In order to understand Caverni's behavior we should perhaps consider that his approach to scientific knowledge was that of an amateur who wished to share his enthusiasm with the reader. If, as we believe, his primary aim was to make everybody able to follow a thread of thought, it is not surprising that he saw nothing blameworthy in composing scattered tessera into a picture without gaps

---

[17] An excerpt of this article is reprinted in this Appendix. In his apologetic introduction to the reprint of *Storia del metodo sperimentale in Italia* Tabarroni omits warning the modern reader about these fakes (Tabarroni 1972).

instead of presenting them as partial and fragmentary as they are. The results are not without scientific interest for the modern reader, as shown by a dialog about the utiliy of catenary for the calculation of projectile trajectories (Caverni 1972, 5:143–53).[18] One of the interlocutors describes a cardboard bearing perforated lines convincingly corresponding to a perforated paper sheet existing among the Galilean Manuscripts at the Biblioteca Nazionale of Florence.[19] The explanations about the possible function of the carboard given in the pseudo-Galilean dialog by Caverni are undoubtedly pertinent, as well as the connections developed in the dialog between the propositions concerning the momentums of falling weights, the catenary and the parabolic trajectory of a projectile. In our opinion, Caverni had indeed understood the reasoning of Galileo. It is a pity that he did not expose his interpretation as such.

Max Planck Institute for the History of Science, Berlin (GC)
Dipartimento di Filosofia e Teoria delle Scienze Umane,
Facoltà di Magistero, Università di Cagliari (MC)

---

[18] The dialog is reprinted in this Appendix.
[19] Ms. Gal. 72, folio 41/42. The document has been published for the first time in Drake 1979, 238–39. It has been studied in Damerow et al. 1996 and in Renn et al. 1998, 9–11.

RAFFAELLO CAVERNI

An Excerpt from *History of the Experimental Method in Italy**

We consider this merit,[1] however, to have almost no value in comparison to one which we wish to acquire from the offended worshippers of Galileo. We announce to them that after having identified and reordered the scattered writings which complete the sixth dialogue as far as percussion is concerned, we were also able to reintegrate the dialogue with regard to the use of a little chain to provide a rule for aiming artillery, without having to resort to laborious calculations.

Toward the end of the Fourth Day[2] Salviati says that little chains held loosely at their extremities hang in a curve very much resembling a parabola. He then alludes to a not unimportant use for such hanging chains which he promises the interlocutors to deal with later, diverting the conversation at first to the demonstration concerning streched rope. After this demonstration Simplicio reminds Salviati of his promise to explain "the utility that may be drawn from the little chain, and afterward give us those speculations made by our Academician about the force of percussion" (Galilei 1842–56, 13:266).[3] But the hour being so late there was not enough time to deal with the mentioned topics, Salviati suggests postponing the meeting to a more opportune time.

Apparently in the next meeting they intended to deal first with the little chain and then with percussion. The program was then changed, for whatever reason, by their discussing the second subject first. This, however, did not relieve Salviati of fulfilling his promise. That he really meant to keep it is evident from the fact that Salviati gauged the time of the conversation very well. When the discussion of the first subject ended, which also covered the theory of collision, it was only nine o'clock. It was then possible to spend what was left of the evening satisfying the curiosity of those who desired to know to what use the little chain might be put.

---

* This is a translation of pages 143–154 of volume 5 of Raffaello Caverni, *Storia del metodo sperimentale in Italia, 1891–1900*, 6 vols. Firenze: Civelli, 1891-1900. All footnotes are by the translator. The list of references can be found in a bibliographical section at the end of the appendix.

[1] In the preceding part of the chapter, Caverni argues that in a truly complete edition of Galileo's works a text written by Torricelli should have been published as a completion of Galileo's fragment concerning the force of percussion. Caverni claims to have the merit of having recognized the connection between the two texts.

[2] Caverni comments here on the fourth part of *Discorsi e dimostazioni matematiche intorno a due nuove scienze* (Galilei 1638).

[3] Caverni's bibliographical references have been standardized. For the accepted edition, see Galilei 1890–1909, 8:312. We adopt the translation by Stillman Drake (Galilei 1974, 259).

Notwithstanding, after more than a century and a half, their curiosity is still not satisfied, and it did not, and does not seem to matter at all to any of the most fervent Galileans. Therefore, we are the first and only ones among them to have searched industriously and finally found that second part of the Galilean dialogue, which, along with the first concerning percussion, gave the good Salviati and his friends subject matter to philosophize upon until evening. We shall abstain from narrating how we made this discovery among certain jumbled manuscripts given to us by a friend for examination because we believe that our readers would rather wish to learn without delay what we copied from it. This reads as follows:

SAGREDO. Your reasoning, Mr. Salviati, has completely convinced me that the forces of natural percussion and collision are infinite, so that you can now save yourself the trouble of further discussion. As far as I'm concerned, you may now keep your other promise, which was to tell us about the usefulness that our Academician hoped to obtain from little chains when applied to plot many parabolic lines on a flat surface. But I see Mr. Aproino looking quite surprised.

APROINO. You have understood me, Mr. Sagredo, because I find your proposal quite new.

SAGREDO. You are right. I had not realized that Your Lordship was not present on the evening of our last meeting when, before leaving, Mr. Salviati gave Mr. Simplicio and me to understand that, following the demonstration on the force of percussion, he would have added the explanation about the little chains. These, when held at their extremities, he said, naturally accommodate the curvatures of parabolic lines.

APROINO. To the first surprise you now add a further one, greatly arousing my curiosity, which is to see the purpose of something that has always been without any meaning to me or anyone else. I therefore join you in requesting Mr. Salviati to begin without further ado this new reasoning.

SALVIATI. Mr. Aproino, who entered too late into our conversation, perhaps does not know that in our previous meeting we read the demonstrations by the Academician concerning the new science of projectiles. This science was founded on the allegation that the projectiles, disregarding any air impediment and any other extrinsic causes, describe in the air a curved line indeed no different from the parabola. Henceforth some absolutely reliable rules concerning the marksmanship of bombardiers were unexpectedly suggested. Having at first established the impetus of an instrument, that is the force with which it shoots a projectile perpendicularly upwards with a given amount of gunpowder, it is possible to determine to what distance the instrument, held at different inclinations, would shoot the same ball merely by using mathematical calculations, presented by the Author in very exact tables for military use. But the use of these tables still required some knowledge of the doctrines, and in any case it was also necessary to consult a book and handle the instruments of learned men, requirements which were not always easily met in a military camp. Therefore the same Academician, having observed that the curvature of a little chain has the shape of a parabola, had the idea to reduce to a simple manual exercise what the Philosopher had written in his books. Suppose Mr. Aproino, you have two pins tacked at the extremities A and B of a horizontal line on a flat surface of wood or cardboard (*figura 46* [fig. 1]). A very thin chain loosely suspended from the pins will hang along lines ACB in the figure of a parabola, the height of which will be CD and the width AB. In order to obtain higher or lower parabolas passing through a given point, for instance E, whilst retaining the same width, you have only to pull one end of the chain. Imagine now that these curves represent the paths traced in the air by a projectile passing through B. You are now easily able to understand how it is possible to measure the angles DBF and DBG by tracing the tangents BF and BG, and in this way find out the elevation of the device required for a certain width and height of a shot. Consequently a quadrant, correctly divided and applied to the board with the center at B, would be enough to resolve this and other similar problems.

APROINO. I understand well how such a device would be very convenient to soldiers. To them it would be of no less service than the proportional compass that the same inventor described and divulged in order to facilitate the geometrical and arithmetic operations for people lacking the endurance needed to follow the rules taught in books, because they are occupied with and distracted by many other tasks. But I have some difficulties in performing the operations conceived above. The first one is how to get the little chain to leave a mark on the surface when it touches it.

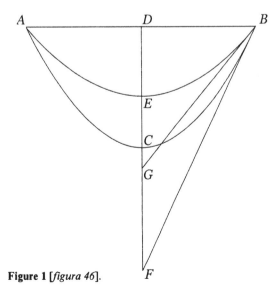

Figure 1 [*figura 46*].

SALVIATI. The easiest way, that does not even depart too much from the required precision, is to dot with a stylus or pen. But our Academician, since he wanted to have a drawing and keep it for use as a print, used to puncture the cardboard with a pin along the contour of the chain. By pouncing he could then reproduce the same drawing elsewhere, as many times as he liked. Do you see this cardboard, so punctured and blackened along the three lines, over which the feather duster full of charcoal powder had been swept? It was prepared in order to determine the degree of elevation for parabolas of various heights but with the same width of 465. On finding the Author in his studio one day, intent on these activities, I asked him for this cardboard. It was no longer of any use to him since he had made another similar and more precise one. Although it is a cheap thing for common people, philosophy and friendship induce me to hold it in high esteem.

APROINO. Mr. Salviati, by reason of friendship I certainly would not hold it in any less esteem. But as far as philosophy is concerned, for my part, I would not be content in prizing the invention's value until it has been demonstrated to me that the line described by the curvature of a chain is really a parabola. And since you assert it with such conviction, I cannot believe that you do not have some demonstrative evidence. I beg you to disclose such evidence to me, so that I am then able to place the same value as you do and I would like to, upon the invention of our common friend.

SALVIATI. The demonstration you are asking for consists of factual evidence. Delineate, with the devices suggested by geometricians and according to the rules they teach, the parabolas ACB and AEB as in the preceding figure or in any other you like, and then place a little chain over them. You will find that the chain coincides to a hair's-breadth with each of the geometric parabolas that you have drawn.

SAGREDO. I made this experiment several times and found that it works, especially in the case of parabolas with an elevation of less than 45 degrees. However, I confess to you Mr. Salviati that I was never convinced of this method of mechanically drawing curves as I would be by a proper mathematical demonstration. Such a demonstration is required, I think, in order to make of the little chain a military instrument perfectly suited to ballistic operations as the compass is suited to arithmetic and geometrical operations. Therefore, I also share Mr. Aproino's perplexities.

SALVIATI. Lucky for me that I am able to amply satisfy both of you, having received from our Academician the mathematical demonstration you are wishing for. Indeed, I will tell you for your relief, that the Academician himself confessed to me several times that he was not content with entrusting such an important conclusion to mere eyesight, which could be suspected of some fallacy. Moreover, matter does not always match the purposes of experimental art. For that reason, our Academician proposed the use of his new military instrument only when he was able to demonstrate that the line along which the chain links arrange themselves is the same as the one traced by projectiles in the air. Likewise I would not have promised to expose you to this matter, was I not scientifically certain about it myself.

SAGREDO. I suppose that this certainty cannot depend on anything but the doctrines concerning the new science of motion which have already been demonstrated.

SALVIATI. It could not be other than as you say. Such doctrines are derived in particular from one of those propositions that, — you will remember —, you heard me reading whilst treating the resistance of solids to break. Imagine that all the links of a chain are threaded through a bar suspended horizontally at both ends. The bar suddenly yields at the points where the weights rest while only its extremities remain immobile. All the other links in the middle are now loosened and will fall. They will not be able to arrange themselves in a new state of equilibrium unless each link has fallen as much as its own momentum requires. It is the disposition of those fallen links, beginning from the second to the middle one, that determines the line of curvature of half of the chain, which is, of course, identical to the other half. You understand that everything depends on knowing with what momentum the links gravitate according to the various distances from each of the supports, presuming the links to be identical along the whole length of the bar.

APROINO. Allow me, Sir, to help my weak intelligence with a bit of drawing. Let CD be the bar resting on its extremities (*figura 47* [fig. 2]); assuming that the weights of two links, one at B and the other at A, are represented by the equally heavy bodies H and F hanging from the bar at those same points B and A, you propose to resolve the question of what is the ratio of the momentum of weight H at B to that of the same weight, or of its equal F, at A. I do not find clear principles to resolve this question in the mathematical science that I have learned up to now through teachers and books. Nevertheless, it seems to me that those principles are not different from the mechanical speculations concerning the balance. Therefore, I would not see what propositions concerning the resistance of solids to breaking have to do with this question even if I had had, like Mr. Sagredo, the luck of attending your past meetings.

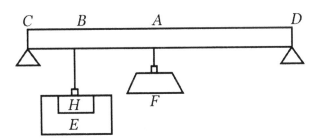

Figure 2 [*figura 47*].

SALVIATI. You need to know, though, that the new science of resistances depends upon nothing but on Archimedes' ancient science of balance, if you consider the geometric line at the extremities of which the weights are added to be a stiff rod that can break. Let the balance AB be supported at C (*figura 48* [fig. 3]). You say that according to the doctrine of weights in equilibrium the balance will be in equilibrium when the weight B resists being lifted adequately to the power of the weight A to heave. But the same ratio of power and resistance can be applied to the instrument, if we consider the line AB to be a stiff rod, which will remain in equilibrium every time the power of A to break equals the resistance of B against breaking. If those two opposite powers of acting and resisting are the strongest in producing their effects, any minimal addition to the one or detraction from the other would be enough to unsettle the equilibrium, that is, to bend the rod by pulling it down and turning it around the center C, as it happens with the simple balance.

SAGREDO. Now, Mr. Salviati, you make me conjecture that the proposition of the treatise on resistances you have just mentioned could be the twelfth, which, if I remember right, you formulated in this way: "If two places are taken on the length of a cylinder at which the cylinder is to be broken, then the resistances at those two places have to each other the inverse ratio [of areas] of rectangles whose sides are the distances of those two places [from the two ends]."[4] However, I must confess that regarding this proposition I am assailed from two sides: the first attack comes from considering the proposition in itself, and the second from applying it to the momentums of the same weight placed at various distances from the middle of the rod. In fact, I never doubted the truth of the mentioned proposition, but the way you demonstrated it. You based your demonstration on the assumption, dubious in my opinion perhaps because I do not understand it well, that the momentums of heavy bodies hanging from a balance are to each other in the ratio compounded from the distances from the support and the weights.[5] So much for

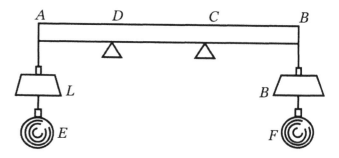

Figure 3 [*figura 48*].

[4] The quoted sentence is taken from *Discorsi e dimostrazioni matematiche intorno a due nuove scienze* (Galilei 1890–1909, 8:176). We adopt the translation by Stillman Drake, who numbered the proposition as the eleventh (Galilei 1974, 133).

[5] In accordance with the context the word "weights" renders Caverni's "moli," although in Galileo's language "mole" means generally "volume."

the proposition itself. As to its intended application to the momentums of weights hanging from a balance supported at its ends, my doubt is due to the consideration that in the twelfth proposition you set the cylinder in which the breakage has to take place as having the supports at middle points.

SALVIATI. Do not doubt, Mr. Sagredo, that I will find the way to satisfy your mind as to both of your doubts. Starting with the first, I will not deny that the ratio of momentums as it shines through the twelfth proposition of the treatise on resistances leaves something to be desired. We could, however, easily compensate for this fault by referring to the definition of momentum given by the authors of mechanical science and to the known laws of weights in equilibrium on a balance. From these laws derives, in fact, that the machine remains in equilibrium when, as in the preceding figure, the weight of A multiplied by the distance AC from the support is the same as the weight of B multiplied by the distance BC. If you give the name *momentum* to the tendency or impetus of going downwards, compounded from gravity and position, you will have already concluded that the momentums in the balance have the ratio compounded from the distances and the weights.

Because of these considerations, the Author of the treatise on resistances did not deem it necessary to demonstrate something that can be so easily concluded from Archimedes' ancient theorems. But then the Academician wanted to expound the propositions he had ultimately demonstrated for serving as fundamental to the new little treatise on the use of little chains. Since he wanted to start by introducing the momentums, according to the ratio of which the links fall down more or less, he thought it better to formulate the proposition that I will read to you from this sheet, in the original form in which it was written. For us, too, this proposition will be the first of all subsequently appearing in our reasoning.

Proposition I.[6] The momentums of weights hanging in the balance have the ratio compounded from the ratio of the weights itself and from the ratio of the distances.

Let the weights DE and F hang at the distances AB and BC (*figura 49* [fig. 4]). I say that the momentum of weight DE have to the momentum of weight F

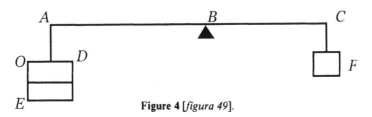

Figure 4 [*figura 49*].

---

[6] Proposition I and its demonstration are in Latin. The original Galilean fragment and figure has been edited by Favaro in 1898 in Galilei 1890–1909, 8:367–68. The figure there differs from that reproduced here by Caverni.

the ratio compounded from the ratio of weight DE to weight F and the ratio of distance AB to distance BC. As AB is to BC, so let weight F be to weight DO. Therefore, since weight F and weight DO have the inversed ratio of the distances AB and BC, the momentum of weight F will be equal to the momentum of weight DE. Thus, whatever the three weights ED, F and DO may be, the ratio of weight ED to weight DO will be compounded from the ratios of ED to F and of F to DO. Moreover, as weight ED is to weight DO, so is the momentum of ED to the momentum of DO, since they hang from the same point. Therefore, since the momentum of DO is equal to the momentum of F, the ratio of the momentum of ED to the momentum of F will be compounded from the ratio of weight ED to weight F and from that of weight F to weight DO. Moreover, weight F has been set to weight DO like distance AB to distance BC. Therefore it follows that the momentum of weight ED has to the momentum of weight F the ratio compounded from the ratios of weights ED and F, and from those of distances AB and BC.

APROINO. I thank you, Mr. Salviati, and at the same time bless Mr. Sagredo's doubts which gave the opportunity to expose a theorem, which I do not remember having ever met when reading what has been written on similar matter by other authors. Moreover, the principles from which the conclusion follows are so clear that they enable me to glimpse many other useful consequences for the doctrine of motion.

SALVIATI. Sir, you will soon see the applications we will make of these principles proving the usefulness that you have shrewdly perceived, but now it is better to proceed in resolving the other doubt of Mr. Sagredo. And it seems to me that on his serene countenance I can read the satisfaction he has already felt regarding the first doubt.

SAGREDO. You should say not just satisfaction but delight because this demonstration of the ratio of momentums is for me like for Mr. Aproino something completely new. And even if I could perhaps be able to understand by myself the reasons for the step from the cylinder sustained in the middle to the cylinder supported at its ends, to the point of yielding because they are both overburdened by the same weights, I am waiting that you alleviate my labor and convince me, better than I can myself, of having seen the truth.

SALVIATI. Very willingly I would leave to you the whole pleasure of finding out how it is true that we have the same conditions of equilibrium in the geometric balance and in the rigid rod near to breaking, whether the supports are in the middle or at the ends, being that indeed very easy to demonstrate. But since you want me to lighten your burden, I will again call your attention to the balance AB just drawn in *figura 48* [fig. 3]. As you well know, it remains in equilibrium around point C when weight A is to weight B as distance BC is to AC. By composition we will find that weights A and B together are to single weight A or to single weight B as BC and AC together, that is AB, are to BC or to AC. Whence it is evident that the balance remains in equilibrium when the support is at C and the weights at A and B as well as when the supports are set at A and B and the sum of those two same weights at C. Proceeding then from the geometric balance to the solid cylinder, you will understand that if A and B are the maximum forces to which the cylinder sustained at C resists without breaking, the sum of the two weights at C will give the measure of the maximum force to which the solid can resist being broken at that same point when it is instead sustained at A and B.

Let us now remember the twelfth proposition about resistances: With it we demonstrated that if forces A and B are the minimum ones for breaking at C, and forces E and F are equally the minimum ones for breaking at D, the forces A and B have to E and F reciprocally the same ratio as the rectangle ADB has to the rectangle ACB. But according to what was already said and agreed upon, it is the same to keep the supports at C or D, and the weights at A and B or at E and F, as to move the supports to A and B, and the weights A and B together to C or the other weights E and F together to D. We will say therefore, and let this be the second proposition, that for a cylinder supported at its ends A and B the weight that can break at C is to the weight that can break at D, that is, the resistance at C is to the resistance at D, as the rectangle ADB is to the rectangle ACB. Thus, the demonstration would now be the same one that was already given and should be repeated only in favor of Mr. Aproino, who was not present then.

APROINO. Sir, with your learned reasoning you have prepared the way for me so well that I do not doubt of being able to trace that demonstration by myself. Anyway, in order not to delay for too long in deducing the rest, — which is the purpose of our conversation, — I will presume as true the proposition that you have put as the second one in the row.

SALVIATI. If so, there is nothing left but to make one step in order to achieve our main purpose which was to know with what various momentums the links gravitate on the bar through which we imagined they were threaded, hence to deduce the ratios of descents to lastly conclude what is the line in which the chain curves. At first I enounce, with reference to the figure drawn for that first purpose,[7] a third proposition that says: The momentum of weight F at A is to the momentum of the same weight or of an equal weight H at B as the rectangle CAD is to the rectangle CBD.

SAGREDO. Therefore, the momentums are to each other in the inverse ratio as the resistances, and the chain's link at B will have less impetus in falling than the link at A because the former encounters in the bar, that resists it more, a greater impediment. Similarly, I understand why the chain from the first link through to middle one diverges more and more from the horizontal arrangement it had when it was threaded through the bar, having been abandoned to its own weight. It seems to me, also, that I can distinctly see the

---

[7] The figure referred to is *figura 47* [fig. 2].

dawning of that light of truth that you will soon reveal to our eager eyes, and since it is unpleasant to wait, proceed, Mr. Salviati, to demonstrate that the momentums of weights F and H have to each other the same ratio as the rectangles whose sides are respectively the distances of those points [from the two ends].

SALVIATI. After all that has been said and agreed upon by you and Mr. Aproino, the demonstration is easy and speedy. For, keeping in sight the same figure, let us suppose that weight F is the measure of the resistance at A, and that the measure of the resistance at B is weight H increased to E. For the second proposition the resistance at A will be to the resistance at B, that is, the weight F will be to weight E as the rectangle CBD is to the rectangle CAD. But since the weights H and E are attached at the same point of the balance the ratio of their momentums is the same, that is, the momentum of H is to the momentum of E (which is equal to the momentum of F for having the same power to break the bar) as the weight F is to weight E. Therefore the momentum of H is to the momentum of F as the rectangle CBD is to the rectangle CAD, which is what I wanted to demonstrate.

Through this ordered series of propositions we came finally to the point where we can find what we were seeking from the beginning of our reasoning and was said to sum up everything. That is, we can learn with what momentum the different links of a chain tend to fall when they are dropped from the bar which held them. Let the bar be represented by the horizontal line HD (*figura 50* [fig. 5]). Let us assume that the link at F because of its impetus or momentum can fall to E for the entire perpendicular line FE and that likewise the link at N can fall for the entire line MN. Since the descents must be proportional to their momentums, and accordingly to what was already demonstrated, FE will be to NM as the rectangle HFD is to the rectangle HND. Now, in order to conclude that the points E, M and all the others corresponding to the links of a chain are really in a parabola, what else is left but to invoke a theorem, that you will not find written by any ancient or modern author but has been demonstrated by our Academician by reason of his treatise on resistances? Now I want to propound to you that theorem which says: The

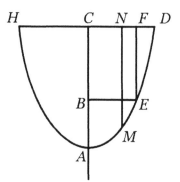

Figure 5 [*figura 50*].

parallels to the diameter of a parabola, whose base they cut perpendicularly, have to each other the same ratio as the rectangles having as sides the segments. Hence, e.g., in the drawn figure the parallel NM and FE to the diameter AC are to each other as the rectangles HND and HFD.

APROINO. When I recently visited father Bonaventura Cavalieri in Bologna and, in relation with my instrument to strengthen hearing, we entered into a conversation on the conics, he told me this very same theorem but I did not quite understand whether he presented it as his own invention or Mr. Galileo's.

SALVIATI. It could very well be that father Bonaventura as well, whom our friend uses to call the Archimedes of our times, had encountered this very same property of the parabola, so useful for many demonstrations of Mechanics and Geometry. But I can assure you that I got news of it during my conversations with the Academician, many years before Cavalieri's genius was ripe to produce such fruits.

SAGREDO. You have now reminded me of having heard the same theorem in Padua when our mathematician was teaching at our University. Now, since truth does not deny itself to anyone who seeks it with desire and along the same right ways, solace us, Mr. Salviati, by showing it anew unveiled to our eyes.

SALVIATI. I am pleased to be able to completely satisfy you this time, too, since you do not, actually, need any other knowledge but that which you had already when from the simple generation of the parabola I immediately concluded that the diameters are to each other like the squares of the ordinates.[8]

APROINO. I remember well the demonstration given by Apollonius in his *Conics* and therefore I also do accept as known that the line AC is to AB as the square of CD is to the square of BE.

SALVIATI. Being it indeed so, let us divide and we will obtain that AC minus AB, that is BC as well as its equal EF is to AC like the square of CD minus the square of BE is to the square of CD. But, as can be easily deduced from the fourth proposition of Euclid's second book, the difference of two squares is equal to the rectangle obtained from the sum and the difference of the roots. Therefore, the square of CD minus the square of BE will be equal to the line CD plus BE, i.e. HF, multiplied by the line CD minus BE, i.e. FD. Said in another way, the difference of the two mentioned squares will be equal to the rectangle HFD. Hence, EF will be to AC as the rectangle HFD is to the square of CD. In the same way we will demonstrate that NM is to AC as the rectangle HND is to square of CD. Hence, since the two ratios have equal consequents and therefore the antecedents must be proportional, we conclude that FE and MN are to each other like the rectangles HFD and HND, as I promised you in order to satisfy your wish.

---

[8] The proposition referred to can be found in Galilei 1890–1909, 8:270–71.

The dialogue breaks off at this point but the treatise on the use of the small chains is, in any case, complete; what we feel might be missing is only the more or less ceremonious farewell of the interlocutors. Anyway, even if our readers agree that the entire argument is included in the transcript, they could ask for the reasons which induced us to ascribe this text to Galileo. In this respect one has to distinguish between form and content. To prove with certainty that the latter is a genuine Galilean one would be enough to adduce the fact that the theorem concerning the momentums compounded of distances and weights read by Salviati is in Galileo's handwriting, in the codex and on the sheet we referred to in the IVth paragraph of chapter VIII of the previous volume;[9] that also in Galileo's handwriting is the proposition, likewise quoted by us at the same place,[10] concerning equal weights that operate on a bar supported at its extremities with momentums homologically proportional to the rectangles having as sides the distances from the supports; and finally that in Galileo's hand is the drawing reproduced by us in the mentioned volume and chapter.[11] As he hinted, Galileo in this drawing wanted to apply the last mentioned proposition to the chain's links, with the manifest intention of concluding that the chain's curvature is parabolic.

And we do not want to carry on our exposition without pointing out that the discovery of the dialogue on small chains, which luckily befell us these last few days, dispelled some of our doubts and made clear some facts still obscure to us when we were relating our story in the mentioned chapter VIII. There we wondered how the handwritings reported herein before could have been left among other useless papers, as if their author, although he could have enriched with these results his treatise on resistances, wanted to leave it with this defect so that later Mssrs. Cavalieri, Torricelli, and Viviani, in order to satisfy the needs of science, could compete in emending it. Now we did understand that the propositions left handwritten were meant for a treatise quite different from that concerning resistances. Far from having been demonstrated only to be then rejected, as it seemed to us when we found them so neglected, these propositons had rather to serve as rich warp upon which the rest of the conversation would have been woven, so leading into evening the day begun with the treatment of percussion.

Let us now resume to tell the reasons proving that the discourse on the use of small chains transcribed by us is shaped upon Galileo's concepts. We can add that a cardboard punctured with a pin along parabolic lines in order to reproduce by pouncing the same figure, this very cardboard with black smudges left by the feather duster and in the conditions described by Salviati, bearing at its opposite corners the repeated handwritten words *amplitudo tota 465*, is still kept sewn in

---

[9] Caverni 1891–1900, 4:484. For the standard edition of the mentioned fragments and figures, see Galilei 1890–1909, 8:367–70. One of the fragments is reproduced there in *facsimile* (ibid., Appendice, car. 43r).

[10] Caverni 1891–1900, 4:485.

[11] Ibid., 495.

place of folio 41 in Volume II of Part V of Galileo's Manuscripts.[12] But the most authoritative confirmation of what we intend to prove is given by Viviani's testimony, to whom we believe we should attribute the writing of the dialogue or rather of the dialogue's fragment, found by us in a copy that must be of that time.

In the margin of page 284 of the Leiden edition, at the place where Sagredo suggests that it is possible to dot with a little chain many parabolic lines and Salviati replies: "*That can be done, and with no little utility, as I am about to tell you*," Viviani appended a similar note: "By means of this small chain Galileo perhaps found the elevations to hit a given target" (Mss. Gal., P. V, T. IX).[13] Furthermore, in one of those notes written by Viviani on folio 23[14] of Volume IV of the same Part V of the collection, he expressed a similar doubt in this other way: "See at page 384,[15] the last sentence, which utility Galileo meant, whether in measuring the parabolic line or in finding the propositions concerning projectile motion."

All Viviani's doubts in this respect were solved when he came upon the hand-written slips bearing the propositions concerning momentums exerted by equal weights on a balance supported at its extremities, propositions from which one can deduce the impetus' strength and the descent's length of each link of a chain. While arranging these dispersed propositions Viviani remembered what he had heard his master saying in the refuge of Arcetri, and then he himself rewrote this little treatise on the use of small chains, about which nothing was known but the allusions made by Salviati in the evening of the Fourth Day. In this way both parts of the last conversation would come to an end, in accordance with the given promises. It is, therefore, natural that Viviani worked out, on the basis of newly found documents, what remained to be said on the use of little chains for military purpose, in order to add this part to the dialogue and so completing it, while he intended to publish the part concerning percussion among the posthumous works that were to be dedicated to the king of France, in attachment to his biography of Galileo. But since he failed in his hopes of gathering in a book the works that his master ultimately considered writing, Viviani was content to satisfy the public on this matter with this information, in the "Summary" he appended to the *Universal Science of Proportions*:

> Now it remains to be said what I know about the use of small chains, concerning which Galileo made a promise at the end of the Fourth Day. I will relate it as Galileo intimated it when, he being present, I was studying his

---

[12] Caverni refers to a document kept in the Galilean Collection at the Biblioteca Nazionale Centrale, Florence, call-nr.: Ms. Gal. 72, folio 42r. See Drake 1979, 238–39; Damerow et al. 1996, 5; Renn et al. 1998, 9–10.

[13] Caverni is referring to a copy of Galilei 1638, with handwritten notes by Viviani, kept in the Galilean Collection, call-nr.: Ms. Gal. 79. For the standard edition, where Viviani's notes are not entirely reported, see Galilei 1890–1909, 8:310. For Salviati's sentence we adopt the translation by Drake (Galilei 1974, 257).

[14] Viviani's note is on folio 33 (Ms. Gal. 74, folio 33r).

[15] Viviani writes "page 284" and not 384, referring to Galileo 1638.

science of projectiles. It seemed to me then that he intended to make use of such very thin chains hanging from their extremities over a flat surface in order to deduce from their different tensions the rule and practice of shooting with artillery at a given target. Our Torricelli, however, wrote adequately and ingeniously about that at the end of his treatise on projectiles, so that the loss is compensated.

If I remember well, Galileo deduced that the natural bend of such small chains always fits with the curvature of parabolic lines from a reasoning similar [to this]: Heavy bodies must naturally fall according to the proportion of the momentums which they have at the places from which they hang, and the momentums of equal heavy bodies, hanging from points of a balance supported at its extremities, are to each other in the same ratio as the rectangles having as sides the parts of that balance, as Galileo himself demonstrated in the treatise on resistances. And, according to the theory of conics, this ratio is the same as the one existing between the straight lines which from the points of that balance, taken as base of a parabola, can be drawn in parallel to the diameter of that parabola. And, finally, since all the links of a small chain, — which are like as many equal weights hanging from points on the straight line that connects the extremities where the chain is attached and serves as base of the parabola, — must fall as much as allowed by their momentums and there must stop, these links must, therefore, stop at those points where their descents are proportional to their momentums at the places from where they hang in the last instant of motion. These then are the points adapting to a parabolic curve [which is] long as much as the chain and whose diameter, which rises from the middle of the said base, is perpendicular to the horizon. (Viviani 1674, 105–06)

It is easy to see summarized in these words the dialogue we have transcribed, the loss of which Viviani thought was compensated for by Torricelli. But Torricelli, in fact, at the end of his treatise on projectiles[16] ingeniously describes a new type of square that could be effectively used by bombardiers; he does not, however, say a word about the instrument conceived by Galileo, nor about the order of the propositions which should give a greater certainty of mechanical science to Galileo than to the instruments to measure the force of percussion [which he] imagined and described.[17] The dialogue as published by Bonaventuri[18] lacks, therefore, its

---

[16] Reference to the last part of "De motu proiectorum" in Torricelli 1644, 204–43. For a modern edition, see Torricelli 1919–1944, 2:197–232.

[17] The sentence is difficult to understand. We interpret it in the sense that Torricelli did not give any information about the logical order of the propositions concerning the alleged connection between catenary and parabola summarized by Viviani. After that Caverni seems, in our opinion, to suggest that Galileo worked on these propositions in order to acquire scientific certainty as far as the mentioned connection, and not in order to give scientific foundation to the use of devices described in "Della forza della percossa" (Galilei 1890–1909, 8:319–46; Galilei 1974, 281–306).

[18] Reference to "Della forza della percossa" in Galilei 1718, 2:693–710.

best part which the devoted public would never have expected could be restored by us, sacrilegious offenders of the Numen. But that is, it seems, how things go in the religion of science as well as in all mundane affairs, whose care we leave to others while we come back to the thread of our previous reasoning.

GIUSEPPE CASTAGNETTI AND MICHELE CAMEROTA

Antonio Favaro and the *Edizione Nazionale* of Galileo's Works

Antonio Favaro was born in Padua on 21 May 1847 to a cultivated family of lower nobility.[1] After having accomplished his studies in mathematics at the University of Padua in 1866, he went to Turin, where he specialized as an engineer at the Scuola d'Applicazione (Polytechnical High School) in 1869. As early as 1872, he was appointed as extra-ordinary professor at the University of Padua. For fifty years he taught graphical statics there.[2] During different periods he also gave courses in infinitesimal calculus and projective geometry. Since 1878 Favaro, as one of the first in an Italian University, also taught history of mathematics. He died in Padua on 30 September 1922, shortly after his retirement.

Despite his academic obligations in geometry and mathematics, Favaro's major focus of interest soon became the history of science in which he published more than 500 articles and several books. Favaro had not received training as an historian. His first, amateurish steps in this field were contributions to priority disputes or historical and bibliographical notes in publications concerning technical subjects (e.g., Favaro 1869 and 1873; Quaranta 1983, 52). He was then encouraged to continue by Baldassarre Boncompagni, who was at that time an authority in the young and not yet academically established discipline of the history of science (Favaro 1878, 800).[3] Favaro took Boncompagni as a model of method thus treading in the wake of an historiographical tradition more interested in the erudite compilation of bibliographies and the recuperation and interpretation of manuscripts, just as the seventeenth-century antiquarians, rather than in explaining the course of history.

---

[1] The references cited in this article can be found in the bibliographical section at the end of the appendix. Our account is based on Favaro G. 1922–23, Bortolotti 1923–24, and Bucciantini 1995. For Favaro's bibliography, see Favaro G. 1922–23, Gabrieli 1925, Baldo Ceolin and Olivieri 1994–95. Evaluations of Favaro's work and his place in the history of science besides the biographical notes mentioned above are Bosmans 1923, Brugnaro 1979, Galluzzi 1983, Lefons 1984, Malusa 1977, Quaranta 1983, Seneca 1995.

[2] Graphical statics was a newly established discipline concerning the use of graphical methods in solving problems of statics (Henneberg 1901–08, 349–51).

[3] In 1868, Boncompagni (1821–1894) had started to publish the probably first worldwide journal exclusively dedicated to the history of science, the *Bullettino di bibliografia e di storia delle scienze matematiche e fisiche*. Among Boncompagni's most important achievements are the history of the medieval translations of Arabic books by Gherardo of Cremona and Platone of Tivoli and the discovery of unknown works of Leonardo Fibonacci. On Boncompagni, see Favaro 1894–95 and Cappelletti 1969; on the *Bullettino*, see Lefons 1984 and Bucciantini 1986.

Favaro soon acquired the standards of accuracy and erudition of his master, although he did not follow him in his nearly maniacal pedantry (e.g., Favaro 1875). On Boncompagni's suggestion, he started to investigate ancient Paduan mathematicians, became interested in the history of the University of Padua, and was finally captured by its major figure, Galileo Galilei. From 1880 until the end of his life, Favaro dedicated himself predominantly to the study of Galileo's life, work, and context. Besides that, he published several works on Padua University, on Niccolò Tartaglia, on Leonardo da Vinci, and occasionally on other historical as well as scientific subjects.[4]

For the scrupulous preparation of a book on Galileo, Favaro at first closely examined the extant Galilean manuscripts and documents in several archives and libraries.[5] He could thus establish that the existing editions of Galileo's works did not satisfy scholarly criteria. Even the most recent one (Galilei 1842–56), although called the "first complete edition" was incomplete and full of errors; in plenty of cases Galileo's text had been manipulated, figures omitted or altered; the matter had been arbitrarily distributed. Furthermore, Favaro complained that the editors "did not consider at all the way followed by Galileo in order to arrive at the formulation of a certain truth." They had disregarded Galileo's preparatory manuscripts and given only the last version of his ideas and production (Favaro' 1883, 1–2; Favaro [1883] 1966, 2:318–25; Favaro 1888, 19–29; Bortolotti 1923–24, 19).

Favaro came to the obvious conclusion that a new edition of Galileo's works was needed.[6] It should include, of course, all of Galileo's writings, also the unpublished and fragmentary ones, Galileo's own marginal notes to other books, and the letters Galileo sent as well as those sent to him. In addition to this, the new edition should include texts of Galileo's disciples commenting on his works, the works of other people to whom Galileo replied or on which he made comments, excerpts from letters between third persons relating to Galileo, and finally archival documents relating to him and his family. Favaro wanted to present not only the "truly complete" works of Galileo, but also his scientific as well as historical context.

With regard to the general partition between scientific texts, literary texts, correspondence, and documents, Favaro opted for a strict chronological order as "being the more suitable in order to faithfully render the natural generation of the ideas" (Favaro 1888, 34). As to the editorial method, all the texts would be collated

---

[4] Major works are Favaro 1877, [1883] 1966, 1886a, 1887a, 1891, 1911–12, 1922a, 1922b. In recent years, some of the journal articles on Galileo and his context have been collected in volumes: Favaro 1968, 1983, 1992a, 1992b.

[5] Articles on Galilean documents are, e.g., Favaro 1880 and 1882. Of paramount importance is the Galilean Collection put together by the former Grand Dukes of Tuscany and preserved at the Biblioteca Nazionale Centrale in Florence. Favaro dealt repeatedly with this Collection (Favaro 1879–80, 1883, 1885, 1886b).

[6] The new edition project is presented in Favaro [1883] 1966, 2:311–31, and in Favaro 1888. For the editorial criteria, see Favaro 1888, 30–42; Favaro 1898–99.

with the published or unpublished sources and should be "reproduced like they have flown out of their author's pen" (Favaro [1883] 1966, 2:328), the language would not be modernized, figures would be exactly reproduced. Favaro shared the rule of strict adherence to the sources with Isidoro Del Lungo, the highly esteemed Italian philologist that he had enlisted as co-editor responsible for textual criticism.[7] Their benchmark was — as Del Lungo wrote — the "testificazione alla tedesca," the establishment of texts following the rules of the German philological school (Seneca 1995, 398).

Having won the financial support of the Italian Government, Favaro started to publish the new edition, the so-called "Edizione Nazionale" of Galileo's works in 1890. The twenty volumes came out in almost regular yearly succession (Galilei 1890–1909). Thanks to Favaro's editorial policy, this edition satisfied scholarly needs for almost a century.[8] Comparable undertakings at the turn of the century were the editions of Descartes (Descartes 1897–1913) and of Huygens (Huygens 1889–1950), indicating the existence of widespread interest in the seventeenth-century scientific revolution. The high level of editorial standards common to all these productions was, furthermore, the sign of a shared awareness of the importance of the sources for their historical reconstruction.

In preparing the Galileo edition and concurrently to it, Favaro concentrated his studies on the historical context, especially on the people whom Galileo came across, scientists as well as relatives, friends as well as foes. The results of this research, for which he systematically explored public and private archives, went into the footnotes to the main opus or into a long series of articles and erudite notes. Favaro was also allowed access to the Vatican Archives and could publish the documents of the trial against Galileo (Favaro 1907b).

Favaro did not cease working on Galileo after the accomplishment of the edition in 1909, his aim being a collection of materials as complete as possible in order to write the long-projected scientific biography. In fact, this major work was never written. Only a short popularization was published (Favaro 1910). The aspiration to completeness probably hid Favaro's hesitation to leave secure documentary basis in order to tackle problems of interpretation of Galileo's scientific development and of the changes he brought about in the natural sciences (Galluzzi 1983, VI). Favaro's major preoccupation was to faithfully present the documents without giving any interpretation. Almost all his papers dealt mainly with persons or institutions, not with topics of natural sciences, and were usually accompanied by rich documentary appendices.

---

[7] On Isidoro Del Lungo (1841–1906), see Strappini 1990. On the collaboration between Favaro and Del Lungo, see Seneca 1967 and 1995.

[8] Only in the last decades has scholarly attention focused on scattered fragments and notes that Favaro was not able to connect with bigger texts and were therefore left unpublished (e.g., Drake 1979). The arrangement given by Favaro to a group of fragments concerning the theory of motion has been criticized by Wisan: "Favaro's treatment of the manuscript seems almost calculated to discourage further investigation of the fragments, and this may be connected with some strong objections Favaro had to Caverni's published interpretation of the manuscript" (Wisan 1974, 126).

It is fair to say that there is a considerable gap between Favaro's ideas on science and its history and his actual historic works. Favaro shared with the scientists of his time the positivistic belief in a steady growth of knowledge (Quaranta 1983, 53, 55). For him science, in the sense of a set of discoveries, was a "transient result" of human activity. In direct polemic with conceptions considering these results as if they were a coherent revelation descending who knows where from, Favaro, in a methodological article, pointed out the way covered in order to get to the present stand. The journey had not been easy and straight-forward, but rather laborious and wandering. The discoveries have been accumulated through the efforts of generations of scientists, genius as well as obscure scholars, all of them worthy of being studied (Favaro 1878; see also Favaro 1874, 456–58; Favaro 1887b; Favaro 1887c, 345). Furthermore, Favaro was aware that intellecual discoveries are not extemporaneous; in certain periods they rather lie everywhere like "germs" that come to light at the same time.[9] Consequently, Favaro stated the need to study the intellectual context in order to detect even minor exchanges and influences (Favaro 1885, 1–2).

In fact, though, Favaro addressed his attention almost exclusively to the biographies of the figures of Galileo's world but wrote little on their scientific contributions. He did not deal with questions concerning the development of a science. As a colleague wrote in the obituary "It seems that Favaro wants to write the *history of mathematicians* in order to collect material for the *history of mathematics*," on which he wrote only rarely. Paradoxically, he omitted dealing with the very object of the history of mathematics, i.e., the mathematics itself, even in publications concerning the works of mathematicians (Bortolotti 1923–24, 12–14). Instead, he limited himself to very short, resumptive notes on their books without trying to understand the intrinsic reason for their achievements (eg., Favaro [1883] 1966, 1:78–105; Favaro 1886a, 73–75).

Similarly, Favaro stated the importance of studying the academic disputes for the history of universities (Favaro 1919, 459) but his own works in this field dealt almost completely with institutional and personal aspects, giving little attention to the ideas. Favaro's work was a determinative factor for the establishment of the interpretation of Galileo as the inventor of the experimental method, but Favaro himself did not take part in any interpretative debate (Malusa 1977, 560; Galluzzi 1983, VI). Besides presenting what he considered the "pure facts," Favaro engaged only in priority disputes in order to clear Galileo from accusations of intellectual dishonesty and to emphasize his originality (Favaro 1907a, 1913–14, 1918, 1919–20a, 1919–20b). His only other interpretative contribution was to depict the anti-Galilean catholic reaction as a curb for scientific progress. Of course, he pursued these targets with sobriety and without polemical excess, relying only on the documentary evidence (e.g., Favaro 1916; Bosmans 1923, 174).

---

[9] The metaphor comparing scientific discoveries with germs that are in the air is expressed in a long quotation from the German historian Alfred Clebsch, with whom Favaro explicitly agrees (Favaro 1874, 457; see also ibid., 565; Lefons 1984, 87).

Favaro has been considered a positivistic historian (Malusa 1977, 552, 555, 560; Galluzzi 1983, VI; Lefons 1984, 89), and indeed he himself declared his agreement with the "positivism indisputably dominating the modern studies." However, he understood positivism in the narrow sense of strict reliance to the sources, highest scrupulosity and exactitude (Favaro 1882, 581). Consequently, Favaro performed the "first part of the positivist programme" in historiography as identified by Collingwood (Collingwood [1946] 1993, 126-27), namely "ascertaining the facts," but he was not interested in the second part, namely "framing laws." Indeed, Favaro delivered an enormous amount of ascertained facts but renounced one of the major tasks of any historian, i.e. trying to give an explanation of why the facts happened as they did. As Micheli remarks, Favaro's approach was essentially an erudite one, as opposite to Caverni, who tried to interpret Galileo's work "in intrinsic terms and in the light of an organic, comprehensive evaluation of science of the Renaissance and the sixteenth century" (Micheli 1980, 611–12).

Nevertheless, Favaro's detailed research on secondary, even anecdotal circumstances of Galileo's life was not carried through only for the sake of accumulating historical facts without connection, as one would understand eruditeness, but with the aim of rendering a broader picture, the scientist's context. Only on the basis of the facts collected by Favaro was it possible to acquire a comprehensive view of science in sixteenth-century Italy. Only thanks to Favaro's reconstruction of Galileo's net of personal, political, and scientific relations does it become possible to understand the scientific world in which he operated.

Max Planck Institute for the History of Science, Berlin (GC)
Dipartimento di Filosofia e Teoria delle Scienze Umane
Facoltà di Magistero, Università di Cagliari (MC)

ANTONIO FAVARO

Apocryphal Galilean Writings*

The historians of science have only recently learned of the existence of a sixth volume of the grandly conceived work of the abbé Raffaello Caverni entitled *Storia del metodo sperimentale in Italia*. Five volumes had already been published between 1891 and 1898. The sixth volume was published, one might say, in an almost furtive way. In fact, the first notice of it was given in a journal by Aldo Mieli just a few months ago. Among materials he is collecting under the heading, "For a Biography of the Italian Historians of Science," in addition to other things already said about Caverni in a previous issue, Mieli has published the following "information received from G. Vacca": "The printing of the sixth volume of the *History of the Experimental Method in Italy* dealing with the history of hydraulics had been started but was interrupted by the author's death. About 450 pages have been printed, the greater part of the copies of the loose sheets have been lost."[1]

I immediately set about trying to procure a copy of these papers through a very influential person in Florence where the first five volumes were printed, but my search was in vain. I had almost given up hope of getting the remaining piece of the sixth volume when its existence was definitely confirmed some weeks later by my colleague Loria.

At almost the same time I received from the bookshop of Oreste Gozzini in Florence the catalog no. 60, dated November 1, 1916, in which I found that a copy of Caverni's work in six volumes was registered under entry number 890. Without losing any time, I approached the National Central Library of Florence, not doubting for a minute that the sixth volume would also be in its possession, at least in virtue of the press laws. In fact, this volume was said to have been printed in Florence in 1900 by the same Civelli Publishing House that had published the other five volumes. To my great surprise, however, I was told that the above-mentioned volume was not in the Library. Finally, my colleague Loria let me know that he had meanwhile been able to buy a copy of the legendary volume from a

* Translator's note: This second version of Favaro's article was published in *Atti e Memorie della Reale Accademia di Scienze, Lettere ed Arti in Padova* (1919–20), 17–30. Reprinted in Favaro 1992a, 141–54. The modernized and standardized list of references can be found in a bibliographical section at the end of the appendix. In this second version, Favaro noted that the article was first published in *Bollettino di bibliografia e di storia delle scienze matematiche*, 19 (1917), 33–43. He added the following comment: "As I did in other cases, I republish this article in the present collection because of its importance for my work on Galileo, adding to the first version the result of subsequent investigations."
[1] *Rivista di storia critica delle scienze mediche e naturali*, 7 (1916), No. 4, 125.

traveling bookseller and so I accepted his kind offer to send it to me on loan. The following is what I should, or rather must, say about it.

All this preamble was not misplaced because I wish to clearly establish that neither myself nor other historians of science, neither bookshops nor public libraries knew about this volume which, according to its front page, was printed in 1900. Therefore, all that I have written up to now on my favorite subject of study has to be considered totally independent from the content of this volume.[2]

This sixth volume, or tome if you will, really appears to be a continuation of the five previously published volumes, the last of which gives, as we said before, 1898 as its date of printing. Volume 6 has been printed with the same fonts and on the same kind of paper, and is therefore in the same format as the other five volumes. The cover is identical to that of the preceding volumes, except that we read: "This volume is incomplete because the author died while he was preparing the work. Everything of the manuscripts left behind by the author has been printed."[3] Actually, the volume ends abruptly on page 464, i.e. at the end of the 29th folio, while it is absolutely impossible that the author's original text ended in this way. The editor, who moreover let pass so many and such serious mistakes, had not enough pity to at least go to the end of the sentence and to add some opportune words, as well as a little index of the chapters contained in this stump of a volume.

First we note that in this sixth volume there is a first part: "On the Experimental Method Applied to the Science of the Motion of Liquids." According to the author's original plan (Caverni 1891–1900, 1:260), volumes 6 and 7 would have dealt with the history of hydraulics. The origins of hydraulics and the progress achieved by Galileo and Castelli was to have been disclosed in the sixth volume, while the seventh volume was reserved to show to what level of perfection the disciples and followers of the two great masters had brought the science of the motion of liquids. Actually, Caverni considered the science of liquids to be divided into three parts, namely hydrostatics, hydrodynamics, and hydraulics. Therefore he declared that he would divide its history into three parts as well, treating the first two parts in this sixth volume while keeping the third part for the seventh one (Caverni 1891–1900, 6:9). In spite of this plan, some questions regarding hydraulics in the proper sense are also discussed from an historical point of view in this uncompleted sixth volume.

It is not our intention, however, to enter into detailed examination of the content of this sixth volume, although it would offer much scope for inquisitive and interesting considerations, particularly regarding the deeds of Leonardo da Vinci and Cardano. One might also examine how the author persists and even degenerates further in pursuing his insane aim of denigrating Galileo at any cost, to the point of representing him as a champion of Peripateticism even in compari-

---

[2] This is meant particularly with regard to my monograph on Mersenne, whose galley proofs I passed for the press on 4 November 1916 (Favaro 1916–17, 92).

[3] Abbé Raffaello Caverni died on 30 January 1900.

son with his Peripatetic opponents. Instead, we want to take this opportunity to draw the scholar's attention to certain passages of this sixth volume, in which we find texts reproduced that are more or less directly attributed to Galileo but which, in our opinion, can by no means be attributed to him.[4]

In the previous volume, Caverni has already made use of his outstanding ability to imitate the Galilean style, particularly in the dialogue form, by taking advantage of or rather exploiting his amazing facility in manipulating language.[5] He had also given a very nice proof of this ability in another occasion.[6] By inserting in the right places typical, authentic phrases from a genuine Galilean text, Caverni had tried to pass off as stuff of Galileo what was indeed nothing but his own very skillful and refined falsification. Such an action is always bad but more so when done by an historian, especially one who has repeatedly been recognized as someone who totally lacks objectivity. We think that we have the right and duty to point this out for two reasons: first, to warn those who have recourse to that work and would not have noticed the falsification on their own; and second, to somehow vindicate the National Edition, which indeed would have been blameworthy of negligence had it omitted such a substantial part of the Galilean work.[7]

Addressing the facts, let's start by noting what Caverni writes in the above-mentioned volume 5, at the point where he mentions the subjects that should have been treated in the sixth of the *Dialogues Concerning the New Sciences*. Caverni reports Salviati's promise to explain the utility of observing how a little chain hanging loose from its extremities bends and assumes a form that in certain cases coincides almost to a hair with the parabola (Caverni 1891–1900, 5:143). At this

---

[4] We do not want to overlook the fact that a short treatise ascribed to Viviani and entitled "De radiis fluidis" is also published in this volume (Caverni 1891–1900, 6:247–64). Since we ourselves published a monograph on Viviani some years ago and on that occasion carefully examined all of his papers, we owe ourselves and the public an explicit declaration that the attribution of this short treatise to Viviani is absolutely arbitrary. This attribution has been patched together by Caverni from some notes of an unknown hand found among the manuscripts of the Accademia del Cimento.

[5] We do not want to go back any further; we shall rather limit ourselves to noting that there are instances since the first volume that reveal Caverni's ease in recognizing strictly Galilean texts in the works of more or less true disciples. In it we read: "Many of the writings of the great master return to life in the handwritten and printed works of Borelli, as would be the tables of the average motions of Jupiter's satellites, the instructions how to use this instrument in the observations of Jupiter, the discourse on the rudder's mechanical function in directing ships, and other Galilean texts whose copy as well as original are lost" (Caverni 1891–1900, 1:190). At this point it would be useful to remember that, as a matter of fact, Gio. Alfonso Borelli was not a disciple of Galileo because he came to Tuscany for the first time in 1656; he was only a pupil of Castelli in Rome and fellow of Torricelli. Galileo knew Borelli only by name in so far as the latter had been suggested to him by Castelli as lecturer for the University of Pisa, at the time when the chair of Mathematics there was about to became vacant due to Dino Peri's hopeless health conditions.

[6] We refer to the dialogue between Galileo and Salviati masterly invented by Caverni and inserted in his *Ricreazioni scientifiche* (Caverni 1882, 204–43).

[7] The same can be said about several other cases in which Caverni ascribes to Galileo writings of other people. However, since it would take too long to enumerate all of these cases, we shall rather limit ourself to pointing out as an example certain handwritten "Notes" made by Viviani about some passages of the *Dialogues Concerning the New Sciences* and contained in the V part, IV vol., fol. 14 f., of the Galilean Manuscripts in the National Library of Florence. Caverni pretends that these notes are handwritten by Galileo and he derives from them what needed to support his arguments (Caverni 1891–1900, 4:303).

point Caverni writes that he himself, as the "first and only one"[8] having searched for where and how Salviati's promise was fulfilled, found the second part of the Galilean dialogue that followed the discussion on percussion and thus completed the Day.

Everyone would now expect to be informed of where and how Caverni found this most valuable manuscript, from which he extracts more than eight pages of dialogue with the same interlocutors as in the authentic one, namely Salviati, Sagredo, and Aproino. In fact, Caverni avoids this question by writing: "We shall abstain from narrating how we made the discovery amongst certain jumbled manuscripts given to us by a friend for examination because we believe that our readers would rather wish to learn without delay what we copied from it" (Caverni 1891–1900, 5:143).[9] But what is presented here as a "Galilean dialogue" is no longer one at the end of the transcript where, on the contrary, doubts are cast upon its authenticity. We can in fact read:

> The dialogue breaks off at this point but the treatise on the use of the small chains is in any case complete; what we feel might be missing is only the more or less ceremonious farewell of the interlocutors. Anyway, even if our readers agree that the entire argument is included in the transcript, they could ask *for the reasons which induced us to ascribe this text to Galileo.* (Caverni 1891–1900, 5:152)

It is then no longer a matter of an authentic text but rather of something attributed to Galileo by Caverni, who then explains that one has to distinguish between form and content. In order to prove that the content is purely Galilean, Caverni appeals to a theorem that he inserted into the text and which does indeed exist written in Galileo's hand among the "Fragments Pertaining to the Discourses and Mathematical Demonstrations Concerning Two New Sciences." These "Fragments" have been published in full for the first time by us in the National Edition, where they constitute one of the most radiant gems (Galilei 1890–1909, 8:363–448. The cited theorem can be read on page 367). Caverni then adduces as evidence for the attribution a drawing, also included in the "Fragments" (ibid., 369–70), in which Galileo looks as if he wants to apply a certain proposition, cited by Caverni, in order to derive the parabolic form of the little chain. Further on, Caverni finally declares that the dialogue, or rather the fragment of a dialogue, which he pretends to have found in a copy of that time, should be attributed to Viviani (Caverni

---

[8] When Caverni wrote that about himself he had probably forgotten the bitter reproach he himself had previously made to Galileo because of a similar boasting (Caverni 1891–1900, 1:128, and 2:374). Furthermore, it is an essential difference that Galileo wished to have been the first and only one (Galilei 1890–1909, 7:540), whereas Caverni claims to have been the first and only one.

[9] In the "List of the documents taken from the Galilean Manuscripts and reported in the order of the chapters," at the end of this volume, we can read about this text: "Treatise on the proprieties of the little chains usable in ballistics, written in dialogue form in order to be added to the treatise on percussion, at last found among the Galilean manuscripts (sic) and here published" (Caverni 1891–1900, 5:653).

1891-1900, 5:153), and adduces as evidence for this attribution a passage from Viviani's "Summary of the last works of Galileo" in which, he writes, "one can easily see summarized the dialogue we have transcribed" (ibid., 154; Viviani 1674, 105–06).

[Text omitted]

And now let us move on to volume six. The source of the new Galilean writings contained in this volume is the same from which Caverni drew, or rather writes that he had drawn in the just mentioned occasion. But even in the case of these manuscripts nothing is said at all about where they were and to whom they belonged at the time when they were offered to Caverni, as he pretends.

One of these new writings is a dialogue rendering in nearly ten whole pages the short Galilean treatise on the small balance (Caverni 1891–1900, 6:111–20). But on this occasion Caverni, in our opinion, relied a bit too heavily on his experience acquired in concocting such fakes and he did not attend to the imitation of Galileo's language and style with the usual diligence. Especially in the objections which he puts into Simplicio's mouth there is too much, let us say, simplicity, which is too far from the relative keenness showed in the genuine Galilean dialogues by this curious character. Our opinion is confirmed also by the insertion in this dialogue of other arguments closely connected with the small balance and extracted from letters and writings of disciples and of the numerous commentators of the original treatise.

Caverni writes that he also drew other things from the same previously mentioned source that are connected with, or at least should be connected with, the studies made by Galileo on the force of percussion. These studies were intended to provide the argument for what Albèri called "Sixth Day," but we have entitled them, with greater adherence to the content as well as appropriate scrupulosity, "Beginning of Day Added to the Discourses and Mathematical Demonstrations Concerning Two New Sciences" (Galilei 1890–1909, 8:319–46). As to the connection, in this fragment of dialogue Galileo puts into Aproino's mouth the account of one of the experiments carried out in his Paduan house (ibid., 323). It was aimed at searching for a way to measure how much the weight of a striking body and the speed with which the body is moved might influence the effect and operation of percussion.

Now Caverni claims to have found among these papers some notes relating to this theme, at the beginning of which Viviani wrote, as he did in some other occasions that we mentioned, "I have the original of this." A dialogue exposing the substance of these notes and carrying the remark "ad mentem Galilaei" can also be found a few pages later, according to Caverni (Caverni 1891–1900, 6:369). Indeed, this remark by Viviani happens to be found on top of some of his other studies on the master's teachings that were scattered among his papers and collected in the Galilean Manuscripts. This dialogue, which Caverni attributes to Viviani, should be inserted, according to the former, at the point of the authentic one where Aproino speaks (Galilei 1890–1909, 8:325). To be more precise, the insertion would take four pages starting from the middle of the first line of Salviati's reply which, in spite of the long variant, would continue unaltered (Caverni 1891–1900, 6:371–75). After this insertion, Caverni writes:

> There is no doubt that Viviani would have reported this part also, together with other parts of perhaps less importance, had it been possible for him to publish the "Dialogue on the Force of Percussion." But the task of publishing the "Dialogue" was to be fulfilled by Bonaventuri and we may assume that Bonaventuri did not include this part in his edition,[10] as would have been desirable, only because Grandi, for whatever reason, did not show it to his editor friend. Grandi should have read this part of the dialogue along with the papers received from Panzanini. (Caverni 1891–1900, 6:375)

By entering into all these details Caverni provides us with the most effective arguments that undermine the foundations of his assertion regarding the origin of the papers which he pretends to have made use of, and he thus enables us to demonstrate that these papers never existed. As we reported above, the correspondence of Panzanini and Buonaventuri with Father Grandi has reached us.[11] These letters provide full and sure evidence for the fact that the only papers Panzanini gave at first to Buonaventuri were the well known bundles of manuscript authenticated by Prince Leopoldo de' Medici on March 2nd, 1667 ab Inc[arnatione Domini], containing Viviani's treatise concerning the resistance of solids. This treatise had given rise to the famous controversy with Marchetti, which went on until the next generation. Buonaventuri was expected to copy the papers and hand the copies over to Grandi but, since the transcription was taking too long, he decided to send Grandi the originals. Grandi then published the treatise with some important additions in the first Florentine edition of Galileo's works.[12] However, all these writings were afterwards returned to Panzanini and are now among the Galilean Manuscripts in the National Library of Florence.

As to what Viviani would have done if he had been able to publish the so-called "Dialogue on the Force of Percussion," we beg to totally differ with Caverni. Granted for a moment that the facts were as Caverni wants us to believe, we firmly maintain that Viviani would never have dared replace the words of his master with his own, no matter how sure he was of correctly interpreting his thought. Our supposition is strongly supported by the meticulous care Viviani exerted in editing

---

[10] Galilei 1718. On this edition, see our article Favaro 1917–18.

[11] University Library of Pisa. Letters to Fr. Grandi. Vol. 4: Letters of Buonaventuri (83 items); vol. 13: Letters of Panzanini (16 items). One should also examine the letters of Benedetto Bresciani in vol. 3 (58 items).

[12] "Trattato delle resistenze principiato da Vincenzio Viviani per illustrare l'opere del Galileo ed ora compiuto, e riordinato colla giunta di quelle dimostrazioni, che vi mancavano dal P. D. Guido Grandi Abate Camaldolese, Mattematico di S. A. R. e dello Studio Pisano." (Galilei 1718, 3:193–305).

the "Beginning of the Fifth Day," to the point that he even marked the paragraphs in which he suspected that something could have been penned by Torricelli.

Indeed, in dealing with the manifestations of Galileo's thought, no care to guarantee perfect authenticity should be considered superfluous. Thus, even if we leave out of consideration the intrinsic reasons we adduced above, we believe that the lack of exact indications as to the sources, whereas the quotations are generally so studiously detailed, justifies the doubts of the most, the suspicion of the many, and the certainty of some that these multiple discoveries of Galilean dialogues are nothing but a literary abuse. Therefore, the greatest caution is recommended when making use of the conclusions of this work, which is otherwise so rich in merits and so worthy of being studied, as should be properly recognized.

[Text omitted]

HANS-WERNER SCHÜTT

Emil Wohlwill, *Galileo and His Battle for the Copernican System**

Galileo Galilei is one of the few figures in the history of science who has attracted the imagination even of laymen to the natural sciences. The battle of this great physicist against the domination of his church, a battle which he ultimately lost, manifests fundamental human interest that extends beyond the individual. Galileo pits the right of the thinking individual against the right of an institution that defends its claim to set norms for individual thinking because it posseses superhuman truths.

Thus it is understandable that the discussion about Galileo continues today and will probably continue on in the foreseeable future. Every year new books and essays appear about Galileo's life or specific aspects of his work; the field of Galileo historiography is expanding and gradually becoming intractable in scope for any individual scholar.

However, the field is divided into categories; it has a number of landmarks that can not be ignored. This applies in particular to the few works written with such expert knowledge and command of the material that they have become indispensable secondary sources for all subsequent Galileo research.

Among these fundamental works is Emil Wohlwill's *magnum opus*, his two-volume Galileo biography. Wohlwill labored on this book forty years with extraordinary energy, but was ultimately unable to complete it. He worked as a rigorous and self-critical historian, under conditions which certainly would have induced others to give up.

Only the rudiments of independent historiography of the natural sciences existed in the second half of the nineteenth century. There were neither special libraries, research institutes, nor professorial chairs at universities.

Not even Wohlwill could make a career of his "unfortunate inclination" (Emil Wohlwill to Hans Zahn, 22 November 1903) toward the history of the natural

* Translator's note: This article was first published as an introduction to Wohlwill 1969. We thank Hans-Werner Schütt and the Sändig Verlag, Wiesbaden, for the kind permission to republish it. The original contains the following information: "A comprehensive portrayal of the personality and the work of Emil Wohlwill is in preparation at the Institute for History of the Natural Sciences of the University of Hamburg (Schütt 1972); to this end a wealth of material of previously unpublished documents has been analyzed." The notes have been standardized. The list of references can be found in a bibliographical section at the end of the appendix.

sciences, even though his background and disposition gave him all the capabilities for it.

Born in Seesen in 1835, he grew up in an atmosphere of intellectual freedom and high demands. Emil's role model was his father Immanuel, despite his early death. He was descended from the sector of Jewish educated bourgeoisie that after the emancipation of the Jews at the beginning of the century had gradually lost its connection to orthodox Judaism. After their bitter historical experience, they regarded intellectual freedom as the only goal in life worthy of human beings.

The son also knew how to defend his intellectual freedom in an often truly tough life struggle. He wavered long between history and the natural sciences, but finally chose chemistry, which he studied from 1855 in Heidelberg, Berlin and Göttingen. In 1860 he received his doctorate from Friedrich Wöhler with a dissertation, "On Isomorphic Mixtures of Selenic Acid Salts."

In Berlin the young student attended a lecture on the history of physics by Johann Christian Poggendorff. As he later declared, his interest in the history of science was already awakening as an inner opprobrium of the dilettantism that accepted fables and legends so uncritically, a phenomenon he found prominent in large sectors of the historiography of his time (Wohlwhill to Hans Zahn, 22 November 1903). *Ad fontes* remained the guiding principle of his activity, and by holding on to this motto he became an expert scholar, even though in his isolation he easily could have become a dilettante himself.

The Hamburg of the 1860s, where Wohlwill settled after his studies, offered no opportunity for a personal exchange of scientific ideas. In any case, the young chemist initially was left with little time for his own research. For a time he pursued three occupations simultaneously: he was a commercial chemist, a physics teacher, and an industrial chemist at the Norddeutsche Affinerie. This non-ferrous metal foundry, one of the world's largest producers of pure copper, benefits even today from the pioneering achievements of its former employee Emil Wohlwill, who brought the electrolysis of non-ferrous metals to technical maturity.

In spite of his isolation and even though he achieved success in his work in other fields, Wohlwill remained true to the history of science. He never expressed his views about why he made Galileo research the most important object of his scientific work, but two reasons appear to be decisive. First of all, Wohwill must have sensed a spiritual relationship to the man who had to struggle for the freedom of his spirit. Secondly, only in Wohlwill's day had it really become possible to access the most important primary sources relating to the Galileo case. The secret files of the Inquisition on Galileo, which Napoleon had brought to Paris, were back in the possession of the Vatican since 1841 (not 1845, as often mistakenly believed) and made accessible for scientific research, initially in part, and later in full (Laemmel 1928, 406–07).

Years of study dedicated to these and other sources — in 1891 in the Vatican itself — made Emil Wohlwill the most important Galileo expert of the late nineteenth century along with the editor of the *Edizione Nazionale*, Antonio

Favaro. As of any historical personality, every period must sketch its own picture of Galileo, but to do this adequately, to find its own perspective, it must know what the other pictures looked like. For the contemporary reader it is of absolutely no disadvantage to see that period — the late nineteenth century — shimmering through the lines of Wohlwill's biography of Galileo.

Wohlwill saw the history of the natural sciences as part of the history of culture. He once wrote, "We take pains for the history of science because from this historical study we may hope for revelation about the nature of all human thought, about the most internal nature of progress in all human development. By following the development of the natural sciences through the centuries, we engage in anthropology, the study of humans in the highest sense of the word" (Wohlwill 1867, Lecture I, 3). These words express a faith in fundamental progress for the entire history of mankind. The history of the natural sciences merely offers the special advantage of being able to recognize and distinctly analyze this progress. The belief in progress, however, forces a certain polarity onto history, since progress, although ultimately accepted, always faces initial opposition from forces that try to obstruct it.

In the Galileo affair, "modern" physics represents the stimulating force; thus the "unmodern" physics of the Peripatetics and the Catholic Church, with its claim to primacy over physics, actually become obstructive forces.

It appears to us today that the contours of this picture are drawn somewhat too sharply. For instance, the Catholic Church and its special — also internal — problems seem today more differentiated than Wohlwill portrayed them. He was thoroughly successful in showing the contradictory aspects of the behavior of the church authorities, but his suspicion that the Roman Inquisition committed crude forgery has not been confirmed (Laemmel 1928, 415). Some indications suggest, however, that during the trial of 1633 there was intent to deceive certain persons. This argues in favor of a very differentiated picture of the Catholic Church, as these persons must have been high church dignitaries themselves.

On the other hand, Wohlwill aptly described Galileo's character and difficult personal situation. The sections about the theory of tides, especially, show that Wohlwill possessed extraordinary sensitivity and empathy for the thought processes of the great physicist.

Two years before his death in 1912 and shortly after the publication of the first volume of his Galileo biography, Wohlwill expressed hope for the book's impact, "that one time or another a young person will discover that something is contained in it which he can use, and in his own interest must use" (Emil Wohlwill to Fritz Wohlwill, 12 August 1910). In view of the echo which his work has found, one can certainly say that this hope was not unjustified.

Institut für Philosophie, Wissenschaftstheorie, Wissenschafts- und Technikgeschichte
Technische Universität Berlin

EMIL WOHLWILL

The Discovery of the Parabolic Shape of the Projectile Trajectory*

In the fourth volume of his *History of the Experimental Method in Italy* (Caverni 1891–1900, 4:506–33), Raffaello Caverni fulfilled the promise he had made four years earlier in an introductory overview of his work (ibid., 1:135–36): to prove on the basis of the history of the discovery of the parabolic shape of the projectile trajectory that Galileo claimed the intellectual property of his eminent contemporaries as his own, and to show how this was accomplished. One had the right to expect that the proponent of such an entirely new opinion in a case which he himself considered to be, in the words of Bacon, an *instantia praerogativa* for the justification of his view, would weigh the value of each individual argument with absolute impartiality and do complete justice to the ambiguity of the given facts. He would have to state his case in compelling logic so that a clear-thinking individual would have no choice, in this case at least, but to believe in the dishonorable theft by a great man. In these expectations we have been disappointed thoroughly by Caverni: his argumentation is in all aspects that of a shrewd lawyer who considers it his task to allow only one side of the question to come to light, to collect everything that might be utilized in favor of his biased reading, to hold back anything which might give rise to the idea that things could be viewed in another way. He believed it permissible for his own purposes to operate with presumptions as if they were facts, and to regard as proven what is at best probable.

Caverni nevertheless serves the cause he champions with comprehensive factual knowledge and an astounding knowledge of the literature. He is especially well-read in the works of Galileo, his correspondence and the handwritten treasures of the Biblioteca Nazionale in Florence. This rare familiarity with the sources of the history of science, with which the friends of Italian science are acquainted from earlier writings by the same author, received extraordinary recognition through an award from the Royal Institute of Venice for Caverni's *History of Experimental Method in Italy*, whose publication was prefaced with a report by the representative of the Academy, Antonio Favaro [Favaro et al. 1889–90]. However, in these introductory remarks Favaro expresses the hope that the author submit his research to repeated scrutiny and correct his antagonistic judgment of Galileo,

---

* This article was originally published as Wohlwill 1899. The notes in square brackets are by Giuseppe Castagnetti. The references have been modernized. The list of references can be found in a bibliographical section at the end of the appendix.

especially his undoubtedly incorrect opinion about the discovery of the parabolic shape of the projectile trajectory. Thus the Academy of Venice and its representatives took precautions in advance so that their recognition would not be interpreted as approval of the attack on the honor and reputation of Galileo, which plays such a great role in Caverni's book. However, this report from such an important source certified the predominant merit of the work by judgment of the Academy of Venice. At the same time it makes it impossible to ignore a contention so emphatically expressed in this work and for which so much evidence is offered. It is not enough to confront the author by fundamentally rejecting his methods of historical research and historiography. In order to refute his case, one must follow the details of his rendering and his treatment of historical sources which were never before available in their entirety. This is what is attempted in the following.

I.

The fact that Galileo's claim to the discovery of the parabolic shape of the projectile trajectory is contestable has to do with the peculiar way in which his theory of motion was published. It is known that the "new science of local motion" became generally accessible only in his masterpiece [Galilei 1638], which appeared four years before Galileo's death although an outline and the main principles of the theory had been completed four decades earlier. In the work of 1638, it is apparent that the theory had been completed long before: its axioms were put forth as components of a separate book in the larger work. An older three-part Latin manuscript in strict scientific form offers the actual substance for the successive discussion in Italian, presented in the form of a dialog. As Galileo generally wrote only in the vernacular after his departure from Padua, one has to consider the mere use of Latin in these sections of the *Discorsi e dimostrazioni* tantamount to a priority claim over any work of similar content written after 1610. Obviously, Galileo's late claim is not definitely binding for the historian even if it is expressed in this form. When Simon Marius says, four years after the discovery of the satellites of Jupiter, that he observed them at least as early as Galileo did, and when Baliani publishes in 1638 that he "discovered" in 1611 that the oscillation periods of two pendula are proportional to the square roots of their lengths, these pretensions expressed by the people concerned cannot be regarded as historical proof for the facts asserted. Similarly, the fact that a pioneering theory of Galileo's is discussed in the Latin text of one of the three books *De motu locali* is not sufficient proof that it was developed in the period before 1610. It is not only possible, but indeed highly probable, that the older text was not adopted into the context of the greater work unchanged, but rather underwent numerous additions and alterations on the basis of later insights.

Proven, however, is that Galileo had already been working for some time on a three-part manuscript on the theory of motion when he moved to Florence in

September 1610, and that large sections of this work had been completed. He mentioned this in his report to the State Secretary Vinta of 7 May 1610 as one of three he was hoping to complete if an appropriate position at the Florentine court were to grant him the necessary leisure [Galilei 1890–1909, 10:348–53]. Galileo spoke here of the three books of *De motu locali* in words very similar to the introduction to the Latin treatises of the *Discorsi* of 1638, which shared the same title. He said,

> It is an entirely new science, since none of the ancients nor any of the new scientists has discovered any of the countless wonderful peculiarities that I am proving for natural and violent motions. Therefore I have every right to call it a new science, and one that I developed from its very beginnings. [Ibid., 10:351–52]

The "new science of local motion," at least the substance of which existed in 1610 according to this claim, evidently also corresponds to the Latin sections of the later work. Its third book, like the "Fourth Day" of the *Discorsi*, concerns "violent" motions, i.e., the theory of *proietti* (projected bodies). Even an entire year earlier (on 23 May 1609) the mathematician Luca Valerio responded by letter [ibid., 10:244–45] to a communication from Galileo concerning his work on naturally moving and projected bodies. Thus, in addition to the research on uniformly accelerating motion, a more or less coherent theory of projectile motion in Galileo's hand must have existed as early as spring 1609.

In a letter addressed to a prince of the Medici House dated 11 February 1609 [ibid., 10:228–30], Galileo spoke of questions which "remained" in regard to the motion of projected bodies. This means that these questions were linked to a series of experiments already completed on the same subject. The additional comments in this letter are of particular interest; until recently they offered the only reference point in the attempt to reconstruct the development of the 1609 theory of projectile motion, without reference to the Latin texts of the *Discorsi*. These remarks are especially important for answering the question of whether Galileo was aware of the parabolic shape of the trajectory at that time.

Galileo reported to the prince of the grand ducal house that he recently found that cannonballs fired horizontally from an elevated place always deviate from the horizontal and approach the earth at the same speed, independent of the amount of powder used, even if it is just enough to cause them to leave the barrel. He deduced further that for all shots directed horizontally, the ball will reach the earth in the same amount of time, independent of the distance of their points of impact. This time is then identical to the time needed for a ball to fall vertically to the earth from the mouth of the cannon. He recognized a similar effect for shots directed diagonally upward: shots which elevate the ball to the same altitude, and thus have trajectories lying between the same horizontal planes, reach the earth or the same lower horizontal plane at different distances at the same time. As a consequence, the descending halves of their trajectories are also covered in the same amount of time, i.e., the time needed for horizontal shots to fall from the same altitude (Galilei 1842–56, 6:69–70 [Galilei 1890–1909, 10:229–30]).

These statements, which may be called for short the "law of equal times of fall," strongly imply that Galileo on discovering them was fully acquainted with the principle of the neutral coexistence of different motions of the same body. For the assertion that horizontally shot cannonballs reach the earth at the same time, despite the greatest difference in their horizontal speeds, like a body falling freely from the same height, is simply another way to express the opinion that the motion in the direction of gravity is in no way influenced by the additional motion in the direction of the shot. The certainty with which Galileo formulated these rules appears to suggest that he derived them from principle rather than from experiments, since experiments could not have yielded certain results in this case.

In order to apply this consideration to shots directed diagonally upward, it must also have been known that bodies projected upward need the same amount of time to rise and fall. According to Paolo Sarpi, Galileo already had recognized this before 9 October 1604 (Galilei 1842–56, 8:29 [Galilei 1890–1909, 10:114]).

Thus when he discovered the law of equal times of fall, Galileo knew everything necessary to determine both components of the trajectory in detail, and therefore, along with the principle of the independent coexistence of motions, everything needed to construct the trajectory. He presented the law of spaces of fall, which determines the changes of the vertical components, as a recent discovery in a letter to Sarpi of 16 October 1604 [Galilei 1890–1909, 10:115–16].[1] Probably even earlier was the discovery of the law of inertia in the form in which it still is used to construct the trajectory in the *Discorsi* of 1638, i.e., in its restriction to motion in the horizontal plane. Direct deduction led from this to the uniformity of motion in the direction of the horizontal shot, and thus everything necessary to determine the horizontal components. To derive this third premise of his construction, Galileo referred in the *De motu proiectorum* of the *Discorsi* to the argumentations of a preceding section of the Latin manuscript [ibid., 8:268]: unequivocal remarks substantiate that at least this part of his reasoning in the publication of 1638 originated in an older text from his days in Padua. It may be sufficient here to point out that Castelli, in an April 1607 letter [ibid., 10:169–71], presented as Galileo's theory: "For motion to begin the moving body is required, but for the continuation of motion it is sufficient that it find no resistance" [ibid., 10:170].[2]

According to this, when he first spoke in 1609 of a book about the theory of

---

[1] This letter was printed in accordance with the original kept in Pisa for the first time in Favaro 1883, 2:226–27 [reprinted in Favaro [1883] 1966, 2:172–73].

[2] This remark, first published in Favaro 1883, 2:268 [reprinted in Favaro [1883] 1966, 2:203], was not known to me during my work on the treatise about the discovery of the law of inertia (Wohlwill 1883–84). It apparently contradicts the opinion conveyed there that Galileo restricted the principle of unchanged conservation of motion to the horizontal motions. In truth, even Castelli's remark merely proves that this limitation, which was determined by individual causes in Galileo's work, was not noted by his students, as I attempted to prove for Cavalieri and Torricelli. For further comments by Galileo about inertia in the horizontal direction, see the treatise cited above.

projectile motion as a component of his work *De motu locali*, Galileo possessed all prerequisites for the correct construction of the trajectory. To see as many trajectories in parabolic form as he liked, he needed only to make exact drawings rendering the law of equal times of fall on the basis of these discovered facts. For the final discovery, it was necessary, of course, that he recognize the parabola in his drawing. It is not thoroughly inconceivable that one could miss this, even if what lies before one's eyes is the long-sought solution to a riddle. But when Galileo said thirty years later, "I saw then what was before my eyes," it would take strong reasons indeed to convince us of the contrary.

However, several facts are known which justify any doubts at first glance. In the *Dialogue Concerning the Two Chief World Systems*, published in 1632 [Galilei 1632], among numerous other propositions of the new theory of motion there is no mention of the parabolic shape of the projectile trajectory. Moreover, in the section concerning the construction of the actual path of a body falling to earth and simultaneously affected by the rotation of the earth, essentially the problem of the projectile trajectory, the procedure imparted for constructing the trajectory would indeed result in a parabola if performed correctly. In the *Dialogue*, however, the result of the combination of the two motions, according to Salviati, is a motion along a semi-circle, not a parabola, and Sagredo rejoices in this unexpected result [Galilei 1890–1909, 7:190–91]. Did Galileo keep his better knowledge secret in this case, or did he not know, when he wrote this part of the *Dialogue* and even later when he published his book, the true form of the trajectory?

A second, perhaps even greater difficulty is presented by the fact that a few months after publication of the *Dialogue*, Father Bonaventura Cavalieri, Galileo's pupil, published the correct solution sought in vain in the *Dialogue*, and made no mention there of the fact that what he imparted was Galileo's discovery. It is true that Cavalieri prefaced the two chapters on the theory of motion added to his book, *Lo Specchio ustorio overo Trattato delle settioni coniche*, the remark that what is presented in these sections originates "in part" from Galileo and Castelli, his two teachers [Cavalieri 1632, 151–53]. He then explained the three propositions he utilized for the construction of the trajectory in close connection to the arguments of the Galilean *Dialogue*, thus eliminating any doubt that his derivation was based on Galileo's thought [ibid., 153–72]. But whether Galileo himself came to the same result long before him using the same or similar means cannot be inferred from the chapter on the trajectory in the *Specchio ustorio*. Rather, the fact that Galileo is not named in the relevant explanations suggests that the author wished to portray them as his own conclusions, as the "part" which did not belong to his two teachers.

A letter Cavalieri wrote to inform Galileo of the contents of his manuscript immediately before publication appears to concur with this interpretation.

> I touched briefly on the motion of projected bodies, by showing that this motion, excluding air resistance, must occur in a parabola if we assume your

principle concerning the motion of ponderable bodies, according to which
their acceleration corresponds to the increment of the odd numbers proceeding from the one. However, I declare that what I mention here I learned in
great measure[3] from you, while at the same time I present my own derivation
for this principle. (Galilei 1842–56, 9:286 [Galilei 1890–1909,14:378])

Here, too, Cavalieri acknowledged that his proof is based on Galileo's theory of
uniformly accelerated motion; but even from this direct pronouncement, the
impartial reader would not receive the impression that Cavalieri was conscious of
directing his words to the discoverer of the parabolic shape of the trajectory.

That Cavalieri in the end believed to have found Galileo's true opinion in the
argumentations of the *Dialogue* mentioned above becomes preeminently probable
through a remark at the close of the penultimate chapter of his *Specchio ustorio*.
There, to the proof that the curvature of a circle of very large diameter would not
diverge essentially from that of a parabola and a hyperbola he added:

This realization should satisfy those who believed that the path described by
a projected body would be circular. For if the circle in question is of
considerable size, and the path of the ponderable body is only a small part of
the entire periphery, its divergence from the parabola would be negligible.
[Cavalieri 1632, 218]

It is hardly arbitrary to seek in this remark a reference to Galileo's considerations
in the *Dialogue*, since there is no other place in which the circular motion of
projected bodies was asserted with similar determination.

II.

The derivation of a circular path in the *Dialogue* and Cavalieri's mention of a
discovery by Galileo, veiled as it may have been, already have been the subject of
many critical discussions. It has not been overlooked that facts exist here which
contradict *per se* the general assumption that the parabolic shape of the trajectory
was discovered before 1610. However, not even those who see this as problematic
have considered calling into question Galileo's discovery. Such a solution appeared
precluded by Galileo's answer to his pupil's preliminary announcement and
through Cavalieri's response to this answer. In 1632 Galileo claimed credit for the
discovery in no uncertain terms; Cavalieri in his response called Galileo's claim to
this discovery one of many well known to contemporaries and one not doubted by
himself. The words expressed by both men on this occasion rule out any doubts
about the truthfulness of their testimony.

Caverni's opposing view is based on his own peculiar conception of the apparent

---

[3] The "*in parte*" in the book has become here "*in gran parte*."

contradictions mentioned here. To him the remarks outlined in the *Dialogue* do not appear to be incompatible with all else that is known about Galileo's theory of projectile motion; rather he sees this explanation about the circular path of a body falling on the rotating earth as Galileo's true opinion about the form of the trajectory as well. Caverni believes that Galileo held to this view for forty years with only minimal changes, and he finds this contradicted only by the dominant opinion that Galileo was the discoverer of the parabolic shape and by a similar claim raised in the Latin section of the *Discorsi*.

Specifically, Caverni believes that the theory to which Galileo professed until the year 1632 is that of Niccolò Tartaglia. According to this, the motion of a projected body consists in part of pure violent motion and, immediately subsequent to this, pure natural motion. The pure violent motion consists of a straight portion and a curved portion; the curve of the latter is circular in shape. As already implied here, in another sentence of his *Scientia nuova* [Tartaglia 1537] Tartaglia expressly denies that the motion of a projected body in any portion of its path could be mixed, i.e., simultaneously violent and natural. This must have been Galileo's view in essence, not only in 1609, but even at the time the *Dialogue* were published. He would have moved beyond this system of Tartaglia's only to the extent that he later acknowledged, under the influence of other researchers, the possibility of both kinds of motion coexisting.

Anyone naive enough to believe that such an interpretation cannot be regarded as proven without supplying literal citations to demonstrate unequivocally that Galileo believed in Tartaglia's teachings over such a long period has got rid of Caverni's argument; for in truth there is no such evidence, and nor does Caverni supply any. However, he does discuss, in an extraordinarily verbose explanation of no fewer than thirteen densely printed, large-sized pages, a number of quotes of the most varied content which bear some resemblance to evidence [Caverni 1891–1900, 4:517–33]. Thanks to Caverni's peculiar way of mixing indistinguishably the words of others with his own commentaries, he manages to create the impression that they confirm his assertion. As they constitute the actual basis of his argumentation, these quotations must be discussed here in turn.

The investigation begins with the treatises written in 1592 at the latest, and with the fragment of a dialog which was printed under the title *De motu* in Volume I of the *Edizione nazionale* [Galilei 1890–1909, 1:243–419]. Anyone who seriously studies these oldest notes preserved in Galileo's hand will recognize no more than the precursor of the actual scientific development of thought during the Padua period. Therefore it would be irrelevant to the question under discussion whether any proof could be found in these older manuscripts that at the time of their creation Galileo thought no differently about the trajectory than Tartaglia. But in a truly honest search, one could not discover even here what Caverni claims to have found. The treatises and dialogs of *De motu* contain not only no mention of a construction according to Tartaglia, but no word at all about the shape of the trajectory, and consequently no indication that Galileo attributes a circular form to its middle section. Neither does the Pisa manuscript indicate, as Caverni avers, that Galileo contradicted Cardano and Benedetti and agreed with Tartaglia that a mixture of violent and natural motion is impossible at any point of the trajectory. Caverni has no other proof for this assertion than the remark that "the rotation of a sphere whose center coincides with that of the world is neither a natural nor a violent motion" [Caverni 1891–1900, 4:513].

I have shown elsewhere (Wohlwill 1883–84, 74–5) that this remark very probably related to Galileo's first efforts to understand the perpetual rotation of the earth. He noted that the parts of a homogenous ball made of ponderable matter, concentric with the sphere of the world, neither can approach nor move away from the center of the world through its rotation. He concluded from this that such a motion, in addition to the natural and violent motion, has a peculiar status which appears to be compatible with the permanent conservation of the impetus imparted.

Needless to say, Galileo explained in detail that the above applied to the rotation to no ball, homogeneous or non-homogeneous, located anywhere other than the center of the world; thus it also could not refer to the circular motion of projected bodies. Caverni is audacious enough to create this connection by saying: "thus it is also for the projected bodies" (così avviene dei proietti) [Caverni 1891–1900, 4:513]. He would have the reader believe that Galileo, like Tartaglia, held that projected bodies move in a circle so that they neither approach the center of the earth nor move away from it!

That Galileo, at this time by no means denied, in contrast to Cardano and Benedetti, the possibility of "mixed" motion, is evident from definitive explanations given in treatises such as the dialog De motu. His examination of the motion of the vertically projected body concludes that this motion, under the simultaneous influences of gravity and the projective force from the start of the climb through the descent, is mixed, i.e., composed of natural and violent motion (Galilei 1890–1909, 1:322).[4] For the descent from horizontal shot, however, the last treatise of the Pisa manuscript contains an observation about the mixture, or the coexistence of the motion in the direction of gravity with that in the direction of the shot. This essentially agrees with the explanation published 46 years later, despite very different wording. In the Discorsi it reads:

> As soon as the horizontally moved ponderable body leaves the fixed support, it will add the downward inclination caused by its own weight to the previously uniform and indestructible motion. Accordingly a compound motion will result, which I call projectile motion. [Ibid., 8:268]

As a further explanation he adds later that the two motions and their velocities, in

---

[4] "Mixed," i.e., composed of natural and violent motion, in his early work (Galilei 1890–1909, 1:373) is also called the motion of the body projected diagonally upward; here it is, however, doubtful whether the expression refers to the double motion of a single point.

mixing with each other, do not change, disturb, or hinder each other. The following explanation in the early work corresponds to this:

> Should the projected body move in a direction nearly parallel to the horizon, it can begin immediately to incline and thus diverge from the straight line of the projection; for it is enough for the violently impelling force that it removes the body from the starting point of the motion, and this removal is not hindered by the inclination.[5]

This assertion hardly suggests opposition to the idea of a mixed movement in Tartaglia's sense; on the contrary, it demonstrates a serious effort to clarify the nature and the cause of the mixture recognized.

In order to support his argument, it was far more important for Caverni to gather evidence from the heyday of Galileo's professorship in Padua, especially from the period between 1602 and 1609. Caverni himself attributes the bulk of the Latin sections of the *Discorsi* to this period because of its main content; nevertheless he believes that he can prove that this is impossible for the main ideas of the theory of projectile motion. In truth, his efforts meet with a lack of evidence in this period, too, which he fills with assumptions and insinuations. A highly unusual role here is played by a sentence about the trajectory published in 1844 [*recte* 1841] by Libri in his *History of the Mathematical Sciences in Italy* [Libri 1838–41] from a manuscript of the Paris Bibliothèque Nationale. According to Libri, the manuscript and the sentence originate from Galileo's friend and mentor, the marquis Guidobaldo del Monte. The sentence, remarkable in any case, reads as follows:

> If one throws a ball with a catapult or with artillery or by hand or by some other instrument above the horizontal line, it will take the same path in falling as in rising, and the shape is that which, when inverted under the horizon, a rope makes which is not pulled, both being composed of the natural and the forced, and it is a line which in appearance is similar to a parabola and hyperbola. And this can be seen better with a chain than with a rope, since [in the case of] the rope *abc* , when *ac* are close to each other, the part *b* does not approach as it should because the rope remains hard in itself, while a chain or a little chain does not behave in this way. The experiment of this movement can be made by taking a ball colored with ink, and throwing it over a plane of a table which is almost perpendicular to the horizontal.
>
> Although the ball bounces along, yet it makes points as it goes, from which one can clearly see that as it rises so it descends, and it is reasonable this way, since the violence it has acquired in its ascent operates so that in falling it overcomes, in the same way, the natural movement in coming down so that the violence that overcame [the path] from *b* to *c*, conserving itself, operates so that from *c* to *d* [the path] is equal to *cb*, and the violence which is gradually lessening when descending operates so that from *d* to *e* [the path] is equal to

---

ba, since there is no reason from *c* towards *de* that shows that the violence is lost at all, which, although it lessens continually towards *e*, yet there remains a sufficient amount of it, which is the cause that the weight never travels in a straight line towards *e*. (Libri 1838 –41, 4: 397–98) (see fig.1)[6]

The author here, like Cardano before him, tells only to which line the trajectory is "similar in its appearance" (*in vista simile*) and in this context he mentions both parabola as well as hyperbola. This mere fact proves that with this reasoning he aimed at nothing less than actually constructing the trajectory. On the other hand, his definitive assertion that the two branches of the trajectory are equal to each other and that no part of it is straight is illustrated in too uncertain terms and not proven at all. Only an arbitrary interpretation would find in these few words that which Caverni reads in them: the opinion that any point on either branch of the trajectory at the same horizontal have the same velocity. The word velocity does not occur in del Monte, who speaks only of the concurrence of the shape at the corresponding points. This concurrence could not be explained if velocity did not increase during descent in the same manner as it decreases during ascent. This does not justify Caverni's extrapolation from the above sentences that del Monte assumes that the "violent motion" remains unchanged in both the horizontal and the vertical because his assertion is true only under this precondition. Such fundamentally new insights should hardly be read between the lines of a sentence that fails to distinguish between horizontal and vertical propulsion and states only that "violence" is conserved, albeit gradually decreasing, until the end of the motion.

The extent to which these handwritten notes by del Monte ever could have had importance for the development of the theory of projectile motion can no longer be determined; it is not known whether any expert ever saw it before Libri. It is not

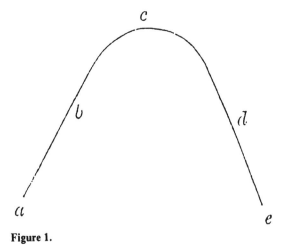

Figure 1.

[6] Prof. Favaro was good enough to check in the *Bibliothèque Nationale* that Libri reproduced the original correctly. Translated from the Italian original by Jürgen Renn.

improbable that its contents were the object of oral discussion or written debate between del Monte and Galileo. Cavalieri relates that around 1622 the engineer Muzio Oddi informed him that Galileo and del Monte had carried out an experiment on the shape of the trajectory [Caverni 1891–1900, 4:516; Galilei 1890–1909, 14:395]. This must have happened before the year 1607, for del Monte died in January of that year. It is conceivable that this experiment was the one described by del Monte, for Galileo cited a similar one in the *Discorsi* of 1638. In Galileo's depiction the experiment was his own, invented as a means of drawing parabolas, whereas del Monte spoke only of a line which looks similar to a parabola and hyperbola. In this passage Galileo also ascribed the parabolic shape to the freely hanging chain fixed at both ends [Galilei 1890–1909, 8:185–86].

If we try without bias to infer from the combination of these late assertions by Galileo together with the remarks by Oddi and the contents of del Monte's handwritten note an insight about a connection between Galileo's and del Monte's research on the theory of projectile motion, we will obtain a number of possible interpretations, but no historical conclusion. If we allow probability to be considered, it is unlikely that in a collaboration of any kind between two men the decisive impulse comes from the older man. Del Monte's achievements lie in another area; the fragmentary note about the trajectory is all we know of del Monte's studies on issues of the theory of motion. In contrast, upon del Monte's death Galileo could look back at nearly two decades of uninterrupted and most successful research dedicated to these issues. In order to believe that Galileo in 1606 would still find del Monte's observations concerning the theory of projectile motion rational, let alone instructive, we would have to dismiss all our inductions from the known data concerning the path of development of his thought.

In his *History* Caverni set forth a completely opposing view of the situation, not as the better founded or more probable but rather as the only possible, and based on it his freely invented reconstruction of an unascertainable course of events. For him it is a certainty that the original or a copy of del Monte's manuscript fell into Galileo's hands soon after 1607, and that Galileo found in it forthwith propositions pertaining to acoustics, the theory of solid bodies and to the theory of motion, which he thought to make his own at an appropriate occasion. Knowing that he might be held accountable for such behavior by those familiar with the contents of the manuscript, Galileo (so Caverni believes) in order to forestall suspicion, spread the tale of the jointly performed experiments.

In keeping with his foible for inventing history, Caverni improves the documents through minor omissions and additions. While Cavalieri heard talk of "an experiment" (*qualche esperienza*), Caverni has him hearing talk of "experiments with cannons." Because of this alteration it is even easier for Caverni to present the experiments as a fabrication, for others also would have heard of experiments with cannons, and yet "we have no certain document, no report which even suggests such events" (Caverni 1891–1900, 4:516).

Caverni even corrected del Monte's decisive text for his purposes by omitting the words *et iperbola* after *parabola*, thus increasing considerably the particle of truth in del Monte's words. In Caverni's portrayal it appears that del Monte left his successor little more than to deliver the proof for a truth which was already established. Galileo's constraint in this earlier period then would be all the more prominent, for despite his supposed tendency to adopt anything useful from the legacy of his friend, he rejected its best part. Thirty years passed before he saw the light and included in his own dialogs the instructions for drawing a parabola from del Monte's manuscript.

"Galileo rejects the similarity to the parabola," which del Monte revealed to him; he even "denied it determinedly when his friend showed him the paint marks of the balls on his polished board," so Caverni relates to the reader (ibid., 4: 524, 531). Yet, such a rejection by Galileo cannot be found in his own outlines from previous or later periods, nor in the report of a contemporary; neither with reference to del Monte's words, nor in any other context. It is a product of the historian's fantasy, who, as on many other occasions, finds it superfluous to enlighten his readers. What he presents in the form of an historical account is nothing more than a report about how he believes events proceeded.

In a similarly misleading reconstruction, Caverni imparts his view of how the inspiration Galileo found in the notebook left behind by del Monte came to bear fruit in his theory of projectile motion. He describes in detail how Galileo's argumentation grew out of del Monte's propositions, culminating in the year 1609 when he conceived, or at least surmised, the law of equal times of fall [ibid., 4:517-19]. For in keeping with the tendency Caverni pursues in this fiction, he does not allow the law, as formulated in Galileo's words, to appear as an inevitable deduction, but rather as an indemonstrable presumption. Only in this way could the law of equal times of fall find room in the same mind which, due to a persistent belief in Tartaglia's theory, dismissed the trajectory's similarity to a parabola.

Caverni did not overlook the fact that the most important insights contained in the letter to the Medici prince could be derived much more simply from the principle of the combination of motions. However, he believes that he can prove that Galileo only came to understand the "mixed motion" significantly later. According to him [ibid., 4:521–23], Galileo only in a letter to Ingoli in 1624 [Galilei 1890–1909, 6:509–61], in reasserting the independence of time of fall from the shot range, adds the remark that this can be proven geometrically; a similar remark cannot be found in the letter of February 1609. In Caverni's opinion, this appears consistent with the fact that the theory of indifferent coexistence of dissimilar motions was advanced clearly for the first time in the letter to Ingoli; therefore Caverni believes that the discovery of the new principle [of mixed motion] must be dated shortly before 1624. The letter to Ingoli links the argumentation on this subject to research on the phenomena of motion on the moving earth; thus Caverni assumes that his study of the Copernican system gave Galileo cause to develop his own principle. Trying to disprove physical objections against the motion of the earth led him to conceive that a body in free fall, which simultaneously takes part in the motion of the space surrounding it, executes each of these motions as if the others did not exist. He then understood that for this reason [the coexistence of motions] a body falling from a crow's nest comes to rest at the foot of the mast, regardless of whether the ship is in motion or at rest. And only now, Caverni thinks, could Galileo also understand the intrinsic necessity of what in 1609 he had only assumed about equal times of fall.

As probable as it is that his intensive occupation with Copernican theory bore fruit for Galileo's general theory of motion,[7] the idea that this fundamental idea emerged in the year 1624 must disturb anyone concerned with his scientific biography. This would mean that it had not occurred to him in the preceding thirty years — during which his primary considerations were the system of the earth's motion and the refutation of its opponents.

There is no need to demonstrate the absurdity of a chronology of Galilean discoveries which assumes that the date of first publication is also the date of discovery. Such a chronology would move the discovery of the laws of fall approximately to the age of 65, had not a letter preserved by chance informed us that the most important of these laws was known to Galileo in 1604.[8] A similar chance allows us to prove that Galileo interpreted the occurrences of motion on the moving earth according to the principle of the indifferent coexistence of the motions at least fourteen years before the letter to Ingoli. In annotations on the treatise by Lodovico delle Colombe contesting the rotation of the earth (ibid., 3:251–90), Galileo answers the author's physical arguments briefly, just as in the later letter to Ingoli and in the *Dialogue*. Colombe's treatise was doubtlessly written in the year 1610 in response to the *Nuncius sidereus*; Galileo's glosses cannot have been much later, for they contain in a few hints an outline of the critique he conducted in July 1611 in the letter to Gallanzoni [ibid., 11:141–55]. Unless one assumes, in accordance with Caverni's method, that by chance Colombe's foolish book first gave Galileo occasion to concern himself with Tycho Brahe's counter-evidence and in particular with the discussion of the phenomena of motion on moving ships, it appears more likely that the answers he gives the Peripatetic [delle Colombe] are the fruits of an essentially completed theory. Thus there appears to be no reason to presume that Galileo in 1609 had no inkling of what was at his disposal a year later, and that the law of equal times of fall could not have been discovered or proven at that time in the way we suggested above (see above, page 582).[9] In any case, neither historical nor psychological reasons can be found for the presumption that Galileo needed del Monte's propositions in order to derive the law.

*****

---

[7] I discussed this view in detail in 1884 (Wohlwill 1883–84).

[8] [Wohlwill refers to Galileo's letter to Paolo Sarpi, 16 October 1604 (Galilei 1890–1909, 10:115–16).]

[9] [Wohlwill refers to the argumentation exposed in section I of this paper.]

The discussion about the letter of 1609 [to the Medici prince] is followed in Caverni by information about a previously unpublished fragment which supposedly originated at approximately the same time [Caverni 1891–1900, 4:519]. The manuscript that is attributed to Galileo contains in the form of a series of chapter headings a plan for a work on artillery problems [Galilei 1890–1909, 8:424]. The fourth of the fourteen planned chapters was to answer the question of whether a ball will move in a straight line if it is not shot vertically; the fifth was to deal with the path described by the ball shot. Had the chapters been preserved along with these chapter headings, and had the problems and their solutions originated at the same time as of those of the theory of projectile motion about which Galileo reported to the mathematician Luca Valerio, then the *instantia exclusiva* to resolve the controversy discussed here would have been found within them. Indeed, there can be no doubt that Galileo had not moved beyond the teachings of Tartaglia by this time, if we accept as proven what Caverni reports about the responses to the fourth and fifth questions. "These two problems," he writes, "were solved by Galileo with Tartaglia's arguments" [Caverni 1891–1900, 4:519]. As regards the fourth, the observation put forward by Tartaglia in the second presupposition of the second book of his *Scientia nuova* [Tartaglia 1537, without pagination] is, indeed, rendered in the words of Simplicio's answer to the question: How long does it take after separation from the hand of the projector for the projected body to begin deviating downward? "I believe," he responded, "that it begins immediately. Since it has nothing supporting it, it is impossible that its own gravity has no effect" [Galilei 1890–1909, 7:221]. Thus Caverni does not report the answer of 1609 here, but rather infers from the *Dialogue* [of 1632] how Galileo would have answered in 1609. What is not taken into consideration, however, is that the question Salviati poses to Simplicio hardly coincides with the fourth chapter heading, and that the Simplicio of the *Dialogue* never represents thoughts and viewpoints new and peculiar to Galileo. There is therefore good reason to doubt that the answer given here is consistent with the entire contents of this fourth chapter. Immediately after the quote Caverni continues,

> In full agreement with these (Tartaglia's) principles, Galileo solves the fifth of the submitted questions by saying (*dicendo*) that the line described by the ball in its motion is in part such that one can consider it a straight line, and in part apparently curved. The curved part will be part of the circumference of a circle, as one reads in Tartaglia's book of the new science.

The above quote is taken verbatim from page 519 of the fourth volume of the Cavernian *History*, where it appears (of course) with neither quotation marks nor reference to a source. Nevertheless, the reference to Galileo's words is so unequivocal that the reader must be a stubborn skeptic indeed to doubt that fragments of these chapters were preserved which give the decisive solution of the controversy. However, the intrinsic improbability of such a solution makes this belief untenable!

Any uncertainty disappeared with the publication in 1898 of the eighth volume of the *Edizione nazionale* of Galileo's works, which includes all remaining and previously unpublished handwritten fragments on the theory of motion. The fragment discussed by Caverni is located on page 424. It contains, like Caverni's transcription, the fourteen chapter headings, but no further indications about the discourse; no second fragment is found that reveals information about the chapter's contents, let alone any that would confirm Caverni's explanation. Thus it is out of the question that a handwritten outline from which the solution of the fifth question could be inferred might exist among the manuscripts preserved at the Biblioteca Nazionale in Florence. Consequently, Caverni, here too, with the words he ascribes to Galileo is merely expressing his own opinion. In view of such argumentation, "no comment" is the best response.

III.

Caverni fails in his attempt to prove that, both during and after the period of his greatest research, Galileo, with respect to the theory of projectile motion, held fast the unrefined ideas of his predecessor and dispensed with any mechanical substantiation of them. The only remark actually preserved from this period is quite simply reconciled with the thesis that the theory of projectile motion of 1609 is in essence that of the "Fourth Day" of the *Discorsi*. In retrospect, it may be worth noting that even the original drawing included in the letter to the Medici prince contradicts the idea that in February 1609 Galileo still believed in the circular trajectory of projected bodies. This drawing is reproduced in relatively good quality in the Albèri edition (Galilei 1842–56, vol. 6, tav. II, fig. 1 [Galilei 1890–1909, 10:229]). For the purpose of this essay, Professor Favaro was kind enough to supply me with a facsimile he himself made in Florence. The following reproduction allows us to recognize distinctly enough that if Galileo had ever found Tartaglia's figure of the trajectory seductive, this was no longer the case by spring 1609 (see fig. 2).

To what extent proof to the contrary can be gathered from the later remarks in which Galileo has a stone falling to the rotating earth describing a circle remains to be investigated. Caverni drew attention to the fact that these statements, presented

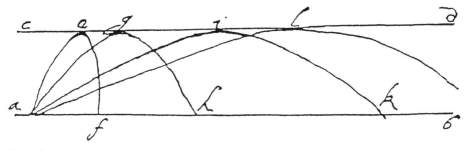

Figure 2.

in the *Dialogue* of 1632, concur with the letter written to Ingoli eight years earlier. "The stone in the crow's nest," is written here,

> moves with the same drive as the ship when it is sailing, and this drive does not disappear just because the person holding the stone opens his hand and lets it fall, rather it remains indelibly in the stone so that this, because of the drive, is able to follow the ship. Because of its own gravity, which is not impeded by the drive, the stone also moves downward combining drive and gravity to a single (and perhaps also circular) transversal motion in the direction in which the ship moves. (Galilei 1890–1909, 6:546)

Apparently the expression "*et forse anco circolare*" indicates at least the possibility of what the *Dialogue* discuss in more detail. This concurrence, however, should not be interpreted as if we have here two independent statements separated by a number of years, and thus proving persistence in the same opinion over a longer period. According to Galileo's testimony to the Inquisition, he began elaborating the *Dialogue* ten to twelve years before April 1633, i.e., between 1621 and 1623. The letter to Ingoli was begun and completed in fall 1624, and thus was written after a number of the subjects in question already had been developed for the *Dialogue*. Therefore it seems preeminently probable that the discussions contained in both writings about the supposed proof against the daily motion of the earth were created practically simultaneously. This is consistent with the fact that in just these sections of both writings there are similar formulations, some even identical to the letter (Strauss 1891, XLIV–XLV). Hence the words "perhaps also circular" also merely render, in a somewhat condensed form, what is at great length, in part expounded and in part outlined, in the *Dialogue*. Thus analysis of the decisive passage in the latter text will be sufficient also to explain the meaning of the remark in the letter to Ingoli.

A contradiction contained in the very first sentences of the discussion in the *Dialogue* (Galilei 1890–1909, 7:189–93) will surely disconcert the attentive reader. Salviati expounds clearly and precisely how to proceed in construction of a trajectory which describes a stone simultaneously falling and participating in the rotation of the earth. He calls special attention to the fact that for the purposes of this construction it is not enough to know that the motion of the falling body is accelerated and that the acceleration is continuous; one has also to know the proportion according to which the acceleration of the falling body takes place. In response to Sagredo's questioning, Salviati then explains that their mutual academic friend (Galileo) discovered this proportion, but that it would require too great a digression to go into details on this occasion. Consequently, the discussion of the relation of acceleration to other factors is put off until a later meeting (ibid., 7:190) and what was just said to be indispensable in order to carry on the construction, and thus solve the problem, is in fact not mentioned. One learns nothing about the law discovered by their academic friend, but in spite of this they take on the construction. Suddenly it appears sufficient to know that the increase of velocity is continuous and that the falling body accordingly will pass through all degrees of slowness lying between the resting position and any given velocity. It is demonstrated that as a result of this, the further the falling body progresses along its trajectory, the more the path of the falling stone must diverge from the circumference parallel to the earth's surface. This fact and the additional condition that the line of motion combined from the earth's rotation and the vertical fall must end at the center of the earth appear to determine the nature of this line. That this is not actually the case emerges directly from Salviati's introductory consideration; for the extent of the increasing deviation from that circular line which would be described by the stone in a position of rest is not determined, and cannot be determined with the given preconditions. Therefore, it is put forward as highly probable, but not as demonstrated, that the line in question, beginning at the top of the tower and ending at the center of the earth, is circular. All that is proven is that, if this were the case, three peculiar consequences would have to result: First, the actual motion of the falling body participating in the earth's rotation is, as a simple circular motion, a motion of exactly the same type as that of the body resting on the tower. Second, the falling body moves neither more nor less than if it had continuously remained on the tower, for the curves which it would have traversed in the latter case are exactly the same as those which it traverses in falling. From this the third miracle follows: the true and actual motion of the stone is not accelerated at all, but rather always steady and uniform, as it takes the same amount of time to traverse equal arcs of both peripheries [ibid. 7:192].

After deriving these propositions geometrically, Salviati reminds us that the value of this proof depends on the truth of his unproven supposition. This, apparently, is the meaning of his closing words:

> I do not wish to claim at this time that things apply in exactly this way for the motion of falling bodies (here one has the *forse circolare* of the letter to Ingoli). I do say, however, that if this is not exactly the line described by a falling body, it certainly approximates it extraordinarily closely. [Ibid., 7:193]

Sagredo, however, overhears the qualification and concludes with great satisfaction that as a consequence of the rotation of the earth, there are no rectilinear motions left in nature at all. Thus even the function which up until that point had been conceded to rectilinear motion, i.e., returning the separated parts back to the whole to which they belong, is attributed to rotation.

Considering that in this conclusion the most important requirement for the combination of the two kinds of motion, the acceleration due to gravity, is ignored intentionally and that therefore from the start any investigation as to whether the presumed circular shape of the resulting motion complies with the odd number rule is renounced,[10] the resulting solution can hardly be regarded as serious; in

---

[10] Strauss took the trouble to prove geometrically that this is not the case (Galilei 1891, 526).

reality, it constitutes the evasion of a solution. Galileo himself called it a pretense. When Pierre Carcavy informed Galileo in 1637 about the reservations of a friend against the mock construction of the *Dialogue*, he responded in a letter of 5 June 1637:

> that from the mixture of the rectilinear motion of the falling body with the uniform circular diurnal motion would result a semi-circle ending at the center of the earth it was said in jest, as is apparent from the fact that it is called a whim and an eccentric idea (*un capriccio e una bizzarria*), that is to say *jocularis quaedam audacia*. Therefore I wish to be granted a dispensation here, all the more since this — might I say — poetic fiction leads to those three unexpected consequences. [Ibid., 17:89]

To this explanation Galileo adds in his letter a further one: As far as the part of the described curve lying over the surface of the earth is concerned, he has no reservations in designating it as a parabolic line, since he asserts that the lines described by projected bodies are parabolae [ibid., 17:90].

According to Caverni, Galileo here dishonestly inserts post-factum as his own ideas what he had since learned from Cavalieri. This interpretation can be rejected with complete conviction. However, the question remains: If Galileo knew this when he concocted his circular line, why did the *Dialogue* not mention the real truth, which surely was not less interesting than his poetic fiction?

The answer to this question, too, almost certainly can be inferred from Galileo's own words. It is established that his *Dialogue* was written over a long period of time, and that Galileo changed the plan for the structure of his work several times over the course of this period. The passage discussed here proves that during its composition, Galileo intended to publish his main ideas on the theory of motion in a separate volume. The *Dialogue*, which was to be published first, was to include just enough about the theory of motion to make it understandable. Therefore the inquisitive Sagredo is referred to the "Treatise of Motion" for details about the law of spaces of fall. If we assume that the treatise, — mentioned here [e.g., ibid., 7:190] as if it were a work already completed —, was identical to or at least, with respect to the main content, congruous with the Latin sections of the *Discorsi* of 1638, then already the well-known introduction [ibid., 8:190] to this section makes it clear that, according to the original plans, the *Dialogue* was to exclude discussions about either the law of fall or about the projectile trajectory for in this introduction the odd number rule and the parabolic shape of the trajectory are strongly emphasized among Galileo's additional new insights as truths "of which nobody previously knew." Had Galileo the intention to publish a second work with such an introduction, he would hardly want to expose in an earlier publication ideas that he later prized as totally new. Therefore he had to do without a clear and correct execution of the construction discussed here.

It seems logical to object against this interpretation that, despite the initial renunciation, the *Dialogue* contains in later passages not only a derivation of the laws of fall, but also detailed considerations about numerous other problems of the theory of motion. A closer examination, however, reveals that these passages were not supplements planned from the beginning, but are rather insertions which clearly contradict the original plan of the work. After thoroughly examining in the "Second Day" of the *Dialogue* the objections to the daily rotation of the earth and preparing the transition to the yearly motion, the dialogists return again to quite a lengthy discussion of the subjects in hand in relation with an extremely sharp critique of the *Disquisitiones mathematicae* by Christoph Scheiner, published as early as 1615 [ibid., 7:244–98]. Doubtlessly it was neither this wretched writing nor the wish not to leave it unchallenged which prompted Galileo to reiterate the arguments already established, but rather Scheiner's violent attacks in the *Rosa Ursina*, which had just appeared in 1631. The apparent purpose of the insertion is to expose his embittered opponent as an ignoramus. Scheiner himself expressed the suspicion that Galileo circumvented the censors to sneak criticism directed against him into the final proofs of the manuscript. Such an insertion long after completion of the edition of the "Second Day" appears to be the only plausible explanation for the multiple repetitions and the contradictions in the final third of this "day."

Apart from the other weaknesses of the *Disquisitiones*, the calculation of the time of fall of a stone traveling from the moon to the earth offered welcome material to discredit Scheiner. The point here was to contrast his completely misguided calculation with the correct one [ibid., 7:250–51].[11] To this end Galileo included a derivation of the laws of free fall [ibid., 7:253–56] and related parts of his theory of motion. He did not deem it necessary, or simply forgot, to reconcile the late derivation with the previous statement that he knew these laws, but did not want to insert them. Consequently, there is no reason for us to consider his subsequent insertion of the laws of fall in our attempt to understand why the previous passage [ibid., 7:191] contains a geometric "joke" instead of the correct derivation of the shape of the trajectory.

Serious people, including Mr. Caverni of course, who find it unbearable to see problems in science taken so lightly, should keep the following in mind. From a superior perspective one may be annoyed with the great man for frolicking through witty games because he does not wish to speak according to his conscience. Those, however, who study history in order to understand people and processes, at least must recognize that joy of wittiness in word and meaning is a significant component of Galileo's cast of mind, and that the remarks about bodies falling in a circular path thus reflected his nature. Similar intellectual games, to be read also as "delightful," but of no value for science, are his hypotheses about the origin of the different speeds of the planets and about the emergence of the acceleration of fall through the interaction of unchanging gravity with the gradually decreasing *vis*

---

[11] The value of this calculation is immaterial here. On this, see the notes by E. Strauss in Galilei 1891, 533–34.

impressa. For the latter case, and possibly also for the two previous ones, writings of earlier years prove with certainty that Galileo here puts forward lines of argumentation from times long past that still seem interesting to him although, having acquired a better insight, he no longer considers them correct. Thus in his letter to Carcavy [ibid., 17:88–93], he still seems to consider his false assumption sufficiently justified by the completely surprising consequences it produces. Anyone wishing to pursue the extent to which witty ideas were tempting — and possibly even dangerous — for Galileo should study thoroughly his letter to the Grand Duchess Christine of Lorraine [ibid., 5:307–48].

IV.

The preceding elaborations serve only to make plausible that Galileo, as in the *Dialogue* of 1632, while speculating about the path of the falling bodies, could still be convinced of the parabolic shape of the trajectory. Anyone willing to concede that the episode of the *Dialogue* might have been created in this manner must thus also concede that no conclusions can be drawn about the contents of the 1609 theory of projectile motion from the arguments exposed there.

Such considerations were alien to the first enthusiastic readers of the *Dialogue*; historical reflections about the creation of the book, the relationship of the author to his work, his motivations for discussing certain details extensively while holding silence about others, in short, all that is of particular interest to anyone who wants to understand the book historically. Such matters were insignificant to those bombarded by new truths and new conclusions about the half-comprehended or misunderstood; at the time, even the peculiar construction of the "Second Day" was doubtlessly understood and admired as a serious teaching of the master. It should come as no surprise, however, that scholars like Carcavy and his friend, who were dissatisfied with Galileo's argumentation, expressed their reservations without taking any notice of what he, in a sense, "secretly sneaked into" this passage.

We have already mentioned the passage of the *Specchio ustorio*, — first pointed out by Caverni —, making it highly probable that Cavalieri also interpreted the elaborations of the *Dialogue* as Galileo's true opinion about the nature of projectile trajectory. In order to judge the extent to which this highly gifted pupil's view allows additional conclusions about Galileo's knowledge in the year 1632, one first must subject to scrutiny how Galileo responded to Cavalieri's announcement about the publication of his *Specchio*, as well as Cavalieri's reaction to this response.

On 11 September 1632 Galileo answered Cavalieri's previously mentioned communication of 31 August [ibid., 14:377–78] indirectly in a letter to their mutual friend Cesare Marsigli, who like Cavalieri lived in Bologna (Galilei 1842–56, 7:5–6 [Galilei 1890–1909, 14:386–87]).

I have letters from Friar Bonaventura (Cavalieri) informing me that he had recently printed a treatise about the burning mirror in which, he says, he inserted at a given occasion the proposition and the demonstration that the path of a projected body is a parabolic line. I cannot conceal from you, my venerable sir, that his report pleased me little; for I see how from over forty years of study, a good part of which I shared with the Father in complete confidence, my first fruits are being taken from me and the blossom of the fame which I had promised myself for my endeavors is to be plucked. For in truth the first thing which caused me to speculate on motion was the effort to find this line. Once it is found, the proof for it is not very difficult; I however, who proved it, know how much trouble I had to find this conclusion. Had Friar Bonaventura informed me of his intention before publishing (as perhaps courtesy would have required), I would have pleaded with him to allow me to publish my book first, and then he could have added as many discoveries as he liked. I will wait and see what he produces; but it must be something great indeed to appease my resentment as well as that of all of my friends who have heard of it and now add insult to injury by accusing me of being too trusting. My destiny is to fight for what is mine and yet to lose.

Marsigli and Cavalieri answered this letter on the same day (21 September 1632) (Galilei 1842–56, 9:290–94 [Galilei 1890–1909, 14:394–96]). The latter writes:

The distress you felt, of which the high illustrious Sir Cesare Marsigli informed me, due to my touching upon the parabolic line described by projected bodies in my *Specchio ustorio*, was certainly not as great as my distress when I heard that you felt offended by my omission, which was due to my excessive reverence more than any other reason. What I said of motion, I said as a pupil of yours and Father Don Benedettos, and as such I declare myself, as you can see from the enclosed sheets, having learned from you, I can say, what little I know. You might surely say that I should have expressed more distinctly that the idea of the parabolic line originated from you, honorable sir. But I wish you to know that my doubt as to whether I was in total agreement with your conclusion kept me from attributing to you with clear words something you might have had to reject as not being your opinion. This fear was the reason for my resorting to the general words on page 152, where I name again Father Don Benedetto, not because I wanted to express that the following originates in part from him, but because he too taught me some of those things, for I saw him performing experiments on them with other pupils from whom I also heard the very same conclusion. In short, it seemed to me that the conclusion is so widely known as you are as its author, that it would be inconceivable that I could claim it as my own property.[12] As I had the courtesy to write to others, like Sir Muzio Oddi,

---

[12] Professor Favaro was kind enough to collate the original of the letter in Florence at my

before I published any subjects we had discussed, much more I would have done the same with you (had I thought that you attach importance to it), since I hold you in such high esteem and both honor and love you for your achievements and for the countless favors you did me. If, whilst teaching me, you had intimated to me that I should not make certain ideas known, I would by no means have done so. However, by explaining them to others and declaring them as yours I believed to do what a good student should, while proving to be someone who at least understands, if not imitates your admirable efforts in the unveiling of nature's secrets.

I must add that I truly thought you had written about this somewhere, since I have not been lucky enough to see all of your works. My opinion was supported by the fact that this theory already is known so widely and for so long, since Oddi had told me ten years ago that you performed experiments about it together with Sir Guidobaldo del Monte. I neglected to write to you earlier also because I truly believed that you would not mind, but rather would be satisfied that one of your pupils at such an opportune occasion showed himself to be a follower of your doctrine while acknowledging that he had learned it from you.

Should you deem this a fault, in spite of what I say in my defense, it was certainly not a malicious one. Think about how I can make it up to you, for I am fully prepared to do so. Here in Bologna I circulated only a few copies, and will not distribute any more until the matter is settled to your satisfaction, if this is possible. Therefore I will either postpone further distribution until you have published your book about motion or can publish a predated edition; reprint the two sheets omitting all you view as disadvantageous to you; if you believe my work is in agreement with yours, add in the margin next to line 22 of page 164, the words: "Conclusion of Sir Galileo Galilei," or finally, burn all copies and thus destroy, as far as possible, the root of the displeasure which I have caused my master Galileo, such that he could despair like Caesar: *Tu quoque Brute, fili!* On the contrary, I always have regarded it as my greatest happiness to have known him, to have been able to honor him and serve him. ....

Tell me freely which of the options listed would grant you the most satisfaction, for I am most prepared to have it carried out immediately.

These letters presented no obstacle for Caverni's interpretation of the context of events [Caverni 1891–1900, 4:526–30]. For him, the entirety of the documents raised an exclusive claim by Cavalieri to the discovery of the parabolic shape of the

---

request. This comparison reveals that in addition to several less important deviations, Albèri's transcription left out an entire line here and through this changed the meaning, not insignificantly. The correct text reads: *da'quali pure ho sentito l'istessa conclusione **parendomi in somma talmente divulgata la conclusione** e ch'ella n'era l'autore, che non potesse cadere etc.* In Albèri the words printed in bold type are missing [Galilei 1842–56, 9:292].

projectile trajectory. He freely fabricates how the reading of the relevant passage of the *Dialogue* distressed Cavalieri and led him after short contemplation to the correct solution both for the problem put forward in the *Dialogue* as well as for the mathematically identical problem of the projectile trajectory. Thus in the *Specchio ustorio* Galileo's tenet is not adopted, but rather disproved. Consequently, Cavalieri's argumentation, according to Caverni's fiction, is focused on just one concern: how Galileo, the strong-willed, passionate tyrant, might react to the fact that he, the pupil, dared to replace his semi-circle with a parabola. Cavalieri is then extremely surprised when, instead of the expected rebuke[13] for this bold deviation, the news arrives that Galileo himself claims the discovery of the parabolic form.

With the brutal pronouncement: "so many propositions, so many lies!" [ibid., 4:529] Caverni disposes of the melancholy, bitter words with which Galileo claims priority over his pupil. He is not even bothered by Cavalieri's willingness to grant Galileo's claims unconditional recognition by this time, in apparent contradiction to his previous comments. Cavalieri is lying, too, but almost unknowingly. Under the spell of the demoniacal or magical influence which Galileo exercises over him, as Caverni interprets his waiver, "the good man" believes he understands and knows what Galileo wants to make him believe. Will-less, the legitimate owner let himself be persuaded to carry home to the thief what this exacts; will-less, he professes to be convinced that he himself was the burglar [ibid., 4:530].

Even disregarding the fact that here, too, the novelist in Caverni takes over from the historian, passing off a combination of possibilities as history — even considered as a simple attempt at explanation, his portrayal of events is absolutely unfounded. His interpretation of known facts and statements would only be permissible if what he promised to prove to us, but failed to demonstrate in his arguments, had been an established truth.

Even with as little authority as one might grant such interpretations, it is indisputable that some points remain open even after Galileo's letter and Cavalieri's response. The only certainty which emerges from Cavalieri's remarks is the wish for reconciliation with the insulted master. To this end, he exhausts himself in listing the reasons for apparently not recognizing Galileo's right of discovery and not requesting his assent for publication, and in proposals for atonement of every possible transgression, culminating in the complete surrender of his own work. These attempts at justification, however, are not entirely convincing, and Cavalieri's boundless willingness to sacrifice for the sake of reconciliation arouses the suspicion that the given reasons might respectfully hide other ones which have more to do with the truth. Cavalieri asserts having been uncertain as to whether Galileo would recognize as his own his formulation of the proposition of the parabola; but if, as he admits unconditionally, he was certain that Galileo recog-

---

[13] He feared, says Caverni, "to become an object of contempt and of anger for Galileo, as Kepler became for similar reasons" [Caverni 1891–1900, 4:528]. It is known that Galileo never agreed with Kepler's discovery of the elliptical orbits of the planets; but it cannot be proven that he reproached the discoverer for giving up the circular shape, as Caverni suggests here.

nized the parabolic shape of the trajectory, where could his uncertainty lie? His thesis reads:

> Excluding the resistance of the medium, ponderable bodies driven from a projector in any direction other than the perpendicular to the horizon describe from the moment of separation from the projector a curved line which is imperceptibly different from a parabola. (Cavalieri 1632, 164–65)

With the caveat of his final words, Cavalieri referred to the fact that because the vertical component is always perpendicular to the earth's curved surface, over short stretches the trajectory deviates negligibly from the parabola. He could hardly doubt that Galileo would agree with his words *insensibilmente differente* understood in this sense. Similarly, in the remaining wording of the thesis nothing expressed could have been rejected by anyone who believed the trajectory of projected bodies to be parabolic. If uncertainty about this truly had led Cavalieri to doubt whether he could name Galileo as the discoverer, he easily could have erased this doubt through the correspondence he neglected to initiate! Since he did not wish to refute, on the contrary, he expressly recognized that the essential contents of the thesis belonged to Galileo, it is incomprehensible that he left Galileo's name unmentioned. Cavalieri believed that an adequate substitute for expressly naming Galileo in connection with the parabolic trajectory was the general remark in which he frankly acknowledged owing "in part" to Galileo his insight into the problems of the theory of motion. Yet, a reader of the *Specchio ustorio* can understand this declaration, at best, in the sense that also the proposition concerning the trajectory may belong to the part of the expounded doctrine for which Cavalieri thanks Galileo, although this is never told explicitly. Furthermore, Cavalieri's subsequent remarks about the general propagation of the theory of the parabola as one originating from Galileo are hardly so definite that they not only would make naming Galileo superfluous but also unthinkable that Cavalieri could have tried to pass himself off as the discoverer. In any case, the theory's presumed wide distribution consisted only of propagation by word of mouth; Cavalieri hardly could doubt, or in any case easily could have found out, that the *Specchio ustorio* was the first published book in which the parabolic shape of the projectile trajectory was openly taught. If the fact that it was discovered by Galileo appeared so apparent that a preliminary inquiry was unnecessary for publication, this is even less of a reason not to attribute the discovery to Galileo unequivocally.

Summarizing these objections, one can say that Cavalieri's defense is insufficient to convince of his total sincerity those who, in both the *Specchio ustorio* and in the August letter, do not find anything more than the explicit attribution to Galileo of the preparatory steps for the discovery of the parabolic shape, but not the discovery itself. In his response to Galileo's letter of 11 September, Cavalieri not only definitively recognized Galileo's priority, but asserted equally decisively that a denial of such a relationship is impossible, neither expressed nor intended in the *Specchio ustorio*.

No matter how the apparent contradiction of remarks before and after 11 September 1632 is reconciled, the *eccesso di reverenza* which Cavalieri himself placed at the head of his defense must have played a role. There is no doubt that for Cavalieri, even if he had discovered the parabolic form himself and not known of Galileo, the words, "I found it!" would have been decisive proof of Galileo's priority. It is even possible that his need to placate the honored master completely was strong enough to induce him to fabricate the story of the theory's general propagation. However, such a fabrication must not necessarily have occurred. Even if Cavalieri in truth had heard talk about Galileo's discovery already in Pisa from Castelli's pupils, and later from Muzio Oddi and many others, the construction in the *Dialogue* renouncing the parabolic form could have made plausible for him that for some reason the discoverer had exchanged his correct insight for that erroneous tenet presented in the *Dialogue* as a highly meaningful truth. In such a case, it is understandable that the brilliant mathematician did not forgo publishing in his book, in contrast to the deceiving argumentations of the *Dialogue*, the simple proof for the parabolic shape based on Galileo's research. No less understandable is that he failed to mention in this particular passage about the projectile trajectory the name of the man who actually discovered the parabolic shape but now dealt with a correlated question in a way as if he would reject his discovery.

This would also explain why Cavalieri speaks of the parabola in the first letter without letting on that it was Galileo's theory he is defending; as a reader of the *Dialogue* he had reason to doubt whether it still was Galileo's theory.

From this point of view, even the otherwise surprising remarks in the second letter are easy to explain, at least in part. A disciple less full of piety and veneration toward Galileo might have ignored the contradiction of the *Dialogue* and been able to attribute to him the discovery of the parabolic shape of the trajectory. Cavalieri, however, could not overcome his doubt as to whether somebody who conceived of the path of a falling body on the moving earth as circular could accept the proposition of the *Specchio ustorio*. This doubt kept him from attributing to Galileo what this had taught in an earlier period.

Even the otherwise almost incomprehensible comment that he believed that Galileo did not trouble himself any more with his theory of the parabola appears justified, if one views the construction of the *Dialogue* as the point of departure of Cavalieri's doubt. Three years later, another letter from Galileo occasioned Cavalieri to justify once more why he neglected to consult Galileo before publishing the *Specchio*. This time, he limited himself to explaining that at the time he believed that Galileo attached little importance to his discovery (Campori 1881, 442).[14]

The *eccesso di riverenza* to which all of Cavalieri's letters to Galileo bear witness explains why neither here nor in his earlier defense he adds that the construction of the *Dialogue* gave him the most urgent cause to take this view. These letters contain not a single word which could be interpreted as critical; it would be

---

[14] [Bonaventura Cavalieri to Galileo Galilei, 24 June 1635, in Galilei 1890–1909, 16:283–84.]

impossible to mention the contradiction between the construction of the *Dialogue* and the correct composition of the trajectory without at least implying criticism.

Instead of this interpretation that limits itself to suppose only the indispensable, consider as more probable that Cavalieri subsequently simulated a knowledge of previous discovery for Galileo's sake and invented fictional facts about this discovery to increase its plausibility. In so doing we would only obtain a condition under which Cavalieri can be seen as an independent discoverer, not at all a sufficient reason to deny Galileo's discovery. For the idea that Galileo could not have discovered and in some way taught something that remained unbeknownst to Cavalieri is in no way substantiated by what we know about the relationship between the two men. If Cavalieri calls himself Galileo's pupil, he certainly meant this in a completely different sense than, for instance, Castelli, Aproino, or Antonini. His relationship to Galileo began nine years after Galileo left Padua and stopped teaching in the usual sense. In 1619 Cardinal Borromeo recommended him to Galileo, calling him a young man of great promise who wanted to dedicate himself to the study of mathematics in Pisa where he then became a pupil of Father Castelli. It can not be established with certainty to what extent Castelli believed himself authorized to initiate Cavalieri into the doctrines which Galileo still reserved for future publication. What Cavalieri's letter of 21 September 1632 relates on this matter suggests more of a coincidental communication than regular teaching. Just as little is known about the extent to which the young mathematician might have enjoyed direct instruction from the man he revered as master during Galileo's occasional presence in Pisa or in Cavalieri's visits to Florence. Galileo speaks unequivocally of such instruction in his letter to Marsigli of 11 September 1632; he does not say so expressly, but he appears certain that Cavalieri, too, has his direct information to thank for the explanation of the trajectory. However, Cavalieri does not confirm this assumption, but rather refutes it with his silence. It was not from Galileo or from Castelli, but from Castelli's pupils and from others that he heard the thesis. This failure to accede to Galileo's allusion is all the more remarkable if one assumes that Cavalieri thoroughly feigns his previous knowledge in order not to contradict Galileo. Not even in this fabrication, where any more or less compliance could make no difference, was he willing to admit that he had Galileo himself to thank for the knowledge of this special theory. In his own opinion, it thus was reconcilable with his Brutus relationship that Galileo did not instruct him personally about a theory to which he attached such great importance.

The *Specchio ustorio* also fails to offer any evidence opposing this interpretation. Though Cavalieri especially emphasized here that he spoke of the problems of the theory of motion as Galileo's pupil, aside from the parabolic shape of the trajectory, his book contains nothing relating to the theory of motion which could not be and indeed was essentially inferred from the published *Dialogue*. In short, there is nothing which would suggest that Cavalieri was preferentially initiated in the Galilean theory.

If he was not, or if it cannot be proven that he was, then his supposed ignorance can not yield any further conclusions about the contents of the older Galilean theory of projectile motion, let alone be suitable as contradictory evidence when Galileo says, "I found it!"

V.

Caverni's novel has a sequel. In the first volume of his *History*, he portrayed Galileo as an unscrupulous tyrant, who would stoop even to fratricide in order to consolidate his dominion, if the brother became a rival and the stolen goods could only be enjoyed in security after his elimination [Caverni 1891–1900, 1:127–28, 135–36]. In our case, how did the rapacious tyrant arrange the theft? How did he deal with the brother, who left it up to him to choose between three ways of securing himself for all time against his claims? Caverni tells of this as well; with disgust the reader sees realized the worst expectations raised by the preceding portrayal. Galileo gave his loyal pupil to understand that of the various means which he proposed for atonement, the destruction by fire of the irksome book would please him most; and Cavalieri does not hesitate: he carries out the work of self-abnegation so completely that today it is difficult to find any copy of his book. As early as 1650 this had become so rare that Daviso, a pupil of the great mathematician, recognized as necessary the publication of a new edition [ibid., 4:533].

The factual basis of this story is nothing more than the publication of a second edition of *Specchio ustorio* in 1650; the remainder is not only unfounded by any document accessible today, but indeed in sharpest contradiction to what one knows about the actual course of events.

Galileo's lively agitation was appeased by Cavalieri's explanations; his letter of 16 October 1632 to Cesare Marsigli documents this. "From the venerable Father Bonaventura," he writes,

> I have received a long letter full of apologies. These were truly unnecessary on his part, for I never doubted his best intentions. Rather, I complained about my own misfortune that caused me grief through something done involuntarily and unintentionally. I can not answer him for now since I am extraordinarily busy, so I ask you, Sir, only to tell him that I do not wish the Father to change anything at all in the published book. Rather, I would like to thank him for the honorable reference. (Galilei 1842–56, 7:14 [Galilei 1890–1909, 14:411])

That Cavalieri received these words as any unbiased reader would today proceeds unequivocally from his response of 7 December, in which he writes: "That you now are satisfied, having seen how I bring up this doctrine, pleases me beyond all measure" (Galilei 1842–56, 9:317 [Galilei 1890–1909, 14:437]).

In the meantime, Cavalieri had sent his book to Florence. In the expectation that Galileo would receive it, he asked him for his opinion, particularly in reference to the thoughts exposed in it about the burning mirror of Archimedes, the main reason he had published the book [ibid., 14:438]. Galileo's answer is directed again to Cesare Marsigli. Here one may expect to hear the fratricide speaking at last; but what Galileo writes on 31 December 1632 to his friend in Bologna, in the middle of the upsetting negotiations about the injunctions of the Roman Inquisition, is as full of warm benevolence and of such ungrudging joy at the successes of his younger colleague as these feelings ever have been expressed by a great researcher. "With you, illustrious Sir," Galileo writes Marsigli,

> and not with the author of the *Specchio ustorio* I wish to delight in the wonderful discovery, because I am certain that he who got to the root of it feels such great joy about it that it cannot bear increasing. I must also delight with you when I see the happy progress and the superhuman success of this genial brain, whom I once recommended to you and was favored by you. And if my judgment still has any value for the gentlemen there, I would advise them to grant to this intellect free run through the broad field of the mathematical sciences, along whatever way his genius takes him. This way will be the best of all and incomparably preferable to the calculation of ephemerides or the tabulation of horoscopes. (Galilei 1842–56, 7:14 [Galilei 1890–1909, 14:444])

After these quotes no more words are necessary to demonstrate that if some kind of secretive machinations obstructed the dissemination of the *Specchio ustorio* in its first edition, or even caused a considerable part of this printing run to be destroyed, Galileo's opinion and will can not have played a part. Moreover, Caverni still owes a proof for what was to be explained by Galileo's intervention: The 1632 edition of the *Specchio ustorio* is to all appearances not any rarer than any other books of similar importance from the same era; it is cited almost exclusively in historical works and is quite generally found, for instance, in larger German libraries. The edition of 1650 does not contain the faintest suggestion that unusual conditions made the appearance of a second printing necessary. One may therefore assume that one of the usual reasons for printing second editions of other books some eighteen years after the first prompted Father Urbano Daviso to re-issue the work of his teacher and fellow monk.

Far from offering support for Caverni's unfounded allegations, there is better reason to presume that Galileo himself, through his extremely warm recommendations, contributed not insignificantly to the dissemination of the *Specchio ustorio* and therefore to the depletion of the first edition. He was not content merely with recommending the book to his friends like Father Fulgenzio Micanzio, and with advising esteemed personalities like Cardinal Capponi to take up personal relations with the author because of his excellent work (Campori 1881, 447, 490); even in public, in his immortal masterpiece [the *Discorsi*], he mentions Father Bonaventura Cavalieri and his work about the concave mirror, "which he read with admiration" (Galilei 1890–1909, 8:86–7). Thus he himself uses expressions reserved only for the greatest of researchers to refer his readers to the book which preceded him by six years in publishing the main proposition of his theory of projectile motion.

In the light of these facts, the story told by Caverni about the *Specchio ustorio* appears to be preposterous and untrue. As such can also be designated what Caverni reports as further proof of Cavalieri's priority, about the emergence of Galileo's theory of projectile motion in the years 1636 and 1637. A series of letters from the year 1637 shows with certainty that Galileo was occupied with the theory of projected bodies during that period, just before the publication of the *Discorsi*. He was, in fact, so distracted that his Dutch publisher had to admonish him to send this part of the manuscript after the previous parts already had been printed. "The documents," says Caverni, referring here to the letters just mentioned, "attest that while the first Latin propositions about accelerated motion go back as far as 1604, those concerning projectiles were written for the most part in 1636 and 1637" [Caverni 1891–1900, 4:531–32]. Here the strict critic confuses the contents of the "documents" with the object of his argumentation: The latter concerns the Latin sections of the "Fourth Day" of the *Discorsi* of 1638; the "documents" speak of the whole content of this "Fourth Day" in general, therefore no decisive evidence can be derived from them that these sections written in Latin arose in the years 1636 and 1637. This can also not be confirmed by the established fact that Galileo in 1637 was still occupied with working out the fourth and final part of his work. The "Fourth Day" contains these Latin passages as a smaller section, but most of it was written in the form of an Italian-language dialog explaining and elaborating on the Latin propositions. These letters hardly imply the doubtful thesis that Galileo worked up to publication on the minor, Latin section; it is much more likely that he was occupied with the subsequent, larger part. They speak generally of *"proietti"* and thus allow us only to recognize that the 73-year old man did what young people are also wont to do in such a case: he worked on the final section of his manuscript up to the last moment before publication.

VI.

Toward the end of his long indictment, Caverni directs the following challenge to Galileo:

> Since you, Mr. Galileo, saw fit to play games with the dignity of your interlocutors and to treat jokingly in your masterpiece a question of such great importance, tell us, in which of the other treatises, letters or notes from your forty years of study of projected bodies you wrote seriously about their parabolic trajectories ... and should you not be in a position to produce a credible document from before September 1632, we can not acquit you of

the charge of having stolen from Cavalieri the discovery of which you were so desirous, in manner unworthy of a philosopher or of any man of honor. [Caverni 1891–1900, 4:531]

As Caverni issued this bold challenge to the grave of the dead hero, he held in his own hands not a single, but rather a number of such documents which would have refuted his charge decisively and forced him to destroy the greater part of his book, had he been capable of escaping from the dense web of self-deception in which he was entangled.

Through precisely this fourth volume of the Cavernian work which treats the controversy discussed here, it became known to the broader public for the first time that the manuscripts of the Biblioteca Nazionale in Florence include remains of an older treatment of the theory of motion intermingled with drafts for the text of the printed *Discorsi* [ibid., 4:338–42]. These fragments written in Galileo's hand apparently belong to the Paduan period and will doubtlessly shed new light on the history of his greatest discoveries. Caverni believed it was possible to reconstruct through these manuscripts parts of an earlier draft of a treatise concerning the "new science" which he dates to the years 1602 and 1604–10. According to his explanation, he used arguments of a formal and a material nature to assist in reconstruction. Among the latter he emphasizes the differences in the handwriting in different periods of his life. "It is known to all," he says,

> how the writing hand is affected over the years in the same way as the movement of all other members; anyone can experience this in himself, if he compares what he wrote at thirty with what he wrote at fifty. The difference would be without doubt more noticeable were the comparison to be drawn between the writing of early youth and that of old age. However, we restricted our range to the twenty years lying between these manuscripts. These were put aside in 1610 and picked up again systematically in 1630, as it will emerge from the most reliable documents when we will deal with this. The theorems proven between 1602 and 1610 are written in an ink lighter in color and with lighter, rounder shapes. In the year 1630 Galileo's sight, which was so weakened that after a few years it was to fail completely, needed clearer signs; therefore the ink is black, the strokes are thick, and the forms square. (Ibid., 4:341)

It must be left to the Italian scholars, the seasoned experts of the manuscripts preserved in Florence, to express themselves about the reliability of these observations and about their utilization for far-reaching conclusions for the history of science. At this juncture only what is relevant for the Galilean theory of projectile motion is to be emphasized. Here, too, Caverni uses extensively Galileo's unprinted and previously unknown notes.[15] These refer to most of the questions treated in the

---

[15] [Wohlwill refers to the detailed examination of Galilean manuscripts pertaining to the theory of motion in vol. IV, chapters VI–IX, of Caverni's *History*.]

Latin texts of the "Fourth Day" of the *Discorsi*, but deviate somewhat from the sections of the printed work with related content both in wording and in the details of argumentation. Yet all of them unquestionably presume the parabolic shape of the trajectory as a given fact. Therefore it was of decisive importance in championing Cavalieri's discovery to remove all doubt that, according to the "material" arguments discovered by Caverni, not a single one of these fragments can be considered as having been written during the Paduan period or at any time before 1632. Caverni negated this eventuality through silence; from the structure of his book, in which the emergence of the scientific theory of projectile motion is linked to the bereavement of Cavalieri, we infer that he considers all the fragments in which the parabolic shape of the trajectory is presumed as written after the theft. By contrast, every single fragment in which the question of the shape of the trajectory is raised, but not answered, is attributed to the period before 1609. Even ignoring Caverni's arbitrary interpolation of a non-existent answer, the resulting chronological separation of the fragments also could be interpreted to support the hypothesis argued here. It appears all the more strange that no attempt whatsoever was made to apply the above material criteria to justify more closely the decisive distinction. There is no pronouncement that the written characters from the two periods exhibit the differences portrayed earlier and thus reveal the different ages of the author. Strictly speaking, the only reason to believe in such dissimilarities is the faith that no reasonable historian would evaluate manuscripts of related content arbitrarily, or merely to suit his purposes, by attributing one to the best years of manhood and the other to the beginning of old age.

Publication of the fragments of the theory of motion in the eighth volume of the *Edizione nazionale* of Galileo's works and the corresponding explanations of the editor [Galilei 1890–1909, 8:33–8, 363–448] have informed us definitively that Caverni used the decisive testimony of the manuscripts against Galileo in the most arbitrary way as evidence supporting his theft hypothesis. Through the publication of the fragments and the explanations by Antonio Favaro the following have been established:

1. The handwriting on those fragments referring to the theory of projectile motion which were found scattered among the sheets of the second part of the fifth section of the Galilean manuscripts indicates that they originate primarily from his youth, but in part from a later period.

2. On several sheets, some of the notes exhibit the youthful handwriting, while other parts in the main text and/or in marginal comments can be recognized through the handwriting as being complements, corrections, or other supplements of a later period.

3. According to their contents, all of the fragments relating to the theory of projectile motion, whether the handwriting attributes them to early or later periods, presuppose in an absolutely unequivocal manner the parabolic shape of the trajectory as an established fact. An exception is the above-mentioned list of questions or chapter headings on the theory of projectile motion, in which the

question about the shape of the trajectory is formulated, but no further information is given about the contents of the answer.

The publication illustrates the parallel occurrence of the different handwriting in a very interesting way through the supplement of three facsimiles [ibid., 8:428–33]. These also demonstrate how Galileo continued work on the same problems over three decades, and bring to mind how in later years Galileo himself criticized the creation of his youth, how he changed formulations, rewrote and corrected proofs, or treated in detail a proposition only jotted down in a previous period. The word "*scritta*" scrawled in the trembling hand of an old man, — it seems to us —, points to an even later period, informing us that the text was incorporated into the treatise on projectile motion of the *Discorsi*.

To the extent possible for a facsimile, the three published in the eighth volume of the *Edizione nazionale* also clarify the question discussed here. The opening sentence of each of the three, written with a relatively young hand, already speak of the parabolic form. What is obvious even to the layman here is elevated to certainty through the explanation, entered above under "3.", given by the editor of the *Edizione nazionale*, the profoundest expert of Galileo's handwriting [ibid., 8:34, note 1]. In the following, it seems to me opportune to complement this general declaration, which is anyway sufficient in itself, through more exact information about the contents of such fragments which with certainty are to be ascribed to the period of youthful research. Professor Favaro again kindly gave me each of the clarifications I requested.

Among the fragments becoming known only now, my attention was captured primarily by a fragment written in Italian (Galilei 1890–1909, 8:373–74), beginning with the words:

> I assume (and perhaps will be able to prove) that the falling ponderable body by nature continuously increases its velocity in proportion to the increases of its distance from the starting point of the motion. ... The principle,

the author continues,

> appears to me very natural and in accordance with all experience we see in instruments and machines which operate by percussion: the percutient causes a greater effect the greater the height is from which it falls. Under presumption of this principle I will prove the rest.

Subsequently, as a necessary consequence of the presupposed principle it is deduced in a very peculiar way that the distances traveled by the falling body in equal times relate to each other like the odd numbers *ab unitate*. To this Galilei adds, "this is in accordance with what I have always said and observed in experiments; and thus all truths are in agreement with one another." Taking the preceding for granted, it is also demonstrated that "the velocity in violent motion decreases in the same ratio as it increases in natural motion along the same straight line."

As is apparent, the fragment essentially agrees with the letter Gaileo wrote to

Paolo Sarpi on 16 October 1604, reporting about this same (later discarded) fundamental idea and its application (Favaro 1883, 2: 226–27 [Galilei 1890–1909, 10:115–16]). This letter appears as an abbreviated rendering of what most probably was written down for the first time in the fragment. A deviation of the wording of the fragment from that of the letter is worthy of being pointed out. In the former the discovery of the law of fall and the discovery of the explanation based on the proportionality of velocities and distances traveled appear separated chronologically in a way which could not be inferred from the letter to Sarpi. The wording of the latter appeared quite reconcilable with the assumption that the discovery of the law of fall, too, occurred in the year 1604. In contrast, the statement in the fragment, in which Galileo designates the odd number rule as something *"che ho sempre detto e con esperienze osservato,"* forces us to consider the possibility that this truth had been recognized much earlier.

The fragment itself, being doubtless an autograph, can be regarded confidently as having been written shortly before the letter to Sarpi; therefore, it belongs in any case to the second half of 1604. If necessary, this established fact can be taken as an indication for the age of the remaining fragments.

The assumption that the fragment discussed here was written in the writing of Galileo's youth is correct with "absolute certainty" according to Favaro. But with equal certainty, the scholar, whose judgment we must view as authoritative in this relation, recognizes the following fragments belonging to the theory of projectile motion as also written by the youthful hand:

1. Pag. 424[16] (Mss. Gal. P. V. T. II. car. 193 r.): This is the compilation, written in Italian, of artillery problems mentioned many times in the preceding and including the two questions: *"se la palla vadia per linea retta, non sendo tirata a perpendicolo"* [whether the ball shot not perpendicularly moves in a straight line] and *"che linea descriva la palla nel suo moto"* [what kind of line the ball describes during its motion].

2. Pag. 427 (Mss. Gal. P. V. T. II. car. 91 t.): Fragment beginning with the words *"Determinetur ergo impetus"* [The impetus therefore is determined]. In this fragment, consisting of only seven lines and a drawing, the problem is formulated how to determine the impetus at individual points of the parabola from the always constant horizontal impetus and the impetus acquired through vertical fall. Here the horizontal velocity is viewed in the manner known from the *Discorsi* as acquired through free fall from a corresponding height.

3. Pag. 428 (Mss. Gal. P. V. T. II. car. 110 t.): Numerical example for the calculation of the impetus at different positions of the parabola.

4. Pag. 428 (Mss. Gal. P. V. T. II. car. 87 t.): *"Datae parabolae elevationem invenire, ex qua decidens mobile parabolam datam describat."* [To find the elevation of a given parabola from which the falling mobile describes the given

---

[16] [This and the following page-numbers refer to vol. 8 of Galilei 1890–1909. They are followed by the call-numbers of the original manuscripts in the Biblioteca Nazionale of Florence.]

parabola.] The solution and the corresponding demonstration coincide completely with the corresponding solution and demonstration imparted in the printed *Discorsi* under *Propositio V* (Galilei 1890–1909, 8:293) with the heading: "*In axe extenso datae parabolae punctum sublime reperire, ex quo cadens parabolam ipsam describit.*" [In the axis of a given parabola extended upward, to find the high point from which a falling body describes this same parabola.] Accordingly, the label "*scritta*" [transcribed] appears at the foot of the fragment. A comparison of the two headings indicates the extent to which the later version with unchanged content strives for heightened distinctness of expression. A deviation in wording only has also to be pointed out: Galileo in the *Discorsi* labels as "*sublimitas*" what is called "*elevatio*" in the fragment. The word "*sublimitas*" appears not to occur in what are certainly the oldest fragments. The solution of the problem is immediately followed on page 429 by the deduction that one-half the base is a mean proportional between the altitude of the parabola and the elevation above the parabola. This deduction coincides with the *Corollarium* to *Propositio V* of the *Discorsi* [ibid., 8:294] except that again the word "*sublimitas*" substitutes in the *Discorsi* the "*elevatio supra parabolam*" of the fragment.

5. Pag. 429–30 (Mss. Gal. P. V. T. II. car. 9a r.): The fragment begins by investigating the correlation between horizontal impetus and range of various semiparabolas, as well as the correlation between the variation of the impetus at the foot and the variation of the horizontal impetus of semiparabolas of equal height. Through inversion, the preceding consideration is then applied to parabolas produced by upwards shots in the direction of the tangent of corresponding semiparabolas; for these parabolas too, the correlation is derived between range and different given start impetuses, at first for individual cases of parabolas with equal height. Through calculation of individual examples it is then illustrated, more than proven, that with a shot at an angle of 45° a greater distance can be reached than with shots of equal force at an higher or lower angle over the horizon. As in the *Discorsi*, this fragment presumes, but does not prove, that a body projected diagonally upward in the direction of the tangent to the semiparabola produced by a horizontal projection must describe the same parabola [as the body horizontally projected].

6. Pag. 431 (Mss. Gal. P. V. T. II. car. 111 r.): Demonstration that for semiparabolas of equal amplitude, the impetus acquired through projection is smaller when the amplitude is twice as great as the altitude than when it is more than twice as great. The demonstration essentially coincides with the argumentation given for *Propositio VII* of the "Fourth Day" of the *Discorsi* [ibid., 8:294–95].

7. Pag. 433 (Mss. Gal. P. V. T. II. car. 111 t.): Beginning of the demonstration for the thesis exposed in *Propositio VIII* of the "Fourth Day" of the *Discorsi* [ibid., 8:297] that the amplitudes of two parabolas are equal when the impetus of the projected bodies is the same and the shots occur at angles which deviate equally above and below half a right angle.

The result of this overview can be summarized briefly. From a period of

Galilean research which surely predated the composition of the *Dialogue Concerning the Two Chief World Systems*, a number of fragments in Galileo's hand relating to the theory of projectile motion are preserved. In them the recognition of the parabolic shape of the trajectory is presupposed and, moreover, the most important principles deriving from this insight are treated in a similar manner as in the Latin sections of the *Discorsi*, partly in literal correspondence with these.

Despite this restriction, the result of the examination of the manuscripts suffices to prove that the discovery of the parabolic shape of the projectile trajectory belongs to Galileo and not to Cavalieri, while doubt is at first possible as to whether the differences in the Galilean handwriting can allow an absolutely certain distinction between what was written before and after 1610. However, according to all of the available information in Galileo's correspondence, he concluded his research on the theory of projectile motion in the year 1609 before the invention of the telescope, and returned to the same area of research two decades later at the earliest. Taking this into consideration, the observation that the fragments in question were written by the young Galileo also furnishes unequivocal evidence that they belong to the Paduan period, and thus constitute fragments of that theory of projectile motion on which Galileo reported to Luca Valerio and Minister Vinta. Thus the Latin sections of the "Fourth Day" of the *Discorsi*, which merely expound that which is essentially in the fragments already, are in truth what they claim to be: they are parts of the manuscript concerning a new theory of motion that had been temporarily finished in Padua. Galileo's theory of projectile motion, as it is presented in the *Discorsi*, is thus a product of the glorious period of his best years to which the majority of his greatest discoveries belong, and not a discovery made in his old age.

Consequently, despite Caverni's bold accusations, any justification for doubting the credibility of the letter of 11 September 1632 is removed. Every part of the letter is true: Galileo's explanation that the search for the shape of the projectile trajectory was the starting point for his research on motion must continue to serve as before and in the future, as an established datum for the historical interpretation of the development of the new theory of motion.

It is certain that the letter of February 1609 can be regarded as written after the discovery of the parabolic shape, and the law of equal times of fall first formulated there as derived from the recognition of the true nature of the projectile trajectory. When Galileo here counts the comparison of times of fall for balls projected with unequal power in the horizontal direction as one of those remaining problems to be studied (*questioni che mi restano intorno al moto dei proietti*), this is in complete accordance with the idea that the main propositions of the theory of projectile motion already had been asserted by this time. The law of equal times of fall was merely a further deduction. This seems to be the reason why Galileo did not include in the Latin sections of the *Discorsi* concerning the theory of projectile motion a conclusion which he often mentioned in other works with particular predilection.

The objections which have been derived from the *Dialogue* and from the *Specchio ustorio* against Galileo's right of discovery are simply reduced to sham proofs when confronted with the evidence given by the manuscripts. This also applies to the many other "proofs" which were constructed and could further be constructed through the interpretation of other passages.

Despite the ultimately decisive importance of the examination of the handwritings, it was nevertheless appropriate that the preceding pages dealt principally with the refutation of apparent counter-evidence, that is, strictly speaking, with the demonstration that what must be considered false on the basis of stronger evidence is at least improbable. The point was not only to answer this one question for which the handwriting evidence would have been sufficient, but rather to interpret the historical context in such a way that even the semblance of a contradiction between the result of the examination of the handwritings and the other established facts vanished. Clear historical insight was only attained with the elimination of this contradiction. There was also a second motive for the exhaustive examination of all adduced causes of suspicion. It was hardly superfluous to show that Caverni's attack would remain untenable even if a happy coincidence had not preserved documents which suffice in and of themselves to refute the charge of theft. The Italian scholar raised other serious insinuations against Galileo's great name and in many cases gave to his insinuations an appearance of reality in the same irresponsible manner as he did in the case discussed above, by combining interpretation and fabrication. In these other cases, though, there are no incontrovertible witnesses to summon to the bar, as we did here, to testify to the hollowness of the insinuations. It was, therefore, important for the examination of these other cases to show in the most complete manner, in a case in which the historical truth is above question, the method according to which all others are constructed.

REFERENCES FOR APPENDIX

Selected Bibliography of Raffaello Caverni

Caverni, Raffaello. 1874. *Problemi naturali di Galileo Galilei e di altri autori della sua scuola*. Firenze: Sansoni.

——. 1877a. *De' nuovi studi della filosofia: Discorsi a un giovane studente*. Firenze: Carnesecchi.

——. 1877b. *Voci e modi nella Divina Commedia dell'uso popolare toscano: Dizionarietto*. Firenze: Giusti [Reprint: Firenze: Pagnini, 1987].

——. 1878. "Notizie storiche intorno all'invenzione del termometro." *Bullettino di bibliografia e di storia delle scienze matematiche e fisiche* 11:531–86.

——. 1879. *Dell'arte dello scrivere: Consigli a un giovinetto*. Firenze: Le Letture di Famiglia.

——. 1881. *Dell'antichità dell'uomo secondo la scienza moderna: Saggio di studi*. Firenze: Cellini.

——. 1882. *Ricreazioni scientifiche*. Firenze: Le Letture di Famiglia Editrici.

——. 1883. "[Review of] *Galileo Galilei e lo Studio di Padova* per Antonio Favaro." *La Rassegna Nazionale* (anno 5) 12:476–80.

——. 1884. *L'estate in montagna: Nozioni di fisica*. Firenze: Le Monnier [Revised edition: 1885].

——. 1886a. "[Review of] *Carteggio inedito di Ticone Brahe, Giovanni Keplero e di altri celebri astronomi e matematici dei secoli XVI, e XVII con Giovanni Antonio Magini, tratto dall'Archivio Malvezzi de' Medici in Bologna*, pubblicato ed illustrato da A. Favaro." *La Rassegna Nazionale* (anno 8) 28:568–75.

——. 1886b. *Fra il verde e i fiori: Nozioni di botanica*. Firenze: Le Monnier [Second edition: 1900].

——. 1888. *Con gli occhi per terra: Nozioni intorno alla natura e alle proprietà di alcune sostanze minerali*. Firenze: Paggi.

——. 1891–1900. *Storia del metodo sperimentale in Italia*. 6 vols. Firenze: Civelli.

——. 1972. *Storia del metodo sperimentale in Italia*. With an Introductory Note by Giorgio Tabarroni. 6 vols. New York/London: Johnson Reprint Corporation [Reprint of Caverni 1891–1900; other reprint Bologna: Forni, 1970].

Selected Bibliography of Antonio Favaro

Favaro, Antonio. 1869. *Studi sul tracciamento della Galleria delle Alpi Cozie preceduti da cenni storici*. Torino: Ceresole e Panizza.

——. 1873. "Beiträge zur Geschichte der Planimeter." *Allgemeine Bauzeitung* 38:68–90, 93–104.

——. 1874. "Notizie storiche sulle frazioni continue dal secolo decimoterzo al decimosettimo." *Bullettino di bibliografia e di storia delle scienze matematiche e fisiche* 7:451–502, 533–89.

——. 1875. *Saggio di cronografia dei matematici dell'antichità.* Padova: Sacchetto.

——. 1877. *Lezioni di statica grafica.* Padova, Sacchetto.

——. 1878. "La storia delle matematiche nella Università di Padova." *Bullettino di bibliografia e di storia delle scienze matematiche e fisiche* 11:799–801.

——. 1879–80. "Ragguaglio dei Manoscritti Galileiani nella Biblioteca Nazionale di Firenze ed annuncio di alcuni frammenti inediti di Galileo." *Atti del Reale Istituto Veneto di Scienze, Lettere ed Arti* 6:847–53.

——. 1880. "Le aggiunte autografe di Galileo al *Dialogo sopra i due massimi sistemi* nell'esemplare posseduto dalla Biblioteca del Seminario di Padova." *Atti della Reale Accademia di Scienze, Lettere ed Arti di Modena* 19:245–75.

——. 1882. "Gli autografi galileiani nell'archivio Marsigli in Bologna." *Bullettino di bibliografia e di storia delle scienze matematiche e fisiche* 15:581–92.

——. 1883. "Alcuni scritti inediti di Galileo Galilei tratti dai manoscritti della Biblioteca Nazionale di Firenze pubblicati ed illustrati." *Bullettino di bibliografia e di storia delle scienze matematiche e fisiche* 16:1–97, 135–210.

——. [1883] 1966. *Galileo Galilei e lo Studio di Padova.* 2 vols. Padova: Editrice Antenore [first edition: Firenze: Le Monnier, 1883].

——. 1885. "Documenti inediti per la storia dei Manoscritti Galileiani nella Biblioteca Nazionale di Firenze pubblicati ed illustrati." *Bullettino di bibliografia e di storia delle scienze matematiche e fisiche* 18:1–112, 151–230.

——. 1886a. *Carteggio inedito di Ticone Brahe, Giovanni Keplero e di altri celebri astronomi e matematici dei secoli XVI e XVII con Giovanni Antonio Magini, tratto dall'Archivio Malvezzi de' Medici in Bologna.* Bologna: Zanichelli.

——. 1886b. "Intorno ad alcuni documenti galileiani recentemente scoperti nella Biblioteca Nazionale di Firenze." *Bullettino di bibliografia e di storia delle scienze matematiche e fisiche* 19:1–54.

——. 1887a. *Miscellanea galileiana inedita. Studi e ricerche.* Venezia: Antonelli [also in *Memorie del Reale Istituto Veneto di Scienze, Lettere ed Arti* 22 (1882):701–1037].

——. 1887b. "Otto anni d'insegnamento di Storia delle Matematiche nella R. Università di Padova." *Bibliotheca Mathematica* 1:49–54.

——. 1887c. "Di Giovanni Tade e di una sua visita a Galileo dal 12 al 15 novembre 1614." *Bullettino di bibliografia e di storia delle scienze matematiche e fisiche* 20:345–71.

——. 1888. *Per la Edizione Nazionale delle opere di Galileo Galilei sotto gli auspicii di S. M. il Re d'Italia. Esposizione e disegno.* Firenze: Barbèra.

——. 1891. "Nuovi studi galileiani." *Memorie del Reale Istituto Veneto di Scienze, Lettere ed Arti* 24:7–430.

——. 1894–95. "Don Baldassarre Boncompagni e la storia delle scienze matematiche e fisiche." *Atti del Reale Istituto Veneto di Scienze, Lettere ed Arti* 6:509–21.

——. 1898–99. "Intorno alle opere scientifiche di Galileo Galilei nella Edizione Nazionale sotto gli auspicii di S. M. il Re d'Italia." *Atti del Reale Istituto Veneto di Scienze, Lettere ed Arti* 58 (part 2):129–204.

——. 1899–1900. "Raffaello Caverni. Nota commemorativa." *Atti del Reale Istituto Veneto di Scienze, Lettere ed Arti* 59 (part 2):377–79.

——. 1907a. "Antichi e moderni detrattori di Galileo." *La Rassegna Nazionale* (anno 29) 153:577–600.

——. 1907b. *Galileo e l'Inquisizione. Documenti del processo galileiano esistenti nell'Archivio del S. Uffizio e nell'Archivio Segreto Vaticano per la prima volta integralmente pubblicati.* Firenze: Barbèra [Reprint: Firenze: Giunti-Barbèra, 1983].

——. [1907–08] 1983. "Amici e corrispondenti di Galileo Galilei XXI. Benedetto Castelli." *Atti del Reale Istituto Veneto di Scienze, Lettere ed Arti* 67 (part 2):1–130 [Reprint in Favaro 1983, 2:737–870].

——. 1910. *Galileo Galilei.* Modena: Formiggini. [New edition: Firenze: Barbèra, 1964].

——. 1911–12. *Atti della Nazione Germanica Artista nello Studio di Padova.* 2 vols. Venezia: Tipografia Emiliana.

——. 1912–13a. "Amici e corrispondenti di Galileo Galilei XXIX. Vincenzio Viviani." *Atti del Reale Istituto Veneto di Scienze, Lettere ed Arti* 72 (part 2):1–155 [Reprint in Favaro 1983, 2:1007–163].

——. 1912–13b. "Emilio Wohlwill. Nota commemorativa." *Atti e Memorie della Reale Accademia di Scienze, Lettere ed Arti in Padova* 29:43–55.

——. [1913–14] 1992b. "Serie ventesimaterza di Scampoli galileiani CXLVI. Galileo e Guidobaldo del Monte." *Atti e Memorie della Reale Accademia di Scienze, Lettere ed Arti in Padova* 30:54–61 [Reprint in Favaro 1992b, 2:716–23].

——. [1914–15] 1983. "Amici e corrispondenti di Galileo Galilei XXXI. Bonaventura Cavalieri." *Atti del Reale Istituto Veneto di Scienze, Lettere ed Arti* 74 (part 2):701–67 [Reprint in Favaro 1983, 3:1245–315].

——. 1916. "La condanna di Galileo e le sue conseguenze per il progresso degli studi." *Scientia* anno 10, vol. 20, no. 51:1–11.

——. 1916–17. "Amici e corrispondenti di Galileo Galilei XXXVIII. Marino Mersenne." *Atti del Reale Istituto Veneto di Scienze, Lettere ed Arti* 76 (part 2):35–92 [Reprint in Favaro 1983, 3:1475–534].

——. 1917. "Scritture Galileiane apocrife", *Bollettino di bibliografia e storia delle scienze matematiche* 19:33–43.

——. 1917–18. "Intorno alla prima edizione fiorentina delle opere di Galileo." *Atti del Reale Istituto Veneto di Scienze, Lettere ed Arti* 77 (part 2):229–42.

——. 1918. "Galileo Galilei *e i Doctores parisienses.*" *Rendiconti della Reale Accademia dei Lincei. Classe di scienze morali, storiche e filologiche* 27:139–50.

——. 1919. "La storia delle scienze e la storia delle università a proposito di un prossimo centenario." *Atti della Società Italiana per il Progresso delle Scienze* 10:457–60.

———. [1919–20a] 1992a. "Adversaria galilaeiana. Serie quinta. XXXIII. Scritture Galileiane apocrife." *Atti e Memorie della Reale Accademia di Scienze, Lettere ed Arti in Padova* 36:17–30 [Reprint in Favaro 1992a, 141–54.]

———. 1919–20b. "Galileo Galilei, Benedetto Castelli e la scoperta delle fasi di Venere." *Archivio di storia della scienza* 1:283–96.

———. 1922a. *Saggio di bibliografia dello Studio di Padova [1500–1920]*. 2 vols. Venezia: Ferrari.

———. 1922b. *L'Università di Padova*. Venezia: Ferrari.

———. 1968. *Galileo Galilei a Padova. Ricerche e scoperte, insegnamento, scolari*. Padova: Editrice Antenore.

———. 1983. *Amici e corrispondenti di Galileo*. A cura e con una nota introduttiva di Paolo Galluzzi. 3 vols. Firenze: Libreria Editrice Salimbeni.

———. 1992a. *Adversaria galilaeiana. Serie I–VII*. Ristampa anastatica dagli "Atti e Memorie" della Accademia Patavina di Scienze, Lettere ed Arti. Edited by Lucia Rossetti and Maria Laura Soppelsa. Trieste: Edizioni Lint.

———. 1992b. *Scampoli galileiani*. Ristampa anastatica dagli "Atti e Memorie" della Accademia Patavina di Scienze, Lettere ed Arti. Edited by Lucia Rossetti and Maria Laura Soppelsa. 2 vols. Trieste: Edizioni Lint.

Favaro, Antonio et al. [1889–90] 1972. "Relazione della Giunta del R. Istituto Veneto deputata all'esame dei lavori presentati al concorso della Fondazione Tomasoni sul tema: Storia del metodo sperimentale in Italia." *Atti del Reale Istituto Veneto di Scienze, Lettere ed Arti* serie 7, vol. 1, 319–43 [Reprinted in Caverni 1972, 1:5–20].

Carli, Alarico; Favaro, Antonio. 1896. *Bibliografia galileiana (1568–1895)*. Roma: Libreria dello Stato [Reprint: Bologna: Brighenti, 1972].

Selected Bibliography of Emil Wohlwill

Wohlwill, Emil. 1863–64. "Bacon von Verulam und die Geschichte der Naturwissenschaft." *Deutsche Jahrbücher für Politik und Literatur* 9 (1863), 382–415; 10 (1864):207–44.

———. 1865. "Zur Geschichte der Erfindung und Verbreitung des Thermometers." *Annalen der Physik und Chemie* 124:163–78.

———. 1866. "Die Entdeckung des Isomorphismus. Eine Studie zur Geschichte der Chemie." *Zeitschrift für Völkerpsychologie und Sprachwissenschaft* 4:1–67.

———. 1867. "Vorlesungen zur Geschichte der Wiederbelebung der Naturwissenschaften im 16. Jahrhundert." Lecture by Dr. Emil Wohlwill in the great auditorium of the Johanneum, January through March (unpublished).

———. 1870. *Der Inquisitionsprocess des Galileo Galilei. Eine Prüfung seiner rechtlichen Grundlage nach den Acten der Römischen Inquisition*. Berlin: R. Oppenheim.

——. 1872a. "Zum Inquisitionsprocess des Galileo Galilei." *Literaturzeitung der Zeitschrift für Mathematik und Physik* 17:9–31.

——. 1872b. "Erwiederung. Zum Inquisitionsprocess des Galileo Galilei." *Literaturzeitung der Zeitschrift für Mathematik und Physik* 17:81–98.

——. 1877a. *Ist Galilei gefoltert worden? Eine kritische Studie.* Leipzig: Duncker & Humblot.

——. 1877b. "Die Fälschung des Protokolls vom 26. Februar 1616." (unpublished; Niedersächsische Staats- und Universitätsbibliothek Göttingen, Hist. lit. biogr. II. 6249).

——. 1879. "Der Original-Wortlaut des päpstlichen Urtheils gegen Galilei." *Historisch-literarische Abtheilung der Zeitschrift für Mathematik und Physik* 24:1–26.

——. 1880. "Erklärung und Abwehr." *Historisch-literarische Abtheilung der Zeitschrift für Mathematik und Physik* 25:185–90.

——. 1883. "[Review of] Antonio Favaro, *Galileo Galilei e lo Studio di Padova.* Florenz, Le Monnier, 1883." *Deutsche Literaturzeitung* 4:204–06.

——. 1883–84. "Die Entdeckung des Beharrungsgesetzes." *Zeitschrift für Völkerpsychologie und Sprachwissenschaft* 14 (1883):365–410; 15 (1884):70–135, 337–87.

——. 1887a. "Die Prager Ausgabe des *Nuncius sidereus.*" *Bibliotheca Mathematica* 1:100–02.

——. 1887b. "Joachim Jungius und die Erneuerung atomistischer Lehren im 17. Jahrhundert. Ein Beitrag zur Geschichte der Naturwissenschaften in Hamburg." In *Festschrift zur Feier des fünfzigjährigen Bestehens des Naturwissenschaftlichen Vereins in Hamburg, 18. November 1887 [= Abhandlungen aus dem Gebiete der Natuwissenschaften,* vol. 10], 1–66. Hamburg: Friederichsen & Co.

——. 1888a. *Joachim Jungius. Festrede zur Feier seines dreihundersten Geburtstages am 22. Oktober 1887. Mit Beiträgen zu Jungius' Biographie und zur Kenntnis seines handschriftlichen Nachlasses.* Hamburg/Leipzig: Voss.

——. 1888b. "Hat Leonardo da Vinci das Beharrungsgesetz gekannt?" *Bibliotheca Mathematica* 2:19–26.

——. 1889. "[Review of] Antonio Favaro: *Per la edizione nazionale delle opere di Galileo Galilei.* Esposizione e disegno. Florenz, Barbèra, 1888." *Deutsche Literaturzeitung* 10:129–31.

——. 1891a. "[Review of] *Le Opere di Galileo Galilei.* Edizione Nazionale. Direttore: Antonio Favaro. Vol. I. Florenz, Barbèra, 1890." *Deutsche Literaturzeitung* 12:823–25.

——. 1891b. "[Review of] Antonio Favaro, *Galileo Galilei e Suor Maria Celeste.* Florenz, Barbèra, 1891." *Deutsche Literaturzeitung* 12:825–27.

——. 1891c. "[Review of] Anton von Braunmühl, *Christoph Scheiner als Mathematiker, Physiker und Astronom.* Bamberg, Buchner, 1891." *Deutsche Literaturzeitung* 12:1247–249.

——. 1894. "Galilei betreffende Handschriften der Hamburger Stadtbibliothek."

Jahrbuch der Hamburgischen Wissenschaftlichen Anstalten 12:147–223. [Also published as a monograhy, Hamburg: Gräfe & Sillem, 1895]

——. 1895. "[Review of] Laurenz Müllner, *Die Bedeutung Galilei's für die Philosophie.* Wien, Selbstverlag der k. k. Universität, 1894." *Deutsche Literaturzeitung* 16:1462–467.

——. 1899. "Die Entdeckung der Parabelform der Wurflinie." *Abhandlungen zur Geschichte der Mathematik* 9 (1899) [= *Zeitschrift für Mathematik und Physik* 44 (1899), Supplement], 577–624.

——. 1902. "Neue Beiträge zur Vorgeschichte des Thermometers." *Mitteilungen zur Geschichte der Medizin und der Naturwissenschaften* 1:5–8, 57–62, 143–58, 282–90.

——. 1904a. "Über einen Grundfehler aller neueren Galileibiographien." In *Verhandlungen der Gesellschaft Deutscher Naturforscher und Ärzte. 75. Versammlung zu Cassel, 20.–26. September 1903. Zweiter Teil, 2. Hälfte,* 100–01. Leipzig: Vogel.

——. 1904b. "Melanchthon und Copernicus." *Mitteilungen zur Geschichte der Medizin und der Naturwissenschaften* 3:260–67.

——. 1905. "Galilei-Studien. I. Die Pisaner Fallversuche." *Mitteilungen zur Geschichte der Medizin und der Naturwissenschaften* 4:229–48.

——. 1906a. "Ein Vorgänger Galileis im 6. Jahrhundert." *Physikalische Zeitschrift* 7:23–32.

——. 1906b. "Galilei-Studien. II. Der Abschied von Pisa." *Mitteilungen zur Geschichte der Medizin und der Naturwissenschaften* 5:230–49, 439–64.

——. 1907. "Galilei-Studien. II. Der Abschied von Pisa (Zweites Nachwort)." *Mitteilungen zur Geschichte der Medizin und der Naturwissenschaften* 6:231–42.

——. 1909. "Zur Geschichte der Entdeckung der Sonnenflecken." *Archiv für die Geschichte der Naturwissenschaften und der Technik* 1:443–54.

——. 1909–29. *Galilei und sein Kampf für die Copernicanische Lehre.* 2 vols. Hamburg/Leipzig: Voss.

——. 1912. "Naturforscher als Historiker der Naturwissenschaften." *Mitteilungen zur Geschichte der Medizin und der Naturwissenschaften* 11:1–5.

——. 1929. "Für eine Veröffentlichung von Jungius' Werken und seine künftige Biographie." In *Beiträge zur Jungius-Forschung. Prolegomena zu der von der Hamburgischen Universität beschlossenen Ausgabe der Werke von Joachim Jungius (1587–1657),* edited by Adolf Meyer, 15–20. Hamburg: Hartung.

——. 1969. *Galilei und sein Kampf für die Copernicanische Lehre.* With an Introductory Note by Hans-Werner Schütt. 2 vols. Wiesbaden: Sändig [Reprint of Wohlwill 1909–29].

Other Works

Baldini, Ugo. 1980. "La scuola galileiana." In *Storia d'Italia. Annali 3: Scienza e tecnica nella cultura e nella società dal Rinascimento a oggi*, edited by Gianni Micheli, 381–463. Torino: Einaudi.

Baldo Ceolin, Milla; Olivieri, Luigi. 1994–95. "Progetto di edizione degli *Studi Galileiani* di A. Favaro." *Atti e Memorie dell'Accademia Patavina di Scienze, Lettere ed Arti* (Parte 3) 107:273–84.

Betti, Umberto; Pagnini, Gian Piero (eds.) 1991. *Raffaello Caverni 1837–1900.* Antologia degli scritti a cura di Umberto Betti. Note biografiche, storico-genealogiche di Gian Piero Pagnini. Firenze: Pagnini.

Bortolotti, Ettore. 1923–24. "Antonio Favaro storico delle scienze matematiche." *Atti e Memorie della Regia Deputazione di Storia Patria per le Provincie di Romagna* 14:1–24.

Bosmans, Henri. 1923. "Antonio Favaro (1847–1922)." *Revue des questions scientifiques* 83:156–75.

Brugnaro, Francesco Giovanni. 1979. "Antonio Favaro, studioso di Galileo Galilei." In *Medioevo e Rinascimento veneto con altri studi in onore di Lino Lazzarini.* vol. II: *Dal Cinquecento al Novecento*, 289–306. Padova: Editrice Antenore.

Bucciantini, Massimo. 1986. "Bibliografia e storia della scienza in Italia (1868–1920)." In *Biblioteche speciali*, edited by Mauro Guerrini, 110–134. Milano: Editrice Bibliografica.

——. 1995. "Favaro, Antonio." *Dizionario biografico degli italiani* 45:441–44. Roma: Istituto della Enciclopedia Italiana.

Campori, Giuseppe (ed.). 1881. *Carteggio Galileano inedito, con note ed appendici.* Modena: Società Tipografica Antica Tipografia Soliani [= *Memorie della Regia Accademia di Scienze, Lettere ed Arti in Modena* 20 (1880–82), part 2].

Cappelletti, Vincenzo. 1969. "Boncompagni Ludovisi, Baldassarre." *Dizionario biografico degli italiani* 11:704–09. Roma: Istituto della Enciclopedia Italiana.

Cappelletti, Vincenzo; Di Trocchio, Federico. 1979. "Caverni, Raffaello." *Dizionario biografico degli Italiani* 23:85–8. Roma: Istituto della Enciclopedia Italiana.

Cavalieri, Bonaventura. 1632. *Lo Specchio ustorio overo Trattato delle settioni coniche, et alcuni loro mirabili effetti intorno al lume, caldo, freddo, suono e moto ancora.* Bologna: Ferroni [Second edition: Bologna, Ferroni, 1650].

Collingwood, R. G. [1946] 1993. *The Idea of History. Revised Edition with Lectures 1926–1928.* Edited with an Introduction by Jan van der Dussen. Oxford: Clarendon Press.

Damerow, Peter; Jürgen Renn, Simone Rieger. 1996. *Pilot Study for a Systematic PIXE Analysis of the Ink Types in Galileo's Ms. 72. Project Report No. 1.* Berlin: Max-Planck-Institut für Wissenschaftsgeschichte [= Max-Planck-Institut für Wissenschaftsgeschichte, *Preprint* 54].

Del Lungo, Carlo. 1919. "Sopra la *Storia del metodo sperimentale in Italia* di R. Caverni." *Atti della Società Italiana per il Progresso delle Scienze* 10:477–78.

——. 1919–20. "La *Storia del metodo sperimentale in Italia* di Raffaello Caverni." *Archivio di storia della scienza* 1:272–82.

——. 1921–22. "Del pendolo e della sua applicazione all'orologio." *Archivio di storia della scienza* 2:147–66.

Descartes, René. 1897–1913. *Oeuvres.* Edited by Charles Adam and Paul Tannery. Paris: Cerf.

Drake, Stillman. 1979. *Galileo's Notes on Motion Arranged in Probable Order of Composition and Presented in Reduced Facsimile.* Firenze: Istituto e Museo di Storia della Scienza.

Duhem, Pierre. 1909. "Sur la mécanique de Léonard de Vinci et les recherches de Raffaello Caverni." In Pierre Duhem, *Études sur Léonard de Vinci. Ceux qu'il a lus et ceux qui l'ont lu,* vol. 2, 361–63. Paris: Hermann [Reprint: Paris: Éditions des Archives Contemporaines, 1984].

Favaro, Giuseppe. 1922–23. "Antonio Favaro. Bio-bibliografia." *Atti del Reale Istituto Veneto di Scienze, Lettere ed Arti* 82 (part 1):221–303.

Gabrieli, Giuseppe. 1925. "A. Favaro e gli studi italiani di storia della scienza." *Isis* 7:456–67.

Galilei, Galileo. 1632. *Dialogo dove ne i congressi di quattro giornate si discorre sopra i due massimi sistemi del mondo tolemaico, e copernicano.* Firenze: Landini. [Reprinted in Galilei 1890–1909, 7:21–520.]

——. 1638. *Discorsi e dimostrazioni matematiche, intorno à due nuove scienze attenenti alla mecanica et i movimenti locali. Con una appendice del centro di gravità d'alcuni solidi.* Leida: Elsevirii. [Reprinted in Galilei 1890–1909, 8:39–318.]

——. 1718. *Opere. Nuova edizione coll'aggiunta di varj trattati dell'istesso autore non più dati alle stampe.* Edited by Tommaso Buonaventuri et al. 3 vols. Firenze: Tartini e Franchi.

——. 1842–56. *Le Opere. Prima edizione completa condotta sugli autentici manos-critti Palatini.* Edited by Eugenio Albèri et al. 16 vols. Firenze: Società Editrice Fiorentina.

——. 1890–1909. *Le Opere. Edizione nazionale sotto gli auspici di Sua Maestà il Re d'Italia.* Edited by Antonio Favaro et al. 20 vols. Firenze: Barbèra [Third edition: Firenze: Barbèra, 1968].

——. 1891. *Dialog über die beiden hauptsächlichsten Weltsysteme, das ptole-mäische und das kopernikanische.* Aus dem Italienischen übersetzt und erläutert von Emil Strauss. Leipzig: Teubner.

——. 1974. *Two New Sciences Including Centers of Gravity and Force of Percussion.* Translated with introduction and notes by Stillman Drake. Madison, Wisconsin: University of Wisconsin Press.

Galluzzi, Paolo. 1983. "[Introductory Note]." In Favaro 1983, 1:V–XII.

Galluzzi, Paolo, et al. 1988. *Galileo: La sensata esperienza*. Cinisello Balsamo (Milano): Pizzi.

Giovannozzi, Giovanni. 1910. "Un tedesco di Montelupo." *La Rassegna Nazionale* (anno 32) 171:25–74.

——. 1920. "Raffaello Caverni e la sua *Storia del metodo sperimentale*." *Archivio di storia della scienza* 1:266–71.

——. 1928. "Un capitolo inedito della *Storia del metodo sperimentale in Italia*." *Memorie della Pontificia Accademia delle Scienze. I Nuovi Lincei* 11:171–90.

Henneberg, Lebrecht. 1901–08. "Die graphische Statik der starren Körper." In *Mechanik*. vol. 1 (= *Encyklopädie der mathematischen Wissenschaften mit Einschluss ihrer Anwendungen*, 4, 1), edited by Felix Klein and Conrad Müller, 345–434. Leipzig: Teubner.

Huygens, Christiaan. 1888–1950. *Oeuvres complètes*. Publiées par la Société Hollandaise des Sciences. La Haye: Nijhoff.

Klug, Josef. 1913. "Die nachgelassenen Schriften Dr. Emil Wohlwills." *Archiv für die Geschichte der Naturwissenschaften und der Technik* 6:216–21.

Laemmel, Rudolf. 1928. "Untersuchung der Dokumente des Galileischen Inquisitionsprozesses." *Archiv für Geschichte der Mathematik, der Naturwissenschaften und der Technik* 10:405–19.

Landucci, Giovanni. 1996. "'Filosofia sperimentale' e immagini di Galileo nell'Ottocento italiano." In Handjaras, Luciano, et al. *Ricerche di filosofia: tra ermeneutica e filosofia analitica*, 151–207. Firenze: Alfani.

Lefons, Chiara. 1984. "Un capitolo dimenticato della storia delle scienze in Italia: Il «Bullettino di bibliografia e di storia delle scienze matematiche e fisiche» di Baldassarre Boncompagni." *Giornale critico della filosofia italiana* 63:65–90.

Libri, Guillaume. 1838–41. *Histoire des sciences mathématiques en Italie, depuis la renaissance des lettres jusqu'à la fin du dix-septième siècle*. 4 vols. Paris: Renouard [Reprint: Sala Bolognese (Bologna): Forni, 1991].

Maffioli, Cesare S. 1985. "Sulla genesi e sugli inediti della *Storia del metodo sperimentale in Italia* di Raffaello Caverni." *Annali dell'Istituto e Museo di Storia della Scienza di Firenze* 10:23–85.

Malusa, Luciano. 1977. "Storiografia filosofica e storiografia scientifica in Antonio Favaro." In Malusa, Luciano. *La storiografia filosofica italiana nella seconda metà dell'Ottocento*. vol. 1: *Tra positivismo e neokantismo*, 551–66. Milano: Marzorati.

Martini, Tito. 1901. "Raffaello Caverni e la sua *Storia del metodo sperimentale in Italia*." *L'Ateneo Veneto* (anno 24) 1:291–321.

Micheli, Gianni. 1980. "Scienza e filosofia da Vico a oggi." In *Storia d'Italia. Annali 3: Scienza e tecnica nella cultura e nella società dal Rinascimento a oggi*, edited by Gianni Micheli, 549–675. Torino: Einaudi.

——. 1987. "La storia della scienza nella cultura italiana." In *La scienza tra filosofia e storia in Italia nel Novecento*, edited by Fabio Minazzi and Luigi Zanzi,

295–308. Roma: Presidenza del Consiglio dei Ministri. Direzione generale delle informazioni e della proprietà letteraria, artistica e scientifica.

——. 1988. "L'idea di Galileo nella cultura italiana dal XVI al XIX secolo." In Galluzzi et al. 1988, 163–86.

Mieli, Aldo. 1919–20. "L'opera di Raffaello Caverni come storico (cenni preliminari)." *Archivio di storia della scienza* 1:262–65.

Procacci, P. 1900. "P. Raffaello Caverni." *La Rassegna Nazionale* (anno 22) 111:804–05.

Quaranta, Mario. 1983. "Antonio Favaro (1847–1922)." In *Galileo e Padova: Mostra di strumenti, libri, incisioni*, 52–55. Padova: Comune di Padova, Assessorato ai beni culturali.

Renn, Jürgen, et al. 1998. *Hunting the White Elephant. When and How Did Galileo Discover the Law of Fall?* Berlin: Max-Planck-Institut für Wissenschaftsgeschichte [= Max-Planck-Institut für Wissenschaftsgeschichte, *Preprint* 97].

Schiaparelli, Giovanni. [1892] 1930. "La *Storia del metodo sperimentale in Italia*. Cenno bibliografico." *Il Pensiero italiano* (anno 2) 4 (1892):405–30 [Reprinted in Schiaparelli, Giovanni V. *Le Opere*. 10:3–23. Milano: Hoepli, 1930. Other Reprint: New York, Johnson Reprint Corporation, 1968].

Schütt, Hans-Werner. 1969. "Emil Wohlwill, *Galilei und sein Kampf für die Copernicanische Lehre*." In Wohlwill 1969, vol. 1, without pagination.

Segre, Michael. 1991. *In the Wake of Galileo*. New Brunswick (New Jersey): Rutgers University Press.

Seneca, Federico. 1967. *Antonio Favaro, Isidoro Del Lungo e l'edizione nazionale delle opere galileiane*. Firenze: Barbèra.

——. 1995 "Antonio Favaro, studioso di Galileo." In *Galileo Galilei e la cultura veneziana*, 381–404. Venezia: Istituto Veneto di Scienze, Lettere ed Arti.

Strappini, Lucia. 1990. "Del Lungo, Isidoro." *Dizionario biografico degli italiani* 38:96–100. Roma: Istituto della Enciclopedia Italiana.

Strauss, Emil. 1891. "Einleitung" in Galilei 1891, VI–LXXIX.

Tabarroni, Giorgio. 1969. "Raffaello Caverni: prete contestatore. A 70 anni dalla morte." *Physis* 11:564–70.

——. 1972. "Raffaello Caverni and his Work: An Introductory Note." In *Caverni* 1972, 1:v–xxii.

Tannery, Paul. 1900. "[Review of] Emil Wohlwill. Die Entdeckung der Parabelform der Wurflinie. *Abhandlungen zur Geschichte der Mathematik*, IX, p. 579–635, Teubner, Leipzig; 1899." *Bulletin des sciences mathématiques* 24:33–7 [Reprinted in: Tannery, Paul. *Mémoires scientifiques*. Edited by J.-L. Heiberg and H.-G. Zeuthen. 12:157–61. Paris: Gauthier-Villars, 1933].

Tartaglia, Niccolò. 1537. *Nova scientia*. Vinegia: Stephano da Sebio.

Timpanaro, Sebastiano. 1957. "Storia del metodo sperimentale in Italia." In *Dizionario Letterario Bompiani delle opere e dei personaggi di tutti i tempi e di tutte le letterature. Opere*. 7:164–65. Milano: Bompiani.

Torricelli, Evangelista. 1644. *Opera geometrica*. Florentiae: Typis Amatoris Massae & Laurentij de Landis.

——. 1919–1944. *Opere*. Edited by Gino Loria and Giuseppe Vassurra. vols. 1–3: Faenza: Montanari, 1919; vol. 4: Faenza: Lega, 1944.

Viviani, Vincenzio. 1674. *Quinto libro degli Elementi d'Euclide ovvero Scienza universale delle proporzioni spiegata colla dottrina del Galileo*. Firenze: Alla Condotta.

Wisan, Winifred Lovell. 1974. "The New Science of Motion: A Study of Galileo's *De motu locali*." *Archive for History of Exact Sciences* 13:103–306.

Index

9 780521 001038